Numerical and Statistical Techniques

Numerical and Statistical Techniques

Dr. Qazi Shoeb Ahmad
Assistant Professor in Department of Mathematics
Integral University, Lucknow

Dr. Zubair Khan
Lecturer in Department of Mathematics
Integral University, Lucknow

Shadab Ahmad Khan
Lecturer in Department of Mathematics
Integral University, Lucknow

Ane Books Pvt. Ltd.

New Delhi ♦ Chennai ♦ Mumbai ♦ Bengaluru
Kolkata ♦ Thiruvananthapuram ♦ Lucknow ♦ Hyderabad

Numerical and Statistical Techniques

Qazi Shoeb Ahmad, Zubair Khan and Shadab Ahmad Khan

© Ane Books Pvt. Ltd.

First Edition : 2009
Reprint : **2010**

Published by

Ane Books Pvt. Ltd.
4821, Parwana Bhawan, 1st Floor, 24 Ansari Road,
Darya Ganj, **New Delhi** - 110 002, India
Tel.: +91(011) 23276843-44, Fax: +91(011) 23276863
e-mail: kapoor@anebooks.com, Website: www.anebooks.com

Branches

- Avantika Niwas, 1st Floor, 19 Doraiswamy Road, T. Nagar,
  Chennai - 600 017, Tel.: +91(044) 28141554, 28141209
  e-mail: anebooks_tn@airtelmail.in

- Plot No. 59, Sector-1, Shirwane, Nerul, **Navi Mumbai** - 400 706,
  Tel.: +91(022) 27720842, 27720851
  e-mail: anebooksmum@mtnl.net.in

- 38/1, 1st Floor, Model House, First Street, Opp. Shamanna Park,
  Basavannagudi, **Bengaluru** - 560 004,
  Tel.: +91(080) 41681432, 26620045
  e-mail: anebang@airtelmail.in

- Flat No. 16A, 220 Vivekananda Road, Maniktalla,
  Kolkata - 700 006, Tel.: +91(033) 23547119, 23523639
  e-mail: anekol@vsnl.net

- # 6, TC 25/2710, Kohinoor Flats, Lukes Lane, Ambujavilasam Road,
  Thiruvananthapuram - 01, Kerala, Tel.: +91(0471) 4068777, 4068333
  e-mail: anebookstvm@airtelmail.in

Representative Office

- C-26, Sector-A, Mahanagar, **Lucknow** - 226 006
  Mobile - +91 93352 29971

- F-413, 'F' Block, Mallapur, May Flower Park, Hyderabad - 76
  Mobile: +91 9618109129

Disclaimer: Please be informed that the author and the publisher have put in their best efforts in producing this book. Every care has been taken to ensure the accuracy of the contents. However, we make no warranties for the same and therefore shall not be responsible or liable for any loss or any commercial damages accruing thereof. Neither the publisher nor the author is engaged in providing services of any professional nature and shall therefore not be responsible for any incidental, consequential, special or any other damages. Please do consult a professional where appropriate.

ISBN : 978-81-8052-259-8

All rights reserved. No part of this book may be reproduced in any form including photocopying, microfilms, photoprints, storage in any retrieval system, transmission in any permanent or temporary form, without the prior written consent of the publisher.

Printed at : A.P. Offset Pvt. Ltd., Delhi

Preface

The authors feel great pleasure in presenting the first edition of the book **"Numerical and Statistical Techniques"**. This book is designed to meet the requirements of B.Tech, B.Tech (Biotech), B.C.A. students of Integral University & its study centres, Lucknow University, Jamia Hamdard and Various other Universities.

The subject matter has been discussed in such a way that the students will find no difficulty to understand it. Each chapter of this book contains complete self-explanatory theory and a large number of solved examples, followed by a collection of good exercises.

The language of the book is simple and easy to understand. The authors hope that the students, teachers and other readers will find the book interesting and to the point covering the whole course. We hope that the students will receive the book warmly.

We have taken great care in eliminating the misprints, but if there are still any, we shall be highly obliged to those who will take trouble of pointing them out. Suggestions for the improvement of the book will be gratefully acknowledged.

Authors

Acknowledgement

All praises and thanks to 'ALLAH', the Almighty, the Merciful, and the Omniscient whose blessings enabled us to complete this work in the present form.

We feel immense pleasure in expressing our gratitude in the honour of Prof. S.W. Akhtar, Vice Chancellor, Prof. S.M. *Iqbal,* Pro-Vice Chancellor and Dr. I.A. Khan, Registrar, Integral University Lucknow for providing us all the necessary facilities to execute this manuscript.

We are also thankful to Prof. Q.H. Ansari, Prof. Mursaleen, Prof. Zafar Ahsan, Prof. Afzal Beg, Prof. Mohd. Imdad, Dr. Rais Ahmad, Dr. Nabiullah Khan, Department of Mathematics and Prof. Abdul Bari, Department of Statistics & Operations Research, A.M.U., Aligarh for their fruitful suggestions and constant encouragement during this work.

We are also greatly indebted to all our colleagues and friends especially Dr. Riyaz Ahmad Khan, HOD, Department of Mathematics, Integral University, Lucknow and Dr. Arshad Khan, Department of Mathematics, J.M.I., New Delhi for creating a healthy environment and sharing of ideas during the preparation of this manuscript.

We are deeply indebted to our parents and family members for their tremendous patience, enthusiastic inspirations in pursuit of this project.

We would like to accord special thanks to Mr. Shoeb Siddiqui, Department of Computer Science, Integral University Lucknow, for his valuable time during the preparation of this manuscript.

In the last but not the least we are also thankful to Mr. Sunil Saxena, Mr. Jai Raj Kapoor, Mr. H. Rahman and their team of **"Ane Books Pvt. Ltd"** for their kind cooperation at every stage, without which it would not have been possible to bring out this book in such a fine format.

Qazi Shoeb Ahmad
Zubair Khan
Shadab Ahmad Khan

Contents

1	**Error and Computer Arithmetic** **1**	
1.1	Introduction	1
1.2	Accuracy of Numbers	1
1.3	Errors	2
1.4	Types of Error	2
1.5	General Formula for Error	5
1.6	Errors in Numerical Computation	5
1.7	Floating Point Representation of Numbers	9
1.8	Arithmetic Operations with Normalized Floating Point Numbers	10
2	**Solution of Algebraic and Transcendental Equations** **17**	
2.1	Solution of Algebraic and Transcendental Equations	17
2.2	Bisection Method or Bolzano Method	17
2.3	Iteration Method or Successive Approxi-mation Method	26
2.4	Condition For Convergence of the Iterative Method	27
2.5	Regula-Falsi Method or Method of False Position	31
2.6	Order and Rate of Convergence of Regula-Falsi Method	33
2.7	Newton-Raphson Method or Newton's Method	38
2.8	Order and Rate of Convergence of Newton-Raphson Method	39
2.9	Geometrical Interpretation of Newton-Raphson method	40
2.10	Solution of Simultaneous Linear Algebraic Equations	47
2.11	Gauss-Elimination Method	48
2.12	Gauss-Elimination Method with Pivoting	52
2.13	Gauss-Jordan Method	54
2.14	Matrix-Inversion Method	59
2.15	Method of Triangularisation or Method of Factorization	62
2.16	Gauss-Jacobi Method or Jacobi Method of Iteration	71
2.17	Gauss-Seidel Method	77
2.18	Lin-Bairstow's method	83

3. Finite Differences 87

3.1	Forward Differences	87
3.2	Backward Differences	89
3.3	Central Differences	89
3.4	Different Types of Operators	90
3.5	Relation Between Operators	91
3.6	Differences of a Polynomial	93
3.7	Factorial Notation	108
3.8	Reciprocal or Negative Factorial Notation	108
3.9	Differences of a Factorial Function	109
3.10	Polynomial in Factorial Notation	110
3.11	Computation of Missing Terms	110
3.12	Finite Integration (or Inverse Operator Δ^{-1})	118
3.13	Summation of Series	119
3.14	Montmort's Theorem	120

4. Interpolation 128

4.1	Interpolation	128
4.2	Newton's-Gregory Forward Interpolation Formula	129
4.3	Error in polynomial Interpolation	131
4.4	Error in Newton's Gregory Forward Interpolation Formula	131
4.5	Newton's Gregory Backward Interpolation Formula	142
4.6	Error in Newton's Gregory Backward Interpolation formula	145
4.7	Central Difference Interpolation Formula	155
4.8	Gauss's Forward Interpolation Formula	156
4.9	Gauss's Backward Interpolation Formula	158
4.10	Stirling's formula	165
4.11	Bessel's formula	166
4.12	Laplace-Everett formula	167
4.13	Relation Between Bessel's and Everett's Formula	168
4.14	Advantages of Central Difference Interpolation Formula	169
4.15	Interpolation with Unequal Intervals	177
4.16	Divided Differences	178
4.17	Properties of Divided Differences	179
4.18	Relation Between Divided Differences and Forward Differences	181
4.19	Newton's General Interpolation Formula (for Unequal Intervals) or Newton's Divided Difference Interpolation Formula	181
4.20	Lagrange's Interpolation Formula (for Unequal Intervals)	188
4.21	Inverse Interpolation	195
4.22	Cubic Spline	197

5. Numerical Differentiation and Integration 203

    Numerical Differentiation 203
- 5.1 Newton's Forward Difference Formula to get the Derivative 203
- 5.1 Newton's Forward Difference Formula to get the Derivative 203
- 5.2 Newton's Backward Difference Formula to get the Derivative 205
- 5.3 Stirling's Interpolation Formula to get the Derivative 207
- 5.4 Bessel's Formula to get the Derivative 209
- 5.5 Numerical Integration 219
- 5.6 Newton-Cote's Quadrature Formula 220
- 5.7 Trapezoidal Rule (for $n = 1$) 220
- 5.8 Simpson's $1/3^{rd}$ Rule (for $n = 2$) 221
- 5.9 Simpson's $3/8^{th}$ Rule (for $n = 3$) 222
- 5.10 Boole's Rule (for $n = 4$) 223
- 5.11 Weddle's Rule (for $n = 6$) 224
- 5.12 Romberg's Method 233
- 5.13 Euler-Maclaurin's Formula 235
- 5.14 Gaussian Quadrature Formula 241

6. Numerical Solution of Ordinary Differential Equations 247

- 6.1 Introduction 247
- 6.2 Inttial and Boundary Value Problems 248
- 6.3 Numerical Methods of Solving Ordinary Differential Equations ... 248
- 6.4 Picard's Method of Successive Approximations 248
- 6.5 Picard's Method for Simultaneous First Order Differential Equations 254
- 6.6 Taylor's Series Method 256
- 6.7 Taylor's Method for Simultateous First Order Differential Equations 256
- 6.8 Euler's Method 263
- 6.9 Modified Euler's Method 267
- 6.10 Runge-kutta Method 272
- 6.11 First Order Runge-kutta Method 272
- 6.12 Second Order Runge-kutta Method 272
- 6.13 Third Order Runge-kutta Method 272
- 6.14 Fourth Order Runge-kutta Method 273
- 6.15 Runge-kutta Method for Simultaneous First Order Differential Equations 284

7. Curve Fitting 288

- 7.1 Introduction 288
- 7.2 Method of Least Squares 288
- 7.3 Fitting of a Straight Line by Method of Least Squares 289
- 7.4 Change of Origin and Scale 290
- 7.5 Normal Equations for Different forms of Curve 290

8. Regression Analysis 303
- 8.1 Regression 303
- 8.2 Linear Regression 303
- 8.3 Lines of Regression 303
- 8.4 Properties of Regression Coefficients 304
- 8.5 Angle Between two Lines of Regression 306
- 8.6 Angle Between two Lines of Regression 307
- 8.7 Linearization 307
- 8.8 Multiple Regression 307

9. Time Series and Forecasting 321
- 9.1 Introduction 321
- 9.2 Analysis of Time Series 321
- 9.3 Applications of Time series 322
- 9.4 Components of Time Series 322
- 9.5 Forecasting 323
- 9.6 Forecasting Models 323
- 9.7 Forecasting Methods 324
- 9.8 Measurement of Trend 325
- 9.9 Measurement of Seasonal Variations 332
- 9.10 Measurement of Cyclical Variations 339
- 9.11 Measurement of Random or Irregular Variations 340

10. Test of Significance and Analysis of Variance 345
- 10.1 Introduction 345
- 10.2 Parameter and Statistic 345
- 10.3 Sampling Distribution 345
- 10.4 Standard Error 346
- 10.5 Uses of standard error 346
- 10.6 Hypothesis Testing 347
- 10.7 Null Hypothesis 347
- 10.8 Alternative Hypothesis 347
- 10.9 Level of Significance 348
- 10.10 Critical Region 348
- 10.11 Critical Value 348
- 10.12 One tailed and Two Tailed Tests 348
- 10.13 Type-I Error and Type-II Error 349
- 10.14 Power of the Test 350
- 10.15 Procedure for Testing of Hypothesis 350

10.16 Student's *t*-Test 350
10.17 Assumptions for Student's *t*-Test 350
10.18 *t*-Test for Single Mean 351
10.19 *t*-Test for Difference of Means 353
10.20 Paired *t*-Test for Difference of Means 347
10.21 Z-Test 361
10.22 Test of Significance for Attributes 361
10.23 Test for Number of Successes 361
10.24 Test for Single Proportion 362
10.25 Test for Difference of Proportions 364
10.26 Test of Significance for Variables 370
10.27 Test of Significance for Single Mean 370
10.28 Test of Significance for Difference of Means 373
10.29 Test of Significance for Difference of Standard Deviations 376
10.30 *F*-Test 380
10.31 Procedure of *F*-Test 381
10.32 Assumptions for *F*-Test 381
10.33 Critical Values of *F*-Distribution 382
10.34 Chi-Square Test 386
10.35 Chi-Square Test to Test the Goodness of Fit 386
10.36 Chi-Square Test to Test the Independence of Attributes .. 392
10.37 Conditions for χ^2 Test 394
10.38 Uses of χ^2 Test 394
10.39 Analysis of Variance 401
10.40 Assumptions in the Analysis of Variance 401
10.41 Technique of Analysis of Variance 401
10.42 The basic Principle of Analysis of Variance 402
10.43 Analysis of Variance in one way Classification 402
10.44 Analysis of Variance in two way Classification 409
Appendix 419-428
Index 429-433

Contents xiii

10.16. Student's t-test	350
10.17. Assumptions for Student's t-test	350
10.18. Test for Single Mean	351
10.19. T-test of Difference of Means	353
10.20. Paired t-test for Difference of Means	357
10.21. F-test	361
10.22. Test of Significance of Attributes	360
10.23. Test for Number of Success	361
10.24. Test of Single Proportion	362
10.25. Test for Difference of Proportions	364
10.26. Test of Significance for Variables	370
10.27. Test of Significance for Single Mean	370
10.28. Test of Significance for Difference of Means	373
10.29. Test of Significance for Difference of Standard Deviations	376
10.30. Chi-square Test	380
10.31. Characteristics of χ^2-test	381
10.32. Assumptions for χ^2-test	381
10.33. General Nature of χ^2-Distribution	382
10.34. Chi-square Test	386
10.35. Chi-square Test for the Goodness of Fit	386
10.36. Chi-Square Test for the Independence of Attributes	392
10.37. Contingency Table	394
10.38. 2×2 Contingency Table	394
10.39. Yates' Correction	401
10.40. Assumptions in the Analysis of Variance	401
10.41. Techniques of Analysis of Variance	401
10.42. One-way Technique of Analysis of Variance	402
10.43. Analytical Methods in one way Classification	402
10.44. Analytical Methods in two way Classification	408
Appendix	419-428
Index	429-432

Chapter 1

Error and Computer Arithmetic

1.1 INTRODUCTION

In this chapter, we examine the sources of various types of errors. A number of different types of errors arise during the process of numerical computation, some are avoidable, and some are not. These errors contribute to the total error in the final result.

Errors, in numerical computation, can be made as small as we please, by taking the number to as many figures as we desired. Therefore, we can assume that the calculations are always carried out in such a manner as to make the errors of calculation negligible.

1.2 ACCURACY OF NUMBERS

  (*i*) **Exact numbers:** The numbers in which, there is no uncertainty and no approximation, are said to be exact numbers.

  e.g. 3, 4, 6, $\frac{5}{2}, \frac{6}{3}, \frac{1}{4}$, 1.45, 8.30,... are exact numbers.

  (*ii*) **Approximate numbers:** The numbers which are not exact are approximate numbers. These numbers contain infinitely many digits.

  e.g. $\sqrt{3}$ = 1.73205 ...., $\frac{1}{3}$ = 0.333333 ...; π = 3.141592 ... are approximate numbers.

  (*iii*) **Significant figures:** The digits used to express a number are called significant digits. Here, we have that all the digits 1, 2, 3, ..., 9 are significant figures and 0 is a significant figure except when it is used to fix the decimal point or to fill the places of unknown digits i.e. 0 may or may not be a significant figure.

Remark: The zeroes used between two non-zero digits are always significant figures.

For example. (*i*) 0.00025, 15000, 4.1000, 0.75 have two significant figures.
(*ii*) 1234, 45.19, 0.6295, contain four significant figures.
(*iii*) 5.23079 have six significant figures.

(*iv*) **Rounding off numbers:** There are numbers which contain large number of digits. Due to limitation of computers and calculators we have to cut them to usable number of figures. This process is called rounding off.

To rounding off a number or digit to n significant figures, discard all digits to the right of the nth place and follow the following rules:

(*i*) If $(n + 1)^{th}$ digit is less than 5, leave the nth digit unaltered.
(*ii*) If $(n + 1)^{th}$ digit is greater than 5, increase the nth digit by unity.
(*iii*) If $(n + 1)^{th}$ digit is exactly 5, increase the nth digit by unity if it is odd and leave it unaltered if it is even.

For example: 2.394 is rounded off to 2.39
11.6936 is rounded off to 11.694
9.13765 is rounded off to 9.1376
9.13775 is rounded off to 9.1378

1.3 ERRORS

Error = True value − Approximate value

1.4 TYPES OF ERROR

Let X_T is the true value and X_A is the approximate value of the given data.

(*i*) **Absolute error:** Denoted by E_a and given by

$$E_a = |X_T - X_A|$$

(*ii*) **Relative error:** Denoted by E_r and given by

$$E_r = \frac{E_a}{X_T} = \left|\frac{X_T - X_A}{X_T}\right|$$

(*iii*) **Percentage error:** Denoted by E_p and given by

$$E_p = E_r \times 100 = \left|\frac{X_T - X_A}{X_T}\right| \times 100$$

Remark: The relative and percentage errors are independent of the units of measurement but absolute errors are expressed in terms of units used.

Note: If a number is correct to n decimal places, then the error $= \dfrac{10^{-n}}{2}$.

For example. If the number 6.789 is correct to 3 decimal places, then the absolute error $= \dfrac{10^{-3}}{2} = \dfrac{0.001}{2} = 0.0005$.

ILLUSTRATIVE EXAMPLES

Example 1. Round off the following numbers to four significant figures and find E_a, E_r, E_p:
  (i) 22.34267   (ii) 0.065739   (iii) 11265.00

Solution. (i) 22.34267 is rounded off to 22.34

$\therefore$ $X_T = 22.34267$ and $X_A = 22.34$

Hence, $E_a = |X_T - X_A| = |22.34267 - 22.34| = 0.00267$
$= 2.67 \times 10^{-3}$

$$E_r = \frac{E_a}{X_T} = \frac{2.67 \times 10^{-3}}{22.34267} = 0.0001195 = 1.195 \times 10^{-4}$$

$E_p = E_r \times 100 = 1.195 \times 10^{-4} \times 100 = 1.195 \times 10^{-2}$

(ii) 0.065739 is rounded off to 0.06574

$\therefore$ $X_T = 0.065739$ and $X_A = 0.06574$

Hence, $E_a = |X_T - X_A| = |0.065739 - 0.06574|$
$= 0.000001 = 1 \times 10^{-6}$

$$E_r = \frac{E_a}{X_T} = \frac{1 \times 10^{-6}}{0.065739} = 0.00001521 = 1.521 \times 10^{-5}$$

$E_p = E_r \times 100 = 1.521 \times 10^{-5} \times 100 = 1.521 \times 10^{-3}$

(iii) 11265 is rounded off to 11260

$\therefore$ $X_T = 11265$ and $X_A = 11260$

Hence, $E_a = |X_T - X_A| = |11265 - 11260| = 5$

$$E_r = \frac{E_a}{X_T} = \frac{5}{11265} = 0.0004444 = 4.44 \times 10^{-4}$$

$E_p = E_r \times 100 = 4.44 \times 10^{-4} \times 100 = 4.44 \times 10^{-2}$

Example 2. Find E_a and E_r if $\frac{1}{9}$ is approximated as 0.111

Solution. Here, $X_T = \frac{1}{9}$ and $X_A = 0.111$

Hence, $E_a = |X_T - X_A| = \left|\frac{1}{9} - 0.111\right| = \left|\frac{1}{9} - 0.111\right|$

$= \left|\frac{0.001}{9}\right| = 0.00011111 = 1.111 \times 10^{-4}$

$$E_p = \left|\frac{X_T - X_A}{X_T}\right| \times 100 = \left|\frac{1.111 \times 10^{-4}}{\frac{1}{9}}\right| \times 100$$

$= 9.999 \times 10^{-4} \times 100 = 0.0999$

Example 3. If $\sqrt{3}$ is approximated as 1.732, find the percentage error.

Solution. Here, $X_T = \sqrt{3} = 1.73205$ and $X_A = 1.732$

Hence, percentage error

$$E_p = \left|\frac{X_T - X_A}{X_T}\right| \times 100 = \left|\frac{1.73205 - 1.732}{1.73205}\right| \times 100$$

$$= \left|\frac{0.00005}{1.73205}\right| \times 100 = 0.00289$$

Example 4. Three approximations to the value of $\sqrt{5}$ are 2.2361, 2.236, 2.24; find which of these three is the best approximation.

Solution. We have, $\sqrt{5} = 2.23607$

Hence, $|\sqrt{5} - 2.2361| = |2.23607 - 2.2361| = 0.00003$

$|\sqrt{5} - 2.236| = |2.23607 - 2.236| = 0.00007$

$|\sqrt{5} - 2.24| = |2.23607 - 2.24| = 0.00393$

$\therefore$ 2.236 is the best approximation.

Example 5. Find the relative error if 235.78659 is approximated to three significant digits.

Solution. We have, $X_T = 235.78659$ and $X_A = 235$

Hence, relative error

$$E_r = \left|\frac{X_T - X_A}{X_T}\right| = \left|\frac{235.78659 - 235}{235.78659}\right| = 0.0033$$

Example 6. Find the percentage error in the sum $\left(\sqrt{11} + \sqrt{19} + \sqrt{2} + \sqrt{8}\right)$, when these numbers are approximated to four significant digits

Solution. We have, $\sqrt{11} = 3.317, \sqrt{19} = 4.359,$

$\sqrt{2} = 1.414$ and $\sqrt{8} = 2.828$

Hence, $\left(\sqrt{11} + \sqrt{19} + \sqrt{2} + \sqrt{8}\right) = 11.9$

Since these numbers are correct to three decimal places.

Therefore absolute error,

$$E_a = \frac{10^{-3}}{2} + \frac{10^{-3}}{2} + \frac{10^{-3}}{2} + \frac{10^{-3}}{2}$$

$$= 4 \times 0.0005 = 0.002$$

and percentage error,

$$E_p = \frac{E_a}{X_T} \times 100 = \frac{0.002}{11.9} \times 100 = 0.017$$

1.5 GENERAL FORMULA FOR ERROR

Let $y = f(x_1, x_2, ..., x_n)$ be a function of n variables $x_1, x_2, ..., x_n$. Suppose Δy be the error in the value of y due to the errors $\Delta x_1, \Delta x_2, ..., \Delta x_n$ in the values of $x_1, x_2, ..., x_n$ respectively. Then the general formula for computing error is given by

$$\Delta y = \frac{\partial y}{\partial x_1} \Delta x_1 + \frac{\partial y}{\partial x_2} \Delta x_2 + ... + \frac{\partial y}{\partial x_n} \Delta x_n$$

where $\frac{\partial y}{\partial x_i}$, $i = 1, 2, 3, ..., n$ is the partial derivative of y with respect to x_i.

1.6 ERRORS IN NUMERICAL COMPUTATION

(i) Error in addition of numbers:

Let $X = x_1 + x_2$

Then maximum absolute error is given by

$$|\Delta X| \leq |\Delta x_1| + |\Delta x_2|$$

and maximum relative error is given by

$$\left|\frac{\Delta X}{X}\right| \leq \left|\frac{\Delta x_1}{X}\right| + \left|\frac{\Delta x_2}{X}\right|$$

(ii) Error in subtraction of numbers:

Let $X = x_1 - x_2$

Then maximum absolute error is given by

$$|\Delta X| \leq |\Delta x_1| + |\Delta x_2|$$

and maximum relative error is given by

$$\left|\frac{\Delta X}{X}\right| \leq \left|\frac{\Delta x_1}{X}\right| + \left|\frac{\Delta x_2}{X}\right|$$

(iii) Error in multiplication of numbers:

Let $X = x_1 \times x_2$

Then maximum absolute error is given by

$$\left|\frac{\Delta X}{X}\right| X = \left|\frac{\Delta X}{X}\right|.(x_1 \times x_2)$$

and maximum relative error is given by

$$\left|\frac{\Delta X}{X}\right| \leq \left|\frac{\Delta x_1}{x_1}\right| + \left|\frac{\Delta x_2}{x_2}\right|$$

(iv) Error in division of numbers:

Let $X = \dfrac{x_1}{x_2}$

Then maximum absolute error is given by

$$\left|\frac{\Delta X}{X}\right| X = \left|\frac{\Delta X}{X}\right| \cdot \left(\frac{x_1}{x_2}\right)$$

and maximum relative error is given by

$$\left|\frac{\Delta X}{X}\right| \leq \left|\frac{\Delta x_1}{x_1}\right| + \left|\frac{\Delta x_2}{x_2}\right|$$

ILLUSTRATIVE EXAMPLES

Example 1. If $\sqrt{29} = 5.385$ and $\sqrt{\pi} = 1.772$ are correct to four significant digits, find the relative error in their sum and difference.

Solution. Since numbers are 5.385 and 1.772 correct to three decimal places.

∴ Maximum absolute error in each case is

$$\frac{10^{-3}}{2} = 0.0005$$

∴ $\Delta x_1 = \Delta x_2 = 0.0005$

Relative error in their sum

$$\left|\frac{\Delta X}{X}\right| \leq \left|\frac{\Delta x_1}{X}\right| + \left|\frac{\Delta x_2}{X}\right| ; \text{ where } X = x_1 + x_2 = 7.157$$

$$\leq \left|\frac{0.0005}{7.157}\right| + \left|\frac{0.0005}{7.157}\right| < 1.397 \times 10^{-4}$$

and relative error in their difference

$$\left|\frac{\Delta X}{X}\right| \leq \left|\frac{\Delta x_1}{X}\right| + \left|\frac{\Delta x_2}{X}\right| ; \text{ where } X = x_1 - x_2 = 3.613$$

$$\leq \left|\frac{0.0005}{3.613}\right| + \left|\frac{0.0005}{3.613}\right| < 2.767 \times 10^{-4}$$

Example 2. If $u = xy^2z^3$ and errors in x, y, z are 0.3, 0.11, 0.016 respectively at $x = 37.1$, $y = 9.87$, $z = 6.052$. Find the maximum absolute error and percentage error in evaluating u.

Solution. We know that,

$$\Delta u = \frac{\partial u}{\partial x}\Delta x + \frac{\partial u}{\partial y}\Delta y + \frac{\partial u}{\partial z}\Delta z$$

$$= y^2 z^3 \Delta x + 2xyz^3 \Delta y + 3xy^2 z^2 \Delta z$$

Now, maximum absolute error
$$(\Delta u)_{max} = |y^2 z^3 \Delta x| + |2xyz^3 \Delta y| + |3xy^2 z^2 \Delta z|$$
$$= |(9.87)^2 (6.052)^3 (0.3)| + |2(37.1)(9.87)(6.052)^3 (0.11)|$$
$$+ |3(37.1)(9.87)^2 (6.052)^2 (0.016)|$$
$$= 6478.17 + 17857.08 + 6354.00 = 30689.25 = 3.0689 \times 10^4.$$

Now, maximum percentage error
$$= \frac{3.0689 \times 10^4}{xy^2 z^3} \times 100 = \frac{3.0689 \times 10^4}{801133.6486} \times 100 = 3.83\%.$$

Example 3. Find the percentage error in the time period $T = 2\pi \sqrt{\frac{l}{g}}$ for $l = 1.5$ m if the error in the measurement of l is 0.01.

Solution. Given, $T = 2\pi \sqrt{\frac{l}{g}}$

$\Rightarrow \qquad \log T = \log 2 + \log \pi + \frac{1}{2} \log l - \frac{1}{2} \log g$

$\Rightarrow \qquad \frac{\delta T}{T} = 0 + 0 + \frac{1}{2} \frac{\delta l}{l} - 0$

$\Rightarrow \qquad \frac{\delta T}{T} = \frac{1}{2} \frac{\delta l}{l}$

$\Rightarrow \qquad \frac{\delta T}{T} \times 100 = \frac{1}{2} \frac{\delta l}{l} \times 100$

$\Rightarrow \qquad \frac{\delta T}{T} \times 100 = \frac{1}{2} \times \frac{0.01}{1.5} \times 100 = 0.33\%.$

Example 4. In a $\triangle ABC$, $b = 4.5$ cm, $c = 8$ cm, $\angle C = 90°$. Find the possible error in the computed value of B, if the errors in the measurement of b and c are 1 mm and 2 mm respectively.

Solution. Given, $a = 4.5$ cm, $b = 8$ cm, $\angle C = 90°$, $\delta b = 0.1$, $\delta c = 0.2$

In $\triangle ABC$, we have
$$\tan B = \frac{b}{c}$$

$\Rightarrow \qquad B = \tan^{-1}\left(\frac{b}{c}\right)$

$\Rightarrow \qquad \delta B = \frac{1}{1+\frac{b^2}{c^2}} \cdot \frac{1}{c} \cdot \delta b + \frac{1}{1+\frac{b^2}{c^2}} \cdot \left(-\frac{b}{c^2}\right) \cdot \delta c$

$\Rightarrow \qquad \delta B = \frac{c}{c^2+b^2} \cdot \delta b - \frac{b}{c^2+b^2} \cdot \delta c$

$$\Rightarrow \quad |\delta B| = \left|\frac{c}{c^2+b^2}\cdot \delta b\right| + \left|\frac{b}{c^2+b^2}\cdot \delta c\right|$$

$$= \left|\frac{8}{8^2+4.5^2}\cdot(0.1)\right| + \left|\frac{4.5}{8^2+4.5^2}\cdot(0.2)\right|$$

$$= 0.00949 + 0.01068 = 0.02017$$

$$\Rightarrow \quad \delta B = 0.02017 \text{ radians.}$$

Example 5. If $r = 3h(h^6 - 2)$, find the percentage error in r at $h = 2.1$, if the error in h is 6.5%.

Solution. We have, $\quad r = 3h(h^6 - 2)$

$$\Rightarrow \quad r = 3h^7 - 6h$$

$$\delta r = (21h^6 - 6)\delta h$$

$$\frac{\delta r}{r}\times 100 = \left(\frac{21h^6 - 6}{3h^7 - 6h}\right)\delta h \times 100$$

$$= \left(\frac{21h^6 - 6}{3h^6 - 6}\right)\left(\frac{\delta h}{h}\times 100\right)$$

$$= \left(\frac{21(2.1)^6 - 6}{3(2.1)^6 - 6}\right)(6.5)$$

$$= \frac{1795.088}{251.298}\times 6.5 = 46.43\%$$

EXERCISE 1.1

1. If $\frac{1}{3}$ is approximated as 0.333, find the absolute and percentage error.
2. Round off the numbers 865250 and 37.46235 to four significant figures and compute the absolute, relative and percentage error.
3. Find the maximum absolute error in the value of $a + b + c + d$, if $a = 10.00 \pm 0.02$, $b = 0.0495 \pm 0.001$, $c = 12391 \pm 3.55$, $d = 31250 \pm 101$.
4. Find the absolute error of the number 2.32, given the percentage error in it is 0.7%.
5. Each edge of a cube measured to within 0.02 cm turned out to be equal to 8 cm. Find the absolute and percentage errors in measuring the volume of the cube.
6. If $u = \dfrac{4x^2 y^3}{z^4}$ and errors in x, y, z are 0.001 at $x = y = z = 1$. Find the maximum relative error.

7. If $u = 10x^3y^2z^2$ and errors in x, y, z are 0.03, 0.01, 0.02 at $x = 3, y = 1, z = 2$. Calculate absolute error and percentage error in evaluating u.
8. The approximate value of $\sqrt[3]{51}$ is 3.708, find absolute and relative errors.
9. In a $\triangle ABC$, $a = 30$ cm, $b = 80$ cm, $\angle B = 90°$. Find the maximum error in the computed value of A, if the possible errors in a and b are $\frac{1}{3}\%$ and $\frac{1}{4}\%$ respectively.
10. Find the percentage error if 0.005988 is approximated to three significant digits.

ANSWERS

1. 0.00033, 0.099
2. 50, 0.0000671, 0.00671 ; 0.00235, 0.0000627, 0.00627
3. 104.571   4. 0.016
5. 3.84 sq.cm.; 0.75%   6. 0.009
7. 75.6, 7%   8. 9.8×10^{-6}, 2.64×10^{-7}
9. 0.00235 radians   10. −0.033%

1.7 FLOATING POINT REPRESENTATION OF NUMBERS

The memory of a digital computer is divided into separate cells called word or location and each word can store a finite number of digits. The numbers in computer can be stored in two ways:

(i) Fixed point mode   (ii) Floating point mode

Let us assume that a hypothetical computer has a memory in which each word can store 6 digits and has provision to store one or more signs. One method of representing real numbers in such a computer would be to assume a 'fixed position for the decimal point' and store all the numbers after appropriate shifting (if necessary) with an assumed decimal point.

Example 1. Let 3978.55 be the number to be stored, then one memory word or location is

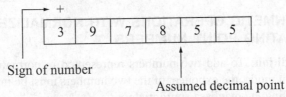

In such a convention, the maximum and minimum numbers in magnitude that may be stored are 9999.99 and 0000.01 respectively. This range is

inadequate in practice and hence a different convention for representing real numbers is adopted. This convention aims to preserve the maximum number of significant digits in a number and also increase the range of values of real numbers stored. This representation is called the normalized floating point mode of representing and storing real numbers.

In this mode, a real number is expressed as a combination of a mantissa and an exponent. The mantissa is the fractional part and is made < 1 or ≥ 0.1 and exponent is the integral part i.e., it is the power of 10 which multiplies the mantissa.

Example 2. The number 12.36×10^5 in normalized floating point mode is represented as 0.1236E5 (where E5 represents 10^5). Here mantissa is 0.1236 and exponent is 5.

The number is stored in a memory location as:

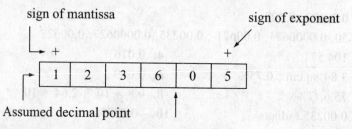

Example 3. The number 0.00004917 (0.4917×10^{-4}) in normalized floating point mode is represented as 0.4917E – 4 (where E – 4 represents 10^{-4}). Here mantissa is 0.4917 and exponent is –4.

The number is stored in a memory location as:

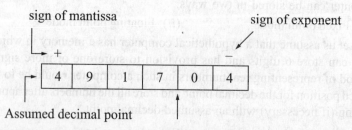

In such a convention, the maximum and minimum numbers in magnitude that may be stored are 0.9999×10^{99} and 0.1000×10^{-99} respectively.

1.8 ARITHMETIC OPERATIONS WITH NORMALIZED FLOATING POINT NUMBERS

(i) **Addition:** To add two numbers represented in normalized floating point mode, the exponent of the two numbers must be made equal by shifting the mantissa appropriately.

Remark: If the exponent is greater than +99, the condition is called overflow condition and if the exponent is less than −99, the condition is called underflow condition.

   (*ii*) **Subtraction:** Like addition, the subtraction operation can be performed on normalized floating point numbers if the exponents of the numbers are equal. If the exponents are not equal then they can be made equal by shifting the mantissa appropriately.

   (*iii*) **Multiplication:** In multiplication of two numbers in normalized floating point mode, the mantissas are multiplied and the exponents are added. After the multiplication of the mantissas, the resulting mantissa is normalized as in the case of addition and subtraction operations and the exponent is adjusted appropriately.

   (*iv*) **Division:** In division, the mantissa of the numerator is divided by the mantissa of the denominator and the denominator exponent is subtracted from the numerator exponent.

ILLUSTRATIVE EXAMPLES

Example 1. Add: (*i*) 0.3592E4 and 0.2213E4

              (*ii*) 0.5291E7 and 0.4860E7

Solution: (*i*) Here, exponents are same, hence the mantissa of the two numbers are added.

$$\begin{array}{r} 0.3592\,E4 \\ +\ 0.2213\,E4 \\ \hline 0.5805\,E4 \end{array}$$

(*ii*) Here, also the exponents are same

$$\begin{array}{r} 0.5291\,E7 \\ +\ 0.4860\,E7 \\ \hline 1.0151\,E7 \end{array}$$

Hence the required sum is 1.0151E7 = 0.1015E8.

Example 2. Add: (*i*) 0.5111E − 3 and 0.0171E − 4

              (*ii*) 0.6100E2 and 0.3965E4

Solution. (*i*) Here, exponents are not equal, so we make them equal

      0.0171E − 4 = 0.0017E − 3

$$\begin{array}{r} 0.5111\,E-3 \\ +\ 0.0017\,E-3 \\ \hline 0.5128\,E-3 \end{array}$$

(*ii*) Here, 0.6100E2 = 0.0061E4

$$0.0061\text{E}4$$
$$+\ 0.3965\text{E}4$$
$$\overline{0.4026\text{E}4}$$

Example 3. Add: (i) 0.6666E99 and 0.4444E99
(ii) 0.3125E – 99 and 0.7592E – 99

Solution. (i)
$$0.6666\text{E}99$$
$$+\ 0.4444\text{E}99$$
$$\overline{1.1110\text{E}99}$$

Here, the required sum is 1.1110E99, which is overflow condition.

(ii)
$$0.3125\text{E}-99$$
$$+\ 0.7592\text{E}-99$$
$$\overline{1.0717\text{E}-99}$$

Here, the required sum is 1.0717E – 99 = 0.1071E – 98.

Example 4. Subtract:
(i) 0.2918E7 from 0.4367E7
(ii) 0.8317E – 4 from 0.6568E – 3
(iii) 0.8363E – 99 from 0.8399E – 99
(iv) 0.3121E5 from 0.0042E7

Solution. (i)
$$0.4347\text{E}7$$
$$-\ 0.2918\text{E}7$$
$$\overline{0.1429\text{E}7}$$

(ii) Here, 0.8317E – 4 = 0.0831E – 3
$$0.6568\text{E}-3$$
$$-\ 0.0831\text{E}-3$$
$$\overline{0.5737\text{E}-3}$$

(iii)
$$0.8399\text{E}-99$$
$$-\ 0.8363\text{E}-99$$
$$\overline{0.0036\text{E}-99}$$

Here, the result is 0.0036E – 99 = 0.3600E – 101, which is underflow condition.

(iv) Here, 0.0042E7 = 0.4200E5
$$0.4200\text{E}5$$
$$-\ 0.3121\text{E}5$$
$$\overline{0.1079\text{E}5}$$

Example 5. Subtract:
  (i) $0.6353E-5$ from $0.9432E-4$
  (ii) $0.5102E57$ from $0.4900E58$

Solution. (i) Here, $0.6353E-5 = 0.0635E-4$

$$\begin{array}{r} 0.9432E-4 \\ -0.0635E-4 \\ \hline 0.8797E-4 \end{array}$$

(ii) Here, $0.5102E57 = 0.0510E58$

$$\begin{array}{r} 0.4900E58 \\ -0.0510E58 \\ \hline 0.4390E58 \end{array}$$

Example 6. Multiply:
  (i) $0.6644E15$ by $0.2311E13$ (ii) $0.1234E75$ by $0.1111E26$
  (iii) $0.2970E19$ by $0.0099E9$

Solution: (i)

$$\begin{array}{r} 0.6644E15 \\ \times\ 0.2311E13 \\ \hline 0.1535E28 \end{array}$$

(ii)

$$\begin{array}{r} 0.1234E75 \\ \times\ 0.1111E26 \\ \hline 0.0137E101 \end{array}$$

Hence, the result is $0.0137E101 = 0.1370E100$, which is overflow condition.

(iii) Here, $0.0099E9 = 0.9900E7$

$$\begin{array}{r} 0.2970E19 \\ \times\ 0.9900E7 \\ \hline 0.2940E26 \end{array}$$

Example 7. Divide:
  (i) $0.8769E8$ by $0.3769E5$ (ii) $0.9998E-6$ by $0.1000E96$
  (iii) 0.6051×10^{22} by 0.0561×10^{19}

Solution: (i)

$$\begin{array}{r} 0.8769E8 \\ \div\ 0.3769E5 \\ \hline 2.3266E3 \end{array}$$

Hence, the result is $2.3266E3 = 0.2326E4$

14 *Numerical and Statistical Techniques*

(ii)
$$0.9998\text{E}-6$$
$$\div 0.1000\text{E}96$$
$$\overline{9.998\text{E}-102}$$

Hence, the result is $9.998\text{E} - 102 = 0.9998\text{E} - 101$, which is underflow condition.

(iii) Here, $0.6051 \times 10^{22} = 0.6051\text{E}22$ and $0.0561 \times 10^{19} = 0.5610\text{E}18$

$$0.6051\text{E}22$$
$$\div 0.5610\text{E}18$$
$$\overline{1.0786\text{E}4}$$

Hence, the result is $1.0786\text{E}4 = 0.1078\text{E}5$.

Example 8. If $x = 0.5665\text{E}1$, $y = 0.5556\text{E} - 1$, $z = 0.5644\text{E}1$, prove that $(x + y) - z \neq (x - z) + y$. (I.U. 2008-09)

Solution. We have, $x + y = 0.5665\text{E}1 + 0.5556\text{E} - 1$
$$= 0.5665\text{E}1 + 0.0055\text{E}1$$
$$= 0.5720\text{E}1$$
$(x + y) - z = 0.5720\text{E}1 - 0.5644\text{E}1$
$$= 0.0076\text{E}1 = 0.7600\text{E} - 1$$
$x - z = 0.5665\text{E}1 - 0.5644\text{E}1$
$$= 0.0021\text{E}1 = 0.2100\text{E} - 1$$
$(x - z) + y = 0.2100\text{E} - 1 + 0.5556\text{E} - 1$
$$= 0.7656\text{E} - 1$$

Hence, $(x + y) - z \neq (x - z) + y$.

Example.9. If $x = 0.1399\text{E} - 1$, find whether $(x + 1)^2$ and $x^2 + 2x + 1$ are equal or not.

Solution. We have, $x + 1 = 0.1399\text{E} - 1 + 0.1000\text{E}1$ $\quad [\because 1 = 0.1000\text{E}1]$
$$= 0.0013\text{E}1 + 0.1000\text{E}1$$
$$= 0.1013\text{E}1$$
$(x + 1)^2 = (x + 1) \times (x + 1)$
$$= 0.1013\text{E}1 \times 0.1013\text{E}1$$
$$= 0.0102\text{E}2 = 0.1020\text{E}1$$
$x^2 = x \times x = 0.1399\text{E} - 1 \times 0.1399\text{E} - 1$
$$= 0.01957\text{E} - 2 = 0.1957\text{E} - 3$$
$2x = 2 \times x = 0.2000\text{E}1 \times 0.1399\text{E} - 1$ $\quad [\because 2 = 0.2000\text{E}1]$
$$= 0.02798\text{E}0 = 0.2798\text{E} - 1$$
$x^2 + 2x + 1 = 0.1957\text{E} - 3 + 0.2798\text{E} - 1 + 0.1000\text{E}1$
$$= 0.0000\text{E}1 + 0.0027\text{E}1 + 0.1000\text{E}1 = 0.1027\text{E}1$$

Hence, $(x + 1)^2 \neq x^2 + 2x + 1$.

Error and Computer Arithmetic

Example 10. If $x = 0.4845$ and $y = 0.4800$, calculate the value of $\dfrac{x^2 - y^2}{x+y}$ using normalized floating point arithmetic. Compare with the value of $x - y$. Indicate the error in the former.

Solution.
$$x + y = 0.4845E0 + 0.4800E0 = 0.9645E0$$
$$x^2 = x \times x = (0.4845E0) \times (0.4845E0) = 0.2347E0$$
$$y^2 = y \times y = (0.4800E0) \times (0.4800E0) = 0.2304E0$$
$$x^2 - y^2 = 0.2347E0 - 0.2304E0 = 0.0043E0 = 0.4300E - 2$$
$$\Rightarrow \quad \dfrac{x^2 - y^2}{x + y} = \dfrac{0.4300E - 2}{0.9645E0} = 0.4458E - 2$$

Also, $\quad x - y = (0.4845E0 - 0.4800E0 = 0.0045E0 = 0.4500E - 2$

$$\text{Error} = \dfrac{(0.4500E - 2) - (0.4458E - 2)}{(0.4500E - 2)}$$
$$= 0.0093 \times 100 = 0.93\%$$

Example 11. Find the solution of the equation $x^2 - 8x + 5 = 0$ using floating point arithmetic with 4 digit mantissa.

Solution. $\quad x^2 - 8x + 5 = 0$
$$\Rightarrow \quad x = \dfrac{8 \pm \sqrt{8^2 - 4 \times 1 \times 5}}{2 \times 1} = \dfrac{8 \pm \sqrt{64 - 20}}{2}$$

Now, $\quad 64 = 0.6400E2$, $20 = 0.2000E2$ and $2 = 0.2000E1$

$\therefore$ $64 - 20 = 0.6400E2 - 0.2000E1 = 0.6400E2 - 0.0200E2 = 0.6200E2$

Hence, $\quad \sqrt{64 - 20} = \sqrt{0.6200E2} = 0.7874E1$

Also, $\qquad 8 = 0.8000E1$

Also,

$\therefore$ Roots are $\left(\dfrac{0.8000E1 + 0.7874E1}{0.2000E1} \right)$ and $\left(\dfrac{0.8000E1 - 0.7874E1}{0.2000E1} \right)$

i.e., $\qquad \left(\dfrac{1.5874E1}{0.2000E1} \right)$ and $\left(\dfrac{0.0126E1}{0.2000E1} \right)$

i.e., $\qquad \left(\dfrac{0.1587E2}{0.2000E1} \right)$ and $\left(\dfrac{0.1260E0}{0.2000E1} \right)$

i.e., $\qquad 0.7935E1$ and $0.6300E0$

EXERCISE 1.2

1. In normalized floating point mode, carry out the following operations:
   (i) $(0.4546E3) + (0.5454E8)$
   (ii) $(0.1111E74) \times (0.2000E80)$
   (iii) $(0.5452E - 99) - (0.5424E - 99)$
   (iv) $(0.1000E5) \div (0.9999E3)$

2. In normalized floating point mode, carry out the following operations:
   (i) $(0.5457E - 3) - (0.9432E - 4)$
   (ii) $(0.7777E8) \div (0.1400E4)$
   (iii) $(0.7259E11) + (0.9361E11)$
   (iv) $(0.4455E6) \times (0.2131E7)$

3. If $x = 0.5555E1, y = 0.4545E1, z = 0.4535E1$, prove that $x(y - z) \neq xy - xz$.

4. Calculate the value of $\sqrt{99} - \sqrt{88}$ using floating point arithmetic.

5. Compute the average value of the numbers 3.795 and 5.154 using the floating point arithmetic. Also compare the result by $a + \left(\dfrac{b-a}{2}\right)$.

ANSWERS

1. (i) 0.5454E8
   (ii) 0.2222E153, overflow condition
   (iii) 0.2800E – 101, underflow condition
   (iv) 0.1000E2

2. (i) 0.4514E – 3 (ii) 0.5555E5
   (iii) 0.1662E12 (iv) 0.9495E12

4. 0.5690E0 5. Both are equal.

Chapter 2

Solution of Algebraic and Transcendental Equations

2.1 SOLUTION OF ALGEBRAIC AND TRANSCENDENTAL EQUATIONS

Consider the equation of the form $f(x) = 0$. If $f(x)$ is an algebraic polynomial of degree two, three or four, we have formulae available for solving them. But, if $f(x)$ is an algebraic polynomial of higher degree or a transcendental function (i.e., a function involving trigonometric function) like $a + b \cos x + c \log x + de^x$, we do not have formulae to find the exact solution.

Here, we will discuss some methods to solve these types of equations.

Corollary: If $f(x)$ is continuous in the interval (a, b) and if $f(a)$ and $f(b)$ are of opposite signs, then the equation $f(x) = 0$ will have at least one real root between a and b.

Note: If $|f(a)| < |f(b)|$, then root lies near to a.

2.2 BISECTION METHOD OR BOLZANO METHOD

This method is also called interval halving method.

Suppose, we have to find the solution of the equation $f(x) = 0$, where $f(x)$ is continuous in the interval (a, b) and may be algebraic or transcendental. Let $f(a)$ be −ve and $f(b)$ be +ve, then at least one real root exist between a and b. We assume $x_1 = \dfrac{a+b}{2}$ as the first approximation to the root.

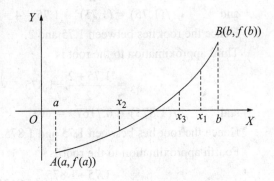

Now calculate $f(x_1)$. If $f(x_1)$ is +ve, the root lies between a and x_1 and if $f(x_1)$ is –ve, the root lies between x_1 and b. Suppose $f(x_1)$ is +ve (as shown in the figure), then the root lies in the interval (a, x_1) and let the second approximation to root be $x_2 = \dfrac{a + x_1}{2}$. Calculate $f(x_2)$, let $f(x_2)$ is –ve (as shown in the figure), then the root lies between x_1 and x_2 and let the third approximation to root be $x_3 = \dfrac{x_1 + x_2}{2}$. Calculate $f(x_3)$, let $f(x_3)$ is +ve, (as shown in the figure), then the root lies between x_2 and x_3 and let the fourth approximation to root be $x_4 = \dfrac{x_2 + x_3}{2}$ and so on. We stop the process when the root start repeating and the root obtained is the exact root.

Note: To find a negative root of the equation $f(x) = 0$, find a positive root of the equation $f(-x) = 0$ and then place negative sign before it.

ILLUSTRATIVE EXAMPLES

Example 1. Find the positive root of the equation $x^3 - x - 4 = 0$ by bisection method.

Solution. Let $\qquad f(x) = x^3 - x - 4$

Here, $\qquad f(0) = -4 = $ –ve; $f(1) = -4 = $ –ve; $f(2) = 2 = $ +ve

Hence a real root lies between 1 and 2.

First approximation to the root is

$$x_1 = \frac{1+2}{2} = 1.5$$

and $\qquad f(1.5) = (1.5)^3 - 1.5 - 4 = -2.125 = $ –ve

Hence the root lies between 1.5 and 2.

Second approximation to the root is

$$x_2 = \frac{1.5 + 2}{2} = 1.75$$

and $\qquad f(1.75) = (1.75)^3 - 1.75 - 4 = -0.39062 = $ –ve

Hence the root lies between 1.75 and 2.

Third approximation to the root is

$$x_3 = \frac{1.75 + 2}{2} = 1.875$$

and $\qquad f(1.875) = 0.71679 = $ +ve

Hence the root lies between 1.75 and 1.875.

Fourth approximation to the root is

$$x_4 = \frac{1.75 + 1.875}{2} = 1.8125$$

and $f(1.8125) = 0.14184 = +\text{ve}$
Hence the root lies between 1.75 and 1.8125.
Fifth approximation to the root is

$$x_5 = \frac{1.75 + 1.8125}{2} = 1.78125$$

and $f(1.78125) = -0.12960 = -\text{ve}$
Hence the root lies between 1.78125 and 1.8125.
Sixth approximation to the root is

$$x_6 = \frac{1.78125 + 1.8125}{2} = 1.79687$$

and $f(1.79687) = 0.00475 = +\text{ve}$
Hence the root lies between 1.78125 and 1.79687.
Seventh approximation to the root is

$$x_7 = \frac{1.78125 + 1.79687}{2} = 1.78906$$

and $f(1.78906) = -0.06275 = -\text{ve}$
Hence the root lies between 1.78906 and 1.79687.
Eighth approximation to the root is

$$x_8 = \frac{1.78906 + 1.79687}{2} = 1.79296$$

and $f(1.79296) = -0.02913 = -\text{ve}$
Hence the root lies between 1.79296 and 1.79687.
Nineth approximation to the root is

$$x_9 = \frac{1.79296 + 1.79687}{2} = 1.79491$$

and $f(1.79491) = -0.012245 = -\text{ve}$
Hence the root lies between 1.79491 and 1.79687.
Tenth approximation to the root is

$$x_{10} = \frac{1.79491 + 1.79687}{2} = 1.79589$$

and $f(1.79589) = -0.003748 = -\text{ve}$
Hence the root lies between 1.79589 and 1.79687.
Eleventh approximation to the root is

$$x_{11} = \frac{1.79589 + 1.79687}{2} = 1.79638$$

and $f(1.79638) = -0.0005043 = -\text{ve}$
Hence the root lies between 1.79638 and 1.79687.
Twelveth approximation to the root is

$$x_{12} = \frac{1.79638 + 1.79687}{2} = 1.796625$$

Hence, the approximate root is 1.796.

Example 2. Find the positive root of the equation $x - \cos x = 0$ by bisection method.

Solution. Let $\quad f(x) = x - \cos x$

Here, $\quad f(0) = -1 = -\text{ve}; f(1) = 0.45970 = +\text{ve}.$

Also $\quad f(0.5) = -0.37758 = -\text{ve}$

Hence a real root lies between 0.5 and 1.

First approximation to the root is

$$x_1 = \frac{0.5 + 1}{2} = 0.75$$

and $\quad f(0.75) = +\text{ve}$

Hence the root lies between 0.5 and 0.75.

Second approximation to the root is

$$x_2 = \frac{0.5 + 0.75}{2} = 0.625$$

and $\quad f(0.625) = -\text{ve}$

Hence the root lies between 0.625 and 0.75.

Third approximation to the root is

$$x_3 = \frac{0.625 + 0.75}{2} = 0.6875$$

and $\quad f(0.6875) = -\text{ve}$

Hence the root lies between 0.6875 and 0.75.

Fourth approximation to the root is

$$x_4 = \frac{0.6875 + 0.75}{2} = 0.71875$$

and $\quad f(0.71875) = -\text{ve}$

Hence the root lies between 0.71875 and 0.75.

Fifth approximation to the root is

$$x_5 = \frac{0.71875 + 0.75}{2} = 0.73438$$

and $\quad f(0.73438) = -\text{ve}$

Hence the root lies between 0.73438 and 0.75.

Sixth approximation to the root is

$$x_6 = \frac{0.73438 + 0.75}{2} = 0.74219$$

and $f(0.74219) = +ve$
Hence the root lies between 0.73438 and 0.74219.
Seventh approximation to the root is
$$x_7 = \frac{0.73438 + 0.74219}{2} = 0.73829$$
and $f(0.73829) = -ve$
Hence the root lies between 0.73829 and 0.74219.
Eighth approximation to the root is
$$x_8 = \frac{0.73829 + 0.74219}{2} = 0.7402$$
and $f(0.7402) = +ve$
Hence the root lies between 0.73829 and 0.7402.
Nineth approximation to the root is
$$x_9 = \frac{0.73829 + 0.7402}{2} = 0.73925$$
and $f(0.73925) = +ve$
Hence the root lies between 0.73829 and 0.73925.
Tenth approximation to the root is
$$x_{10} = \frac{0.73829 + 0.73925}{2} = 0.7388$$
Hence, the root of the given equation is 0.7388.

Example 3. Find the positive root of the equation $x \log_{10} x - 1.2 = 0$ by bisection method.

Solution. Let $f(x) = x \log_{10} x - 1.2$
Here, $f(1) = \log_{10} 1 - 1.2 = -1.2 = -ve$
$f(2) = 2 \log_{10} 2 - 1.2 = -0.598 = -ve$
$f(3) = 3 \log_{10} 3 - 1.2 = 0.2313 = +ve$
Hence the root lies between 2 and 3.
First approximation to the root is
$$x_1 = \frac{2+3}{2} = 2.5$$
and $f(2.5) = 2.5 \log_{10} 2.5 - 1.2 = -0.2052 = -ve$
Hence the root lies between 2.5 and 3.
Second approximation to the root is
$$x_2 = \frac{2.5 + 3}{2} = 2.75$$
and $f(2.75) = 2.75 \log_{10} 2.75 - 1.2 = 0.0081 = +ve$

Hence the root lies between 2.5 and 2.75.
Third approximation to the root is

$$x_3 = \frac{2.5 + 2.75}{2} = 2.625$$

and $f(2.625) = -0.0999 = -ve$

Hence the root lies between 2.625 and 2.75.
Fourth approximation to the root is

$$x_4 = \frac{2.625 + 2.75}{2} = 2.6875$$

and $f(2.6875) = -0.0463 = -ve$

Hence the root lies between 2.6875 and 2.75.
Fifth approximation to the root is

$$x_5 = \frac{2.6875 + 2.75}{2} = 2.71875$$

and $f(2.71875) = -0.019 = -ve$

Hence the root lies between 2.71875 and 2.75.
Sixth approximation to the root is

$$x_6 = \frac{2.71875 + 2.75}{2} = 2.73437$$

and $f(2.73437) = -0.00546 = -ve$

Hence the root lies between 2.73437 and 2.75.
Seventh approximation to the root is

$$x_7 = \frac{2.73437 + 2.75}{2} = 2.74218$$

and $f(2.74218) = +ve$

Hence the root lies between 2.73437 and 2.74218.
Eighth approximation to the root is

$$x_8 = \frac{2.73437 + 2.74218}{2} = 2.73828$$

and $f(2.73828) = -ve$

Hence the root lies between 2.73828 and 2.74218.

$$x_9 = \frac{2.73828 + 2.74218}{2} = 2.74023$$

and $f(2.74023) = -ve$

Hence the root lies between 2.74023 and 2.74218.
Tenth approximation to the root is

$$x_{10} = \frac{2.74023 + 2.74218}{2} = 2.74121$$

and $f(2.74121) = +ve$
Hence the root lies between 2.74023 and 2.74121.

$$x_{11} = \frac{2.74023 + 2.74121}{2} = 2.74072$$

and $f(2.74072) = +ve$
Hence the root lies between 2.74023 and 2.74072.

$$x_{12} = \frac{2.74023 + 2.74072}{2} = 2.74048$$

Since x_{11} and x_{12} are same upto three decimal places, hence the approximate root of the given equation is 2.740.

Example 4. Find a positive real root of the equation $8x^3 - 2x = 1$ in the interval (0,1) using bisection method upto two decimal places.

Solution. Let $f(x) = 8x^3 - 2x - 1 = 0$
Here, $f(0) = 0 - 0 - 1 = -1 = -ve$
          $f(1) = 8 - 2 - 1 = 5 = +ve$
Also, $f(0.5) = 8(0.5)^3 - 2(0.5) - 1 = -1 = -ve$
Hence the root lies between 0.5 and 1.
First approximation to the root is

$$x_1 = \frac{0.5 + 1}{2} = 0.75$$

and $f(2.74121) = 8(0.75)^3 - 2(0.75) - 1 = 0.875 = +ve$
Hence the root lies between 0.5 and 0.75.
Second approximation to the root is

$$x_2 = \frac{0.5 + 0.75}{2} = 0.625$$

and $f(0.625) = 8(0.625)^3 - 2(0.625) - 1 = -0.297 = -ve$
Hence the root lies between 0.625 and 0.75.
Third approximation to the root is

$$x_3 = \frac{0.625 + 0.75}{2} = 0.688$$

and $f(0.688) = 0.229 = +ve$
Hence the root lies between 0.625 and 0.688.
Fourth approximation to the root is

$$x_4 = \frac{0.625 + 0.688}{2} = 0.657$$

and $f(0.657) = -0.045 = -ve$
Hence the root lies between 0.657 and 0.688.

Fifth approximation to the root is

$$x_5 = \frac{0.657 + 0.688}{2} = 0.673$$

and $f(0.673) = 1.093 = +\text{ve}$

Hence the root lies between 0.657 and 0.673.

Sixth approximation to the root is

$$x_6 = \frac{0.657 + 0.673}{2} = 0.665$$

and $f(0.665) = 0.023 = +\text{ve}$

Hence the root lies between 0.657 and 0.665.

Seventh approximation to the root is

$$x_7 = \frac{0.657 + 0.665}{2} = 0.661$$

Since x_6 and x_7 are same upto two decimal places, hence the approximate root of the given equation is 0.66.

Example 5. Find a negative root of the equation $x^3 + x + 1 = 0$ using bisection method.

Solution. Let $f(x) = x^3 + x + 1 = 0$

The negative root of $f(x)$ is the positive root of $f(-x) = 0$.

$$f(-x) = -x^3 - x + 1$$

Let $\phi(x) = x^3 + x - 1 = 0$

Here, $\phi(0) = 0^3 + 0 - 1 = -1 = -\text{ve}$

$\phi(1) = 1^3 + 1 - 1 = 1 = +\text{ve}$

Also, $\phi(0.5) = (0.5)^3 + 0.5 - 1 = -0.375 = -\text{ve}$

Hence the root lies between 0.5 and 1.

First approximation to the root is

$$x_1 = \frac{0.5 + 1}{2} = 0.75$$

and $f(0.75) = (0.75)^3 + (0.75) - 1 = 0.17188 = +\text{ve}$

Hence the root lies between 0.5 and 0.75.

Second approximation to the root is

$$x_2 = \frac{0.5 + 0.75}{2} = 0.625$$

and $f(0.625) = (0.625)^3 + (0.625) - 1 = -0.13086 = -\text{ve}$

Hence the root lies between 0.625 and 0.75.

Third approximation to the root is

$$x_3 = \frac{0.625 + 0.75}{2} = 0.688$$

and $f(0.688) = (0.688)^3 + (0.688) - 1 = 0.01366 = +ve$
Hence the root lies between 0.625 and 0.688.
Fourth approximation to the root is
$$x_4 = \frac{0.625 + 0.688}{2} = 0.657$$
and $f(0.657) = (0.657)^3 + (0.657) - 1 = -0.05941 = -ve$
Hence the root lies between 0.657 and 0.688.
Fifth approximation to the root is
$$x_5 = \frac{0.657 + 0.688}{2} = 0.673$$
and $f(0.673) = (0.673)^3 + (0.673) - 1 = -0.02218 = -ve$
Hence the root lies between 0.673 and 0.688.
Sixth approximation to the root is
$$x_6 = \frac{0.688 + 0.673}{2} = 0.6805$$
and $f(0.6805) = (0.6805)^3 + (0.6805) - 1 = -0.00437 = -ve$
Hence the root lies between 0.688 and 0.6805.
Seventh approximation to the root is
$$x_7 = \frac{0.688 + 0.6805}{2} = 0.68425$$
and $f(0.68425) = (0.68425)^3 + (0.68425) - 1 = 0.00461 = +ve$
Hence the root lies between 0.6805 and 0.68425.
Eighth approximation to the root is
$$x_8 = \frac{0.68425 + 0.6805}{2} = 0.682375$$
and $f(0.682375) = (0.682375)^3 + (0.682375) - 1 = 0.000113 = +ve$
Hence the root lies between 0.6805 and 0.682375.
Nineth approximation to the root is
$$x_9 = \frac{0.682375 + 0.6805}{2} = 0.681438$$
and $f(0.681438) = (0.681438)^3 + (0.681438) - 1 = -0.00213 = -ve$
Hence the root lies between 0.681438 and 0.682375.
Tenth approximation to the root is
$$x_{10} = \frac{0.682375 + 0.681438}{2} = 0.681907$$
and $f(0.681907) = (0.681907)^3 + (0.681907) - 1 = -0.001008 = -ve$
Hence the root lies between 0.681907 and 0.682375.

Eleventh approximation to the root is

$$x_{11} = \frac{0.682375 + 0.681907}{2} = 0.682141$$

and $f(0.682141) = (0.682141)^3 + (0.682141) - 1 = -0.000448 = -\text{ve}$

Hence the root lies between 0.682141 and 0.682375.

Twelfth approximation to the root is

$$x_{12} = \frac{0.682375 + 0.682141}{2} = 0.682258$$

The approximate root of the equation $\phi(x) = 0$ is 0.682.

Hence, the root of the given equation is -0.682.

EXERCISE 2.1

Use Bisection method to find a real positive root of the following equations upto three decimal places:

1. $x^3 + x^2 - 1 = 0$
2. $x^3 - 4x = 9$
3. $3x = \sqrt{1 + \sin x}$
4. $3x - \cos x - 1 = 0$
5. $e^x = 3x$
6. $x = \tan x$
7. $xe^x = 1$
8. $x^3 - 18 = 0$
9. Find a positive root of the equation $x^3 - 7x^2 + 14x - 6 = 0$ in the interval (3, 3.5) using bisection method.
10. Find a negative root of the equation $x^3 + x^2 + x + 7 = 0$ using bisection method.

ANSWERS

1. 0.7548
2. 2.687
3. 0.3918
4. 0.666
5. 0.619
6. 4.493
7. 0.567
8. 2.621
9. 3.419
10. -2.105

2.3 ITERATION METHOD OR SUCCESSIVE APPROXIMATION METHOD

Suppose, we have to find the solution of the equation

$$f(x) = 0 \quad \ldots(1)$$

Write the given equation in the form

$$x = \phi(x) \quad \ldots(2)$$

Find an interval (a, b) such that $f(a)$ and $f(b)$ are of opposite signs. Assume $x_0 \in (a, b)$ as the initial approximation to the root.

Now put $x = x_0$ in the R.H.S. of equation (2), we get the first approximation
$$x_1 = \phi(x_0)$$
Again putting $x = x_1$ in the R.H.S. of (2), we get the second approximation
$$x_2 = \phi(x_1)$$
Continuing this process, we get the successive approximations
$$x_3 = \phi(x_2)$$
$$x_4 = \phi(x_3)$$
....

....
$$x_n = \phi(x_{n-1})$$

The sequence of approximate roots $x_1, x_2, x_3, ..., x_n$ converges to the root α of the given equation (1).

2.4 CONDITION FOR CONVERGENCE OF THE ITERATIVE METHOD

Theorem: Let α be the actual root of the equation $f(x) = 0$ and

(i) I be the interval containing the root $x = \alpha$,

(ii) $|\phi'(x)| < 1$ for all x in I,

then the sequence of approximations $x_0, x_2, x_3, ..., x_n$ will converge to the root α, provided the initial approximation is taken in the interval I.

Note: The sufficient condition for the convergence of the iterative method is
$$|\phi'(x)| < 1 \text{ for all } x \text{ in } I.$$

ILLUSTRATIVE EXAMPLES

Example 1. Solve the equation $x^3 - 2x^2 - 4 = 0$ by iteration method correct to three decimal places.

Solution. Let $\quad f(x) = x^3 - 2x^2 - 4 = 0$

Now, $\quad f(1) = 1 - 2 - 4 = -5 = -\text{ve};$

$\quad f(2) = -4 = -\text{ve}; f(3) = 5 = +\text{ve}$

Also $\quad f(2.5) = -\text{ve}$

Hence one real root lies in the interval (2.5, 3).

Now, the given equation can be written as
$$x = (2x^2 + 4)^{\frac{1}{3}} = \phi(x)$$

$\Rightarrow \quad \phi'(x) = \dfrac{4}{3}x(2x^2 + 4)^{\frac{-2}{3}} \Rightarrow |\phi'(x)| = \dfrac{4}{3}x(2x^2 + 4)^{\frac{-2}{3}}$

$\therefore \quad |\phi'(2.5)| < 1$ and $|\phi'(3)| < 1$

We have, $\quad |\phi'(x)| < 1$ in the interval (2.5, 3).

28 *Numerical and Statistical Techniques*

Hence, the iteration method may be applied.

Let $x_0 = 2.5$ be the initial approximation of the root.

Then the successive approximations to the root are

$$x_1 = \phi(x_0) = (2x_0^2 + 4)^{\frac{1}{3}} = \left[2(2.5)^2 + 4\right]^{\frac{1}{3}} = 2.5458$$

$$x_2 = (2x_1^2 + 4)^{\frac{1}{3}} = \left[2(2.5458)^2 + 4\right]^{\frac{1}{3}} = 2.56937$$

$$x_3 = (2x_2^2 + 4)^{\frac{1}{3}} = \left[2(2.56937)^2 + 4\right]^{\frac{1}{3}} = 2.58149$$

$$x_4 = (2x_3^2 + 4)^{\frac{1}{3}} = \left[2(2.58149)^2 + 4\right]^{\frac{1}{3}} = 2.58772$$

$$x_5 = (2x_4^2 + 4)^{\frac{1}{3}} = \left[2(2.58772)^2 + 4\right]^{\frac{1}{3}} = 2.59092$$

$$x_6 = (2x_5^2 + 4)^{\frac{1}{3}} = \left[2(2.59092)^2 + 4\right]^{\frac{1}{3}} = 2.59211$$

$$x_7 = (2x_6^2 + 4)^{\frac{1}{3}} = \left[2(2.59211)^2 + 4\right]^{\frac{1}{3}} = 2.59342$$

$$x_8 = (2x_7^2 + 4)^{\frac{1}{3}} = \left[2(2.59342)^2 + 4\right]^{\frac{1}{3}} = 2.59385$$

Hence the root of the equation correct to three decimal places is 2.593.

Example 2. Use iteration method to find a positive root of the equation $3x - \log_{10} x = 6$ correct to four decimal places.

Solution. Let $f(x) = 3x - \log_{10} x - 6 = 0$

Now, $\qquad f(2) = 3(2) - \log_{10} 2 - 6 = -0.3010 = -\text{ve}$

$\qquad\qquad f(3) = 3(3) - \log_{10} 3 - 6 = 2.5228 = +\text{ve}$

Hence one real root lies in the interval (2,3).

Now, the given equation can be written as

$$x = \frac{1}{3}[\log_{10} x + 6] = \phi(x)$$

$\Rightarrow \qquad\qquad \phi'(x) = \frac{1}{3}\left(\frac{1}{x}\log_{10} e\right)$

∴ $|\phi'(2)| = 0.07238 < 1$ and $|\phi'(3)| = 0.048255 < 1$

$$[\because \log_{10} e = 0.4343]$$

We have, $|\phi'(x)| < 1$ in the interval $(2, 3)$.
Hence, the iteration method may be applied.
Let us take $x_0 = 2.1$
Then the successive approximations to the root are

$$x_1 = \frac{1}{3}[\log_{10} x_0 + 6] = \frac{1}{3}[\log_{10} 2.1 + 6] = 2.1074$$

$$x_2 = \frac{1}{3}[\log_{10} 2.1074 + 6] = 2.1079$$

$$x_3 = \frac{1}{3}[\log_{10} 2.1079 + 6] = 2.10795$$

$$x_4 = \frac{1}{3}[\log_{10} 2.10795 + 6] = 2.107953$$

Here x_3 and x_4 are same upto five decimal places, hence the approximate positive root of the given equation upto five places of decimals is 2.10795.

Example 3. Using iteration method find a real root of the equation $\cos x - 3x + 1 = 0$ correct to three decimal places. **[I.U. 2008-09]**

Solution. Let $f(x) = \cos x - 3x + 1 = 0$
Now, $f(0) = \cos 0 - 3(0) + 1 = 2 = +ve$

$$f\left(\frac{\pi}{2}\right) = \cos\frac{\pi}{2} - 3\left(\frac{\pi}{2}\right) + 1 = -ve$$

Hence one real root lies in the interval $\left(0, \frac{\pi}{2}\right)$

Rewriting the given equation as follows

$$x = \frac{1}{3}(1 + \cos x) = \phi(x)$$

$\Rightarrow$ $\phi'(x) = \frac{1}{3}(-\sin x) = -\frac{\sin x}{3} \Rightarrow |\phi'(x)| = \frac{\sin x}{3}$

∴ $|\phi'(0)| = \frac{\sin 0}{3} = 0 < 1$ and $\left|\phi'\left(\frac{\pi}{2}\right)\right| = \frac{\sin\frac{\pi}{2}}{3} = \frac{1}{3} < 1$

Hence, $|\phi'(x)| < 1$ in the interval $\left(0, \frac{\pi}{2}\right)$

Let $x_0 = 0$ be the initial approximation of the root.
Then the successive approximations to the root are

30 *Numerical and Statistical Techniques*

$$x_1 = \phi(x_0) = \frac{1}{3}(1+\cos x_0) = \frac{1}{3}(1+\cos 0) = \frac{2}{3} = 0.6667$$

$$x_2 = \phi(x_1) = \frac{1}{3}(1+\cos x_1) = \frac{1}{3}(1+\cos 0.6667) = 0.5953$$

$$x_3 = \phi(x_2) = \frac{1}{3}(1+\cos x_2) = \frac{1}{3}(1+\cos 0.5953) = 0.6093$$

$$x_4 = \phi(x_3) = \frac{1}{3}(1+\cos x_3) = \frac{1}{3}(1+\cos 0.6093) = 0.6067$$

$$x_5 = \phi(x_4) = \frac{1}{3}(1+\cos x_4) = \frac{1}{3}(1+\cos 0.6067) = 0.6072$$

$$x_6 = \phi(x_5) = \frac{1}{3}(1+\cos x_5) = \frac{1}{3}(1+\cos 0.6072) = 0.6071$$

Hence the approximate real root of the given equation correct to three places of decimals is 0.607.

Example 4. Find the cube root of 48 correct to four decimal places by using iteration method.

Solution. Let $\quad x = \sqrt[3]{48} = (48)^{\frac{1}{3}} \quad \Rightarrow \quad x^3 = 48$

Let $\quad f(x) = x^3 - 48 = 0$

$\quad f(3) = 3^3 - 48 = 27 - 48 = -21 = -\text{ve}$

and $\quad f(4) = 4^3 - 48 = 64 - 48 = 16 = +\text{ve}$

Hence, the cube root of 48 lies in the interval (3, 4).

Now, $48 - x^3 + 50x = 50x \quad \Rightarrow \quad x = \dfrac{48 - x^3 + 50x}{50}$

Let $\quad \phi(x) = \dfrac{48 - x^3 + 50x}{50}$

$\Rightarrow \quad \phi'(x) = \dfrac{50 - 3x^2}{50} \quad \Rightarrow \quad |\phi'(x)| = \dfrac{50 - 3x^2}{50}$

$\therefore \quad |\phi'(3)| = \dfrac{50-27}{50} = \dfrac{23}{50} < 1 \text{ and } |\phi'(4)| = \dfrac{50-48}{50} = \dfrac{1}{25} < 1$

$\therefore \quad |\phi'(x)| < 1$ in the interval (3, 4).

Let us consider $x_0 = 3.6$

Now, $\quad x_1 = \dfrac{48 - x_0^3 + 50x_0}{50} = \dfrac{48 - (3.6)^3 + 50(3.6)}{50} = 3.62688$

$$x_2 = \dfrac{48 - (3.62688)^3 + 50(3.62688)}{50} = 3.632702$$

$$x_3 = \frac{48-(3.632702)^3 + 50(3.632702)}{50} = 3.633922$$

$$x_4 = \frac{48-(3.633922)^3 + 50(3.633922)}{50} = 3.63418$$

$$x_5 = \frac{48-(3.63418)^3 + 50(3.63418)}{50} = 3.634228$$

$$x_6 = \frac{48-(3.634228)^3 + 50(3.634228)}{50} = 3.634238$$

Hence, the cube root of 48 correct to four decimal places is 3.6342.

EXERCISE 2.2

Solve the following equations using the Iteration method:
1. $2x - \log_{10} x - 7 = 0$
2. $\cos x = xe^x$
3. $x^3 - x^2 - 1 = 0$
4. $x^3 + 2x^2 + 10x - 20 = 0$
5. $x^3 - 9x + 1 = 0$
6. $2 \sin x = x$
7. $xe^x = 2$
8. Find the reciprocal of 41 correct to four decimal places by using iterative method.
9. Find the square root of 24 correct to 3 decimal places by using the iterative formula

$$x_{n+1} = \frac{x_n^2 + 20}{2x_n}$$

10. Using iteration method find a real root of the equation $x = (5-x)^{\frac{1}{3}}$ correct to four decimal places.

ANSWERS
---

1. 3.7892 2. 0.5177 3. 1.466
4. 1.3688 5. 2.94   6. 1.8955
7. 0.853   8. 0.0244 9. 4.472
10. 1.516

2.5 REGULA-FALSI METHOD OR METHOD OF FALSE POSITION

This is the oldest method of finding the real roots of the equation $f(x) = 0$ and is a modification of bisection method.

Consider the equation $f(x) = 0$. Let a and b be two points such that $a < b$ and $f(a)$ and $f(b)$ are of opposite signs. Then the graph of the curve $y = f(x)$ crosses the x-axis at some point between a and b.

The equation of the chord joining the two points $A(a, f(a))$ and $B(b, f(b))$ is

$$y - f(a) = \frac{f(b) - f(a)}{b - a}(x - a)$$

Since the above equation intersects the x-axis between A and B, where $y = 0$, we have

$$-f(a) = \frac{f(b) - f(a)}{b - a}(x - a)$$

$$\Rightarrow \quad x - a = \frac{(b - a)f(a)}{f(a) - f(b)}$$

$$\Rightarrow \quad x = \frac{af(b) - bf(a)}{f(b) - f(a)}$$

Hence $\quad x_1 = \dfrac{af(b) - bf(a)}{f(b) - f(a)} \quad$ or $\quad x_1 = a - \dfrac{(b - a)}{f(b) - f(a)} \cdot f(a)$

which is first approximation to the root.

Now, if $f(a)$ and $f(x_1)$ are of opposite signs, then the root lies between a and x_1.

Hence, the second approximation to the root is

$$x_2 = \frac{af(x_1) - x_1 f(a)}{f(x_1) - f(a)} \quad \text{and so on.}$$

This process is repeated, until we get the root to the desired accuracy.

In general, the iterative formula for successive approximations is given by

$$x_{n+1} = \frac{x_{n-1} f(x_n) - x_n f(x_{n-1})}{f(x_n) - f(x_{n-1})}$$

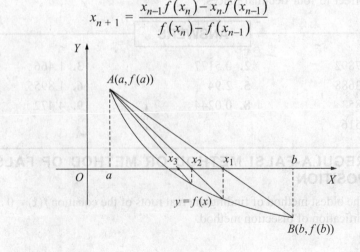

2.6 ORDER AND RATE OF CONVERGENCE OF REGULA-FALSI METHOD

We have, from Regula-Falsi method

$$x_{n+1} = \frac{x_{n-1}f(x_n) - x_n f(x_{n-1})}{f(x_n) - f(x_{n-1})} \qquad \ldots(1)$$

Let α be the root of the equation $f(x) = 0$ such that $f(\alpha) = 0$.

Let e_{n-1}, e_n, e_{n+1}' be the errors in $(n-1)^{th}, n^{th}, (n+1)^{th}$ approximations, then

$$e_{n-1} = x_{n-1} - \alpha \quad \text{or} \quad x_{n-1} = e_{n-1} + \alpha$$
$$e_n = x_n - \alpha \quad \text{or} \quad x_n = e_n + \alpha$$

and
$$e_{n+1} = x_{n+1} - \alpha \quad \text{or} \quad x_{n+1} = e_{n+1} + \alpha$$

Putting these values in equation (1), we have

$$e_{n+1} + \alpha = \frac{(e_{n-1}+\alpha)f(e_n+\alpha) - (e_n+\alpha)f(e_{n-1}+\alpha)}{f(e_n+\alpha) - f(e_{n-1}+\alpha)}$$

$$\Rightarrow \quad e_{n+1} = \frac{(e_{n-1}+\alpha)f(e_n+\alpha) - (e_n+\alpha)f(e_{n-1}+\alpha)}{f(e_n+\alpha) - f(e_{n-1}+\alpha)} - \alpha$$

$$\Rightarrow \quad e_{n+1} = \frac{e_{n-1}f(e_n+\alpha) - e_n f(e_{n-1}+\alpha)}{f(e_n+\alpha) - f(e_{n-1}+\alpha)}$$

Now, expanding $f(\alpha + e_n)$ and $f(\alpha + e_{n-1})$ by Taylor's series, we have

$$e_{n+1} = \frac{e_{n-1}\left[f(\alpha)+e_n f'(\alpha)+\frac{e_n^2}{2!}f''(\alpha)+\ldots\right] - e_n\left[f(\alpha)+e_{n-1}f'(\alpha)+\frac{e_{n-1}^2}{2!}f''(\alpha)+\ldots\right]}{\left[f(\alpha)+e_n f'(\alpha)+\frac{e_n^2}{2!}f''(\alpha)+\ldots\right] - \left[f(\alpha)+e_{n-1}f'(\alpha)+\frac{e_{n-1}^2}{2!}f''(\alpha)+\ldots\right]}$$

$$\Rightarrow \quad e_{n+1} = \frac{(e_{n-1}-e_n)f(\alpha) + \frac{e_{n-1}e_n}{2!}(e_n - e_{n-1})f''(\alpha) + \ldots}{(e_n - e_{n-1})f'(\alpha) + \frac{(e_n - e_{n-1})(e_n + e_{n-1})}{2!}f''(\alpha) + \ldots}$$

$$\Rightarrow \quad e_{n+1} = \frac{\frac{e_{n-1}e_n}{2}f''(\alpha) + \ldots}{f'(\alpha) + \frac{(e_n+e_{n-1})}{2}f''(\alpha) + \ldots} \qquad [\because f(\alpha) = 0]$$

or
$$e_{n+1} = \frac{e_{n-1}e_n f''(\alpha)}{2f'(\alpha)} \qquad \ldots(2)$$

(By neglecting higher powers of e_{n-1} and e_n)

Let $e_{n+1} = c e_n^k$, where c is a constant and $k > 0$.

$$\Rightarrow e_n = c e_{n-1}^k \quad \text{or} \quad e_{n-1} = c^{-\frac{1}{k}} e_n^{\frac{1}{k}}$$

From (2), we get

$$c e_n^k = \frac{c^{-\frac{1}{k}} e_n^{\frac{1}{k}} \cdot c e_n^k f''(\alpha)}{2 f'(\alpha)} = c^{-\frac{1}{k}} e_n^{1+\frac{1}{k}} \cdot \frac{f''(\alpha)}{2 f'(\alpha)}$$

Comparing the two sides, we get

$$k = 1 + \frac{1}{k} \quad \text{and} \quad c = c^{-\frac{1}{k}} \cdot \frac{f''(\alpha)}{2 f'(\alpha)}$$

Now, $k^2 - k - 1 = 0 \Rightarrow k = 1.618, -0.618$

But $k \neq -0.618$, hence $k = 1.618$

Also, $c = c^{-\frac{1}{k}} \cdot \frac{f''(\alpha)}{2 f'(\alpha)} \Rightarrow c^{1+\frac{1}{k}} = \frac{f''(\alpha)}{2 f'(\alpha)}$

$$\Rightarrow c^{1+\frac{1}{1.618}} = \frac{f''(\alpha)}{2 f'(\alpha)} \Rightarrow c^{1.618} = \frac{f''(\alpha)}{2 f'(\alpha)}$$

$$\Rightarrow c = \left[\frac{f''(\alpha)}{2 f'(\alpha)}\right]^{0.618}$$

Hence, order of convergence of Regula-Falsi method is 1.618 and rate of convergence is $\left[\dfrac{f''(\alpha)}{2 f'(\alpha)}\right]^{0.618}$

ILLUSTRATIVE EXAMPLES

Example 1. Using Regula-Falsi method find a real root of the equation $x^3 - x^2 - 2 = 0$ correct to three decimal places.

Solution. Let $f(x) = x^3 - x^2 - 2 = 0$

Now, $f(1) = 1 - 1 - 2 = -2 = -\text{ve}$

and $f(2) = 2^3 - 2^2 - 2 = 2 = +\text{ve}$

Therefore at least one real root lies in the interval (1, 2).

By Regula-Falsi method, we have

$$x_1 = \frac{a f(b) - b f(a)}{f(b) - f(a)}$$

Here $a = 1$, $b = 2$, $f(a) = -2$, and $f(b) = 2$

First approximation:

$$x_1 = \frac{1 \times 2 - 2 \times -2}{2 - (-2)} = 1.5$$

$$f(x_1) = f(1.5) = -0.875 = -\text{ve}$$

Hence the root lies between 1.5 and 2.

Second approximation:
$$x_2 = \frac{1.5 \times 2 - 2 \times -0.875}{2 - (-0.875)} = 1.652$$
$$f(x_2) = f(1.652) = -0.2207 = -\text{ve}$$

Hence the root lies between 1.652 and 2.

Third approximation:
$$x_3 = \frac{1.652 \times 2 - 2 \times -0.2207}{2 - (-0.2207)} = 1.6865$$
$$f(x_3) = f(1.6865) = -0.0474 = -\text{ve}$$

Hence the root lies between 1.6865 and 2.

Fourth approximation:
$$x_4 = \frac{1.6865 \times 2 - 2 \times -0.0474}{2 - (-0.0474)} = 1.6938$$
$$f(x_4) = f(1.6938) = -0.00951 = -\text{ve}$$

Hence the root lies between 1.6938 and 2.

Fifth approximation:
$$x_5 = \frac{1.6938 \times 2 - 2 \times -0.00951}{2 - (-0.00951)} = 1.6953$$
$$f(x_5) = f(1.6953) = -0.00168 = -\text{ve}$$

Hence the root lies between 1.6953 and 2.

Sixth approximation:
$$x_6 = \frac{1.6953 \times 2 - 2 \times -0.00168}{2 - (-0.00168)} = 1.6955$$

Since x_5 and x_6 are equal, hence the required root is 1.695.

Example 2. Find a real root of the equation $x^2 - \log_e x - 12 = 0$ correct to three decimal places.

Solution. Let $\quad f(x) = x^2 - \log_e x - 12 = 0$
Now, $\quad\quad\quad f(3) = 3^2 - \log_e 3 - 12 = -4.0986$
and $\quad\quad\quad f(4) = 4^2 - \log_e 4 - 12 = 2.6137$

Therefore a real root lies in the interval (3, 4).

By Regula-Falsi method, we have
$$x_1 = \frac{af(b) - bf(a)}{f(b) - f(a)}$$

Here $a = 3$, $b = 4$, $f(a) = -4.0986$ and $f(b) = 2.6137$
First approximation:

$$x_1 = \frac{3 \times 2.6137 - 4 \times -4.0986}{2.6137 - (-4.0986)} = 3.6106$$

$f(x_1) = f(3.6106) = -0.2475 = -\text{ve}$

Hence the root lies between 3.6106 and 4.
Second approximation:

$$x_2 = \frac{3.6106 \times 2.6137 - 4 \times -0.2475}{2.6137 - (-0.2475)} = 3.6443$$

$f(x_2) = f(3.6443) = -0.0123 = -\text{ve}$

Hence the root lies between 3.6443 and 4.
Third approximation:

$$x_3 = \frac{3.6443 \times 2.6137 - 4 \times -0.0123}{2.6137 - (-0.0123)} = 3.6459$$

$f(x_3) = f(3.6459) = -0.001 = -\text{ve}$

Hence the root lies between 3.6459 and 4.
Fourth approximation:

$$x_4 = \frac{3.6459 \times 2.6137 - 4 \times -0.001}{2.6137 - (-0.001)} = 3.6460$$

$f(x_4) = f(3.6460) = -0.0003 = -\text{ve}$

Hence the root lies between 3.6460 and 4.
Fifth approximation:

$$x_5 = \frac{3.6460 \times 2.6137 - 4 \times -0.0003}{2.6137 - (-0.0003)} = 3.6461$$

Hence, the root correct to three decimal places is 3.646.

Example 3. Find a real root of the equation $x \tan x + 1 = 0$ correct to three decimal places in the interval (2.5, 3) using Regula-Falsi method.

Solution. Let $\quad f(x) = x \tan x + 1 = 0$
Now, $\quad f(2.5) = 2.5 \tan(2.5) + 1 = -0.867556$
and $\quad f(3) = 3 \tan(3) + 1 = 0.572360$
Hence, First approximation is

$$x_1 = \frac{2.5 \times 0.57236 - 3 \times -0.867556}{0.57236 - (-0.867556)} = 2.801252$$

$f(x_1) = f(2.801252) = 0.00802 = +\text{ve}$

Hence the root lies between 2.5 and 2.801252.
Second approximation is

$$x_2 = \frac{2.5 \times 0.00802 - 2.801252 \times -0.867556}{0.00802 - (-0.867556)} = 2.806513$$

$f(x_2) = f(2.806513) = 0.022743 = +\text{ve}$
Hence the root lies between 2.5 and 2.806513.
Third approximation is

$$x_3 = \frac{2.5 \times 0.022743 - 2.806513 \times -0.867556}{0.022743 - (-0.867556)} = 2.798683$$

$f(x_3) = f(2.798683) = 0.000831 = +\text{ve}$
Hence the root lies between 2.5 and 2.798683.
Fourth approximation is

$$x_4 = \frac{2.5 \times 0.000831 - 2.798683 \times -0.867556}{0.000831 - (-0.867556)} = 2.7981$$

Hence the root is 2.798.

Example 4. Find the real root of the equation $x^3 - 9x + 1 = 0$ correct to four decimal places.

Solution. Let $\quad f(x) = x^3 - 9x + 1 = 0$
Now, $\quad f(2) = 2^3 - 9(2) + 1 = -9 = -\text{ve}$
and $\quad f(3) = 3^3 - 9(3) + 1 = 1 = +\text{ve}$
Therefore a real root lies in the interval (2, 3).
By Regula-Falsi method, we have

$$x_1 = \frac{af(b) - bf(a)}{f(b) - f(a)}$$

Here $a = 2, b = 3, f(a) = -9$ and $f(b) = 1$
First approximation is

$$x_1 = \frac{2 \times 1 - 3 \times -9}{1 - (-9)} = 2.9$$

$f(x_1) = f(2.9) = -0.711 = -\text{ve}$
Hence the root lies between 2.9 and 3.
Again $a = 2.9, b = 3, f(a) = -0.711$ and $f(b) = 1$
Second approximation is

$$x_2 = \frac{2.9 \times 1 - 3 \times -0.711}{1 - (-0.711)} = 2.9416$$

$f(x_2) = f(2.9416) = -0.0207 = -\text{ve}$
Hence the root lies between 2.9416 and 3.
Again $a = 2.9416, b = 3, f(a) = -0.0207$ and $f(b) = 1$

Third approximation is

$$x_3 = \frac{2.9416 \times 1 - 3 \times -0.0207}{1-(-0.0207)} = 2.9428$$

$$f(x_3) = f(2.9428) = -0.0003 = -\text{ve}$$

Hence the root lies between 2.9428 and 3.
Again $a = 2.9428$, $b = 3$, $f(a) = -0.0003$ and $f(b) = 1$
Fourth approximation is

$$x_4 = \frac{2.9428 \times 1 - 3 \times -0.0003}{1-(-0.0003)} = 2.94281$$

Hence, the root is 2.9428 correct to four places of decimals.

EXERCISE 2.3

Find a root of the following equations using the Method of False Position:
1. $x^3 - 2x = 5$
2. $2x = 6 + \log x$
3. $x^6 - x^4 - x^3 = 1$
4. $e^{-x} - \sin x = 0$
5. $e^x - 4x = 0$ in the interval (2, 2.5)
6. $x^3 - 4x - 9 = 0$
7. $(x - 5)e^x - 5 = 0$
8. Find a negative root of the equation $x^3 - 3x + 4 = 0$ using the method of False Position in the interval (−2, −3).
9. Find a positive root of the equation $3x - 1 = \cos x$ using the method of False Position.

ANSWERS

1. 2.094
2. 3.257
3. 1.4036
4. 0.5885
5. 2.1533
6. 2.7065
7. 4.964
8. −2.198
9. 0.6071

2.7 NEWTON-RAPHSON METHOD OR NEWTON'S METHOD

This is the most widely used methods of solving the equations. Given an approximate root of an equation, a better and closer approximation to the root can be obtained by this method.

Consider the equation $f(x) = 0$. Let x_n be an approximation to the root of the equation. Let h be a very small interval such that

$$h = x_{n+1} - x_n \qquad \ldots(1)$$

Now, $f(x_{n+1}) = f(x_n + h)$

$$= f(x_n) + hf'(x_n) + \frac{h^2}{2!}f''(x_n) + \ldots \text{ (By Taylor's expansion)}$$

Since h is small, we can neglect h^2, h^3, ... and hence we have
$$f(x_{n+1}) = f(x_n) + hf'(x_n) \qquad ...(2)$$
If x_{n+1} is a root of the equation $f(x) = 0$, then
$$f(x_{n+1}) = 0$$
$\therefore$ (2) gives, $f(x_n) + hf'(x_n) = 0$
$$\Rightarrow \qquad h = -\frac{f(x_n)}{f'(x_n)}$$
Using (1), we get
$$x_{n+1} - x_n = -\frac{f(x_n)}{f'(x_n)}$$
$$x_{n+1} = x_n - \frac{f(x_n)}{f'(x_n)}, \text{ where } n = 0, 1, 2, ... \qquad ...(3)$$
Equation (3) is the iterative formula of Newton's Method.

2.8 ORDER AND RATE OF CONVERGENCE OF NEWTON-RAPHSON METHOD

Let α be the exact root of the equation $f(x) = 0$ such that $f(\alpha) = 0$.

Let x_n and x_{n+1} be the nth and $(n+1)^{th}$ approximations respectively, then
$$x_n = e_n + \alpha$$
and
$$x_{n+1} = e_{n+1} + \alpha$$
By Newton-Raphson method
$$x_{n+1} = x_n - \frac{f(x_n)}{f'(x_n)} \qquad ...(1)$$
$$\Rightarrow \quad \alpha + e_{n+1} = e_n + \alpha - \frac{f(e_n + \alpha)}{f'(e_n + \alpha)}$$
$$\Rightarrow \qquad e_{n+1} = e_n - \frac{f(e_n + \alpha)}{f'(e_n + \alpha)}$$
$$\Rightarrow \qquad e_{n+1} = e_n - \frac{f(\alpha) + e_n f'(\alpha) + \frac{e_n^2}{2!} f''(\alpha) + ...}{f'(\alpha) + e_n f''(\alpha) + \frac{e_n^2}{2!} f'''(\alpha) + ...}$$
$$\text{(By Taylor's expansion)}$$
$$\Rightarrow \qquad e_{n+1} = e_n - \frac{e_n f'(\alpha) + \frac{e_n^2}{2!} f''(\alpha) + ...}{f'(\alpha) + e_n f''(\alpha) + \frac{e_n^2}{2!} f'''(\alpha) + ...} \qquad [\because f(\alpha) = 0]$$

$$\Rightarrow \quad e_{n+1} = \frac{\dfrac{e_n^2}{2} f''(\alpha) + \ldots}{f'(\alpha) + e_n f''(\alpha) + \dfrac{e_n^2}{2!} f'''(\alpha) + \ldots}$$

$$\Rightarrow \quad e_{n+1} = \frac{\dfrac{e_n^2}{2} f''(\alpha)}{f'(\alpha) + e_n f''(\alpha)} \quad \text{neglecting higher powers of } e_n.$$

$$\Rightarrow \quad e_{n+1} = \frac{e_n^2 f''(\alpha)}{2\left[f'(\alpha) + e_n f''(\alpha)\right]}$$

$$\Rightarrow \quad e_{n+1} = \frac{e_n^2 f''(\alpha)}{2 f'(\alpha)\left[1 + e_n \dfrac{f''(\alpha)}{f'(\alpha)}\right]}$$

$$\Rightarrow \quad e_{n+1} = \frac{e_n^2 f''(\alpha)}{2 f'(\alpha)}\left[1 + e_n \dfrac{f''(\alpha)}{f'(\alpha)}\right]^{-1}$$

$$\Rightarrow \quad e_{n+1} = \frac{e_n^2 f''(\alpha)}{2 f'(\alpha)}\left[1 - e_n \dfrac{f''(\alpha)}{f'(\alpha)} + \ldots\right]$$

$$\Rightarrow \quad e_{n+1} = \frac{e_n^2}{2} \frac{f''(\alpha)}{f'(\alpha)} \quad \text{neglecting higher powers of } e_n.$$

$$\Rightarrow \quad e_{n+1} \propto e_n^2$$

This shows that the subsequent error at each step is proportional to the square of the previous error. Hence the convergence is of order 2 i.e., the convergence of the Newton-Raphson method is quadratic.

2.9 GEOMETRICAL INTERPRETATION OF NEWTON-RAPHSON METHOD

Consider the graph of the curve $y = f(x)$.

Let x_0 be the initial approximation to the root of the equation $f(x) = 0$. Draw a tangent at any point $P\{x_0, f(x_0)\}$ on the curve $y = f(x)$. Then the equation of the tangent at P is

$$y - f(x_0) = f'(x_0)(x - x_0)$$

Let the tangent intersects the x-axis at the point Q where $x = x_1$ and $y = 0$, then

$$-f(x_0) = f'(x_0)(x_1 - x_0)$$

$$\Rightarrow \quad x_1 = x_0 - \frac{f(x_0)}{f'(x_0)}$$

Solution of Algebraic and Transcendental Equations

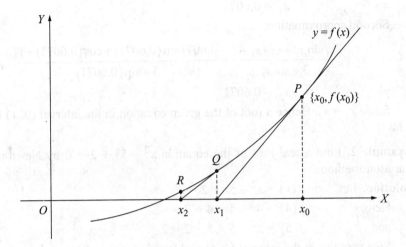

Now, taking x_1 as the initial approximation to the root, draw a tangent at the point $Q\{x_1, f(x_1)\}$ on the curve $y = f(x)$. The intersection of this tangent with x-axis is at the point $R\{x_2, f(x_2))\}$ and it gives the next approximation. Repeating this process, we will get the desired root of the given equation.

ILLUSTRATIVE EXAMPLES

Example 1. Find a real root of the equation $3x = \cos x + 1$ by Newton-Raphson method in between 0 and 1.

Solution: Let $f(x) = 3x - \cos x - 1 = 0$
Here, $f(0) = 3 \times 0 - \cos 0 - 1 = -2$
and $f(1) = 3 \times 1 - \cos 1 - 1 = 1.4597$
So a root lies in the interval (0, 1) and let us take $x_0 = 0.6$.
Also, $f'(x) = 3 + \sin x$
Therefore, by Newton-Raphson method, we have

$$x_{n+1} = x_n - \frac{f(x_n)}{f'(x_n)}$$

$$\Rightarrow \quad x_{n+1} = x_n - \frac{3x_n - \cos x_n - 1}{3 + \sin x_n}$$

$$\Rightarrow \quad x_{n+1} = \frac{x_n \sin x_n + \cos x_n + 1}{3 + \sin x_n} \qquad \ldots(1)$$

Equation (1) gives the iterative formula of Newton-Raphson Method.
First approximation:

$$x_1 = \frac{x_0 \sin x_0 + \cos x_0 + 1}{3 + \sin x_0} = \frac{0.6 \sin(0.6) + \cos(0.6) + 1}{3 + \sin(0.6)}$$

$$\Rightarrow \qquad x_1 = 0.6071$$

Second approximation:

$$x_2 = \frac{x_1 \sin x_1 + \cos x_1 + 1}{3 + \sin x_1} = \frac{0.6071 \sin(0.6071) + \cos(0.6071) + 1}{3 + \sin(0.6071)}$$

$$\Rightarrow \qquad x_2 = 0.6071$$

Since $x_1 = x_2$. Hence a root of the given equation in the interval (0, 1) is 0.6071.

Example 2. Find a real root of the equation $x^2 - 5x + 2 = 0$ by Newton-Raphson method.

Solution. Let $\qquad f(x) = x^2 - 5x + 2 = 0$
Here, $\qquad f(4) = 4^2 - 5 \times 4 + 2 = -2$
and $\qquad f(5) = 5^2 - 5 \times 5 + 2 = 2$

So a root lies in the interval (4, 5) and let us take $x_0 = 4$.
Also, $\qquad f'(x) = 2x - 5$
Therefore, by Newton-Raphson method, we have

$$x_{n+1} = x_n - \frac{f(x_n)}{f'(x_n)}$$

$$\Rightarrow \qquad x_{n+1} = x_n - \frac{x_n^2 - 5x_n + 2}{2x_n - 5}$$

$$\Rightarrow \qquad x_{n+1} = \frac{x_n^2 - 2}{2x_n - 5}, \ n = 0, 1, 2, 3, \ldots \qquad \ldots(1)$$

Equation (1) gives the iterative formula of Newton-Raphson Method.

First approximation ($n = 0$):

$$x_1 = \frac{x_0^2 - 2}{2x_0 - 5} = \frac{4^2 - 2}{2 \times 4 - 5} = \frac{14}{3} = 4.6667$$

Second approximation ($n = 1$):

$$x_2 = \frac{x_1^2 - 2}{2x_1 - 5} = \frac{(4.6667)^2 - 2}{2 \times 4.6667 - 5} = \frac{19.7781}{4.3334} = 4.5641$$

Third approximation ($n = 2$):

$$x_3 = \frac{x_2^2 - 2}{2x_2 - 5} = \frac{(4.5641)^2 - 2}{2 \times 4.5641 - 5} = \frac{18.8310}{4.1282} = 4.5616$$

Fourth approximation ($n = 3$):

$$x_4 = \frac{x_3^2 - 2}{2x_3 - 5} = \frac{(4.5616)^2 - 2}{2 \times 4.5616 - 5} = \frac{18.8082}{4.1232} = 4.5616$$

Now since, $x_3 = x_4$. Hence a root of the given equation is 4.5616.

Example 3. Use Newton-Raphson method to find a real root of the equation $\log x - \cos x = 0$ correct to three places of decimals.

Solution. Given $f(x) = \log x - \cos x = 0$

Here, $\quad f(1) = \log 1 - \cos 1 = -0.5403$
and $\quad f(2) = \log 2 - \cos 2 = 1.1092$
Also, $\quad f(1.3) = \log 1.3 - \cos 1.3 = -0.0051$
$\quad f(1.4) = \log 1.4 - \cos 1.4 = 0.1665$

Thus a real root lies in the interval (1.3, 1.4) and let us take $x_0 = 1.3$.

Also, $\quad f'(x) = \dfrac{1}{x} + \sin x$

Therefore, by Newton-Raphson method, we have

$$x_{n+1} = x_n - \frac{f(x_n)}{f'(x_n)}$$

$\Rightarrow \quad x_{n+1} = x_n - \dfrac{\log x_n - \cos x_n}{\dfrac{1}{x_n} + \sin x_n}$ ...(1)

First approximation ($n = 0$):

$$x_1 = x_0 - \frac{\log x_0 - \cos x_0}{\dfrac{1}{x_0} + \sin x_0} = 1.3 - \frac{\log 1.3 - \cos(1.3)}{\dfrac{1}{1.3} + \sin(1.3)} = 1.3029$$

Second approximation ($n = 1$):

$$x_2 = x_1 - \frac{\log x_1 - \cos x_1}{\dfrac{1}{x_1} + \sin x_1} = 1.3029 - \frac{\log 1.3029 - \cos(1.3029)}{\dfrac{1}{1.3029} + \sin(1.3029)} = 1.30295$$

Similarly, third approximation ($n = 2$):

$$x_3 = 1.30295$$

Now since, $x_2 = x_3$. Hence a root of the given equation is 1.30295.

Example 4. Use Newton-Raphson method to find a root of the equation $x^3 + 2x^2 + 10x - 20 = 0$.

Solution. Let $\quad f(x) = x^3 + 2x^2 + 10x - 20 = 0$
Here, $\quad f(1) = 1^3 + 2 \times 1^2 + 10 \times 1 - 20 = -7 = -\text{ve}$
and $\quad f(2) = 2^3 + 2 \times 2^2 + 10 \times 2 - 20 = 16 = +\text{ve}$

$\therefore$ a root lies in the interval (1, 2).

Let us take the initial approximation $x_0 = 1.2$

Also, $\quad f'(x) = 3x^2 + 4x + 10$

Therefore, by Newton-Raphson method, we have

$$x_{n+1} = x_n - \frac{f(x_n)}{f'(x_n)}$$

$$\Rightarrow \quad x_{n+1} = x_n - \frac{x_n^3 + 2x_n^2 + 10x_n - 20}{3x_n^2 + 4x_n + 10}$$

$$\Rightarrow \quad x_{n+1} = \frac{2x_n^3 + 2x_n^2 + 20}{3x_n^2 + 4x_n + 10} \qquad \ldots(1)$$

∴ First approximation:

$$x_1 = \frac{2x_0^3 + 2x_0^2 + 20}{3x_0^2 + 4x_0 + 10} = \frac{2(1.2)^3 + 2(1.2)^2 + 20}{3(1.2)^2 + 4(1.2) + 10} = 1.37741$$

Second approximation:

$$x_2 = \frac{2x_1^3 + 2x_1^2 + 20}{3x_1^2 + 4x_1 + 10} = \frac{2(1.37741)^3 + 2(1.37741)^2 + 20}{3(1.37741)^2 + 4(1.37741) + 10} = 1.36883$$

Third approximation:

$$x_3 = \frac{2x_2^3 + 2x_2^2 + 20}{3x_2^2 + 4x_2 + 10} = \frac{2(1.36883)^3 + 2(1.36883)^2 + 20}{3(1.36883)^2 + 4(1.36883) + 10} = 1.36881$$

Fourth approximation:

$$x_4 = \frac{2x_3^3 + 2x_3^2 + 20}{3x_3^2 + 4x_3 + 10} = \frac{2(1.36881)^3 + 2(1.36881)^2 + 20}{3(1.36881)^2 + 4(1.36881) + 10} = 1.36881$$

Hence, the required root is 1.36881.

Example 5. Find an iterative formula to find $\sqrt{N}$, where N is a positive integer and hence find the square root of 12 correct to four decimal places.

Solution. Let $\quad x = \sqrt{N} \Rightarrow x^2 = N$ i.e., $x^2 - N = 0$
Let $\quad f(x) = x^2 - N = 0$
∴ $\quad f'(x) = 2x$
Therefore, by Newton-Raphson method, we have

$$x_{n+1} = x_n - \frac{f(x_n)}{f'(x_n)} \Rightarrow x_{n+1} = x_n - \frac{x_n^2 - N}{2x_n}$$

$$\Rightarrow \quad x_{n+1} = \frac{x_n^2 + N}{2x_n}$$

$$\Rightarrow \quad x_{n+1} = \frac{1}{2}\left[x_n + \frac{N}{x_n}\right] \qquad \ldots(1)$$

which is the required iterative formula.

Now, put $N = 12$, we have from (1)
$$x_{n+1} = \frac{1}{2}\left[x_n + \frac{12}{x_n}\right]$$

Also, the value of $\sqrt{12}$ lies between 3 and 4.
Let the initial approximation be $x_0 = 3.1$
$\therefore$ First approximation:
$$x_1 = \frac{1}{2}\left[x_0 + \frac{12}{x_0}\right] = \frac{1}{2}\left[3.1 + \frac{12}{3.1}\right] = 3.48548$$

Second approximation:
$$x_2 = \frac{1}{2}\left[x_1 + \frac{12}{x_1}\right] = \frac{1}{2}\left[3.48548 + \frac{12}{3.48548}\right] = 3.46417$$

Third approximation:
$$x_3 = \frac{1}{2}\left[x_2 + \frac{12}{x_2}\right] = \frac{1}{2}\left[3.46417 + \frac{12}{3.46417}\right] = 3.46410$$

Hence, $\sqrt{12} = 3.4641$.

Example 6. Find an iterative formula to find the reciprocal of a given number N and hence evaluate $\frac{1}{24}$.

Solution. Let $x = \frac{1}{N} \Rightarrow N = \frac{1}{x}$

Let $f(x) = \frac{1}{x} - N = 0$ $\therefore$ $f'(x) = -\frac{1}{x^2}$

By Newton-Raphson method, we have
$$x_{n+1} = x_n - \frac{f(x_n)}{f'(x_n)}$$

$$\Rightarrow \quad x_{n+1} = x_n - \frac{\dfrac{1}{x_n} - N}{-\dfrac{1}{x_n^2}} = \frac{-\dfrac{2}{x_n} + N}{-\dfrac{1}{x_n^2}} = x_n(2 - Nx_n)$$

$$\Rightarrow \quad x_{n+1} = x_n(2 - Nx_n) \qquad \ldots(1)$$

which is the required iterative formula.

Now, put $N = 24$, we have from (1)
$$x_{n+1} = x_n(2 - 24x_n) \Rightarrow x_{n+1} = 2x_n(1 - 12x_n)$$

Also, the value of $\frac{1}{24}$ lies between 0.01 and 0.05.

Let the initial approximation be $x_0 = 0.03$
First approximation:
$$x_1 = 2x_0(1 - 12x_0) = 2(0.03)[1 - 12(0.03)] = 0.0384$$
Second approximation:
$$x_2 = 2x_1(1 - 12x_1) = 2(0.0384)[1 - 12(0.0384)] = 0.04141$$
Third approximation:
$$x_3 = 2x_2(1 - 12x_2) = 2(0.04141)[1 - 12(0.04141)] = 0.041665$$
Fourth approximation:
$$x_4 = 2x_3(1 - 12x_3) = 2(0.041665)[1 - 12(0.041665)] = 0.041667$$
Fifth approximation:
$$x_5 = 2x_4(1 - 12x_4) = 2(0.041667)[1 - 12(0.041667)] = 0.041667$$
Hence $\dfrac{1}{24} = 0.041667$

Example 7. Find an iterative formula to find the pth root of a given number N and hence evaluate the cube root of 10.

Solution. Let $\qquad x = \sqrt[p]{N} \implies x^p = N$
Let $\qquad f(x) = x^p - N = 0$
$\therefore \qquad f'(x) = px^{p-1}$
By Newton-Raphson method, we have
$$x_{n+1} = x_n - \frac{f(x_n)}{f'(x_n)}$$
$$\implies \qquad x_{n+1} = x_n - \frac{x_n^p - N}{px_n^{p-1}} = \frac{(p-1)x_n^p + N}{px_n^{p-1}}$$
$$\implies \qquad x_{n+1} = \frac{(p-1)x_n^p + N}{px_n^{p-1}} \qquad \text{...(1)}$$

which is the required iterative formula.
Now, put $N = 10$ and $p = 3$, we have from (1)
$$x_{n+1} = \frac{2x_n^3 + 10}{3x_n^2}$$
The value of $(10)^{1/3}$ lies between 2.1 and 2.2.
Let the initial approximation be $x_0 = 2.1$
First approximation:
$$x_1 = \frac{2x_0^3 + 10}{3x_0^2} = \frac{2(2.1)^3 + 10}{3(2.1)^2} = \frac{28.522}{13.23} = 2.15586$$

Second approximation:

$$x_2 = \frac{2x_1^3 + 10}{3x_1^2} = \frac{2(2.15586)^3 + 10}{3(2.15586)^2} = \frac{30.03972}{13.94319} = 2.154436$$

Third approximation:

$$x_3 = \frac{2x_2^3 + 10}{3x_2^2} = \frac{2(2.15444)^3 + 10}{3(2.15444)^2} = \frac{30.00015}{13.92483} = 2.154435$$

Hence $\sqrt[3]{10} = 2.15443$

EXERCISE 2.4

Find a root of the following equations using the Newton-Raphson Method:
1. $x^2 + x - \cos x = 0$
2. $x^4 - x - 10 = 0$ [I.U.2008-09]
3. $2x - 5 = 3 \sin x$
4. $2x^3 - 3x - 6 = 0$
5. $2x - \log_{10} x = 7$
6. $x^3 - 3x - 5 = 0$ [I.U. 2007-08]
7. Find an iterative formula to find the reciprocal of a given number N and hence evaluate $\dfrac{1}{13}$.
8. Find the values of $\sqrt[3]{17}$ and $\sqrt{37}$
9. Find an iterative formula to find the reciprocal of the pth root of given number N and hence evaluate $\dfrac{1}{\sqrt{11}}$.
10. Find a root of the equation $x^3 - 10 = 0$ by Newton-Raphson Method correct to 4 decimal places.

ANSWERS

1. 0.5500
2. 1.856
3. 2.7985
4. 1.78377
5. 3.7893
6. 2.279
7. 0.07692
8. 2.571; 6.08276

9. $x_{n+1} = \dfrac{x_n(p+1 - Nx_n^p)}{p}$; 0.30151
10. 2.1547

2.10 SOLUTION OF SIMULTANEOUS LINEAR ALGEBRAIC EQUATIONS

For solving a system of simultaneous linear equations, we have studied Cramer's rule or matrix method. These methods become tedious when number of unknowns is large. In this case we use numerical methods, some of which are discussed here.

2.11 GAUSS-ELIMINATION METHOD

This method reduces the system of equations into an upper triangular matrix which can be solved by back substitution.

Let the system of equations in three unknowns x, y, z is

$$a_1 x + b_1 y + c_1 z = d_1$$
$$a_2 x + b_2 y + c_2 z = d_2$$
$$a_3 x + b_3 y + c_3 z = d_3$$

The above system of equations can be written as

$$\begin{bmatrix} a_1 & b_1 & c_1 \\ a_2 & b_2 & c_2 \\ a_3 & b_3 & c_3 \end{bmatrix} \begin{bmatrix} x \\ y \\ z \end{bmatrix} = \begin{bmatrix} d_1 \\ d_2 \\ d_3 \end{bmatrix}$$

$$\Rightarrow \qquad AX = B \qquad \ldots(1)$$

Now, $\quad (A, B) = \begin{bmatrix} a_1 & b_1 & c_1 & \vdots & d_1 \\ a_2 & b_2 & c_2 & \vdots & d_2 \\ a_3 & b_3 & c_3 & \vdots & d_3 \end{bmatrix}$

Now, reduce this matrix to upper triangular matrix; i.e., to a matrix whose all the elements below the principal diagonal are zero. We get

$$(A, B) = \begin{bmatrix} A_1 & B_1 & C_1 & \vdots & d_1 \\ 0 & B_2 & C_2 & \vdots & d_2 \\ 0 & 0 & C_3 & \vdots & d_3 \end{bmatrix}$$

Now by (1), we have

$$\begin{bmatrix} A_1 & B_1 & C_1 \\ 0 & B_2 & C_2 \\ 0 & 0 & C_3 \end{bmatrix} \begin{bmatrix} x \\ y \\ z \end{bmatrix} = \begin{bmatrix} d_1 \\ d_2 \\ d_3 \end{bmatrix}$$

which can be written as,

$$A_1 x + B_1 y + C_1 z = d_1$$
$$B_2 y + C_2 z = d_2$$
$$C_3 z = d_3$$

Solving the above equations, we get values of x, y, z.

ILLUSTRATIVE EXAMPLES

Example 1. Solve the following system of equations by Gauss-Elimination method:

$$x + 2y + 3z = 6$$
$$2x + 4y + z = 7$$
$$3x + 2y + 9z = 14$$

Solution. Let $(A, B) = \begin{bmatrix} 1 & 2 & 3 & 6 \\ 2 & 4 & 1 & 7 \\ 3 & 2 & 9 & 14 \end{bmatrix}$

$\sim \begin{bmatrix} 1 & 2 & 3 & 6 \\ 0 & 0 & -5 & -5 \\ 0 & -4 & 0 & -4 \end{bmatrix} \quad R_2 - 2R_1; \; R_3 - 2R_1$

$\sim \begin{bmatrix} 1 & 2 & 3 & 6 \\ 0 & -4 & 0 & -4 \\ 0 & 0 & -5 & -5 \end{bmatrix} \quad R_2 \leftrightarrow R_3$

$\therefore \quad x + 2y + 3z = 6$
$\qquad -4y = -4$
$\qquad -5z = -5$
$\Rightarrow \qquad y = 1 \text{ and } z = 1$
and $\qquad x + 2 + 3 = 6 \Rightarrow x = 1$
Hence, $x = 1, y = 1, z = 1$.

Example 2. Solve by Gauss-Elimination method
$\qquad 2x + 2y + z = 12$
$\qquad 3x + 2y + 2z = 8$
$\qquad 5x + 10y - 8z = 10$

Solution: Let $(A, B) = \begin{bmatrix} 2 & 2 & 1 & 12 \\ 3 & 2 & 2 & 8 \\ 5 & 10 & -8 & 10 \end{bmatrix}$

$\sim \begin{bmatrix} 2 & 2 & 1 & 12 \\ 0 & -1 & \dfrac{1}{2} & -10 \\ 0 & 5 & -\dfrac{21}{2} & -20 \end{bmatrix} \quad R_2 - \dfrac{3}{2}R_1 \; ; \; R_3 - \dfrac{5}{2}R_1$

$\sim \begin{bmatrix} 2 & 2 & 1 & 12 \\ 0 & -1 & \dfrac{1}{2} & -10 \\ 0 & 0 & -8 & -70 \end{bmatrix} \quad R_3 + 5R_2$

$\sim \begin{bmatrix} 2 & 2 & 1 & 12 \\ 0 & -2 & 1 & -20 \\ 0 & 0 & -8 & -70 \end{bmatrix} \quad 2 \times R_2$

$\therefore \quad 2x + 2y + z = 12$ ...(1)
$\quad\quad\quad -2y + z = -20$ ...(2)
$\quad\quad\quad\quad\quad -8z = -70$ ...(3)

By (3), we have $z = \dfrac{70}{8} = 8.75$

(2) $\Rightarrow \quad -2y + 8.75 = -20 \quad\quad\quad\quad\quad \Rightarrow y = 14.375$

and (1) $\Rightarrow 2x + 2 \times 14.375 + 8.75 = 12 \quad\quad \Rightarrow x = -12.75$

Hence, $x = -12.75, y = 14.375, z = 8.75$.

Example 3. Solve by Gauss-Elimination method

$0.19x + 0.22y + 0.42z = 0.25$
$0.27x + 0.34y + 0.56z = 0.18$
$0.52x + 0.41y + 0.17z = 0.69$

Solution. Let $(A, B) = \begin{bmatrix} 0.19 & 0.22 & 0.41 & 0.25 \\ 0.27 & 0.34 & 0.56 & 0.18 \\ 0.52 & 0.41 & 0.17 & 0.69 \end{bmatrix}$

$\sim \begin{bmatrix} 0.19 & 0.22 & 0.41 & 0.25 \\ 0 & 0.0274 & -0.0227 & -0.1753 \\ 0 & -0.1921 & -0.9521 & 0.0058 \end{bmatrix} R_2 - 1.4211 R_1; \; R_3 - 2.7368 R_1$

$\sim \begin{bmatrix} 0.19 & 0.22 & 0.41 & 0.25 \\ 0 & 0.0274 & -0.0227 & -0.1753 \\ 0 & 0 & -1.1112 & -1.2231 \end{bmatrix} R_3 + 7.0101 R_2$

$\sim \begin{bmatrix} 0.19 & 0.22 & 0.41 & 0.25 \\ 0 & 1 & -0.8285 & -6.3978 \\ 0 & 0 & 1 & 1.1007 \end{bmatrix} \quad R_2 \div 0.0274; \; R_3 \div -1.1112$

$\therefore \; 0.19x + 0.22y + 0.41z = 0.25$ ...(1)
$\quad\quad\quad y - 0.8285z = -6.3978$ ...(2)
$\quad\quad\quad\quad\quad z = 1.1007$ ...(3)

By (2), we have $y - 0.8285 \times 1.1007 = -6.3978$

$\Rightarrow \quad\quad\quad y = -5.4859$

(1) $\Rightarrow 0.19x + 0.22 \times -5.4859 + 0.41 \times 1.1007 = 0.25$

$\Rightarrow \quad\quad\quad x = 5.2927$

Hence, $\quad\quad x = 5.2927, y = -5.4859, z = 1.1007$.

Example 4. Solve by Gauss-Elimination method

$x + y + 2z = 4$
$3x + y - 3z = -4$
$2x - 3y - 5z = -5$

Solution: Let $(A, B) = \begin{bmatrix} 1 & 1 & 2 & 4 \\ 3 & 1 & -3 & -4 \\ 2 & -3 & -5 & -5 \end{bmatrix}$

$\sim \begin{bmatrix} 1 & 1 & 2 & 4 \\ 0 & -2 & -9 & -16 \\ 0 & -5 & -9 & -13 \end{bmatrix} \quad R_2 - 3R_1; R_3 - 2R_1$

$\sim \begin{bmatrix} 1 & 1 & 2 & 4 \\ 0 & 2 & 9 & 16 \\ 0 & 5 & 9 & 13 \end{bmatrix} \quad -1 \times R_2; -1 \times R_3$

$\sim \begin{bmatrix} 1 & 1 & 2 & 4 \\ 0 & 2 & 9 & 16 \\ 0 & 0 & -13.5 & -27 \end{bmatrix} \quad R_3 - 2.5R_2$

$\therefore \quad x + y + 2z = 4$ ...(1)
$\quad\quad 2y + 9z = 16$ ...(2)
$\quad\quad -13.5z = -27$ ...(3)

By (3), we have $z = \dfrac{27}{13.5} = 2$

(2) $\Rightarrow \quad 2y + 9 \times 2 = 16 \quad\Rightarrow\quad y = -1$
and (1) $\Rightarrow x - 1 + 2 \times 2 = 4 \quad\Rightarrow\quad x = 1$
Hence, $x = 1, y = -1, z = 2$.

EXERCISE 2.5

Solve the following system of equtions by Gauss– Elimination Method
1. $x - y + z = 1, -3x + 2y - 3z = -6, 2x - 5y + 4z = 5$
2. $2x + 2y + z = 6, 4x + 2y + 3z = 4, x + y + z = 0$
3. $8x - 6y + z = 13.67, 3x + y - 2z = 17.59, 2x - 6y + 9z = 29.29$
4. $x + y + z = 9, 2x - 3y + 4z = 13, 3x + 4y + 5z = 40$
5. $x + 3y + 10z = 24, 2x + 17y + 4z = 35, 28x + 4y - z = 32$
6. $x + 4y - z = -5, x + y - 6z = -12, 3x - y - z = 4$

ANSWERS

1. $-2, 3, 6$
2. $5, 1, -6$
3. $2.45, 1.62, 3.79$
4. $1, 3, 5$
5. $0.99, 1.50, 1.84$
6. $2.085, -1.141, 1.648$

2.12 GAUSS-ELIMINATION METHOD WITH PIVOTING

In gauss-elimination method, the elements of the principal diagonal (*i.e.*, the element a_{ij}, when $i = j$) is known as pivot element.

The process of pivoting involves the following steps:

(*i*) Search and locate the largest absolute value in the first column.

(*ii*) Interchange the first row with the row containing that element.

(*iii*) Then eliminate the first element in the second row.

(*iv*) Now search for the largest element in the column from second column to the last. Interchange the second row with the row containing the largest element.

(*v*) Continue this process.

---| **ILLUSTRATIVE EXAMPLES** |---

Example 1. Solve the following system of equations with pivoting:

$$2x + y + z = 10$$
$$3x + 2y + 3z = 18$$
$$x + 4y + 9z = 16$$

Solution: Let $(A, B) = \begin{bmatrix} 2 & 1 & 1 & | & 10 \\ 3 & 2 & 3 & | & 18 \\ 1 & 4 & 9 & | & 16 \end{bmatrix}$

$\sim \begin{bmatrix} 3 & 2 & 3 & | & 18 \\ 2 & 1 & 1 & | & 10 \\ 1 & 4 & 9 & | & 16 \end{bmatrix}$ $R_1 \leftrightarrow R_2$

$\sim \begin{bmatrix} 3 & 2 & 3 & | & 18 \\ 0 & \frac{-1}{3} & -1 & | & -2 \\ 0 & \frac{10}{3} & 8 & | & 10 \end{bmatrix}$ $R_2 - \frac{2}{3} R_1, R_3 - \frac{1}{3} R_1$

$\sim \begin{bmatrix} 3 & 2 & 3 & | & 18 \\ 0 & \frac{10}{3} & 8 & | & 10 \\ 0 & \frac{-1}{3} & -1 & | & -2 \end{bmatrix}$ $R_2 \leftrightarrow R_3$

$$\sim \begin{bmatrix} 3 & 2 & 3 & | & 18 \\ 0 & \dfrac{10}{3} & 8 & | & 10 \\ 0 & 0 & \dfrac{-1}{5} & | & -1 \end{bmatrix} \quad R_3 + \dfrac{1}{10} R_2$$

$$\sim \begin{bmatrix} 3 & 2 & 3 & | & 18 \\ 0 & \dfrac{10}{3} & 8 & | & 10 \\ 0 & 0 & \dfrac{1}{5} & | & 1 \end{bmatrix} \quad -1 \times R_3$$

$\Rightarrow \quad 3x + 2y + 3z = 18$

$\qquad \dfrac{10}{3}y + 8z = 10$

$\qquad \dfrac{1}{5}z = 1$

$\therefore \qquad z = 5$

$\qquad \dfrac{10}{3}y + 8 \times 5 = 10$

$\Rightarrow \qquad y = -9$

$\qquad 3x + 2 \times -9 + 3 \times 5 = 18$

$\Rightarrow \qquad x = 7$

Hence, $\qquad x = 7, y = -9, z = 5$.

Example 2. Solve the following system of equations with pivoting:

$\qquad 10x + y + z = 12$
$\qquad 2x + 10y + z = 13$
$\qquad 2x + 2y + 10z = 14$

Solution: Let $\quad (A,B) = \begin{bmatrix} 10 & 1 & 1 & | & 12 \\ 2 & 10 & 1 & | & 13 \\ 2 & 2 & 10 & | & 14 \end{bmatrix}$

$$\sim \begin{bmatrix} 10 & 1 & 1 & | & 12 \\ 0 & 9.8 & 0.8 & | & 10.6 \\ 0 & 1.8 & 9.8 & | & 11.6 \end{bmatrix} \quad R_2 - \dfrac{1}{5} R_1, R_3 - \dfrac{1}{5} R_1$$

$$\sim \begin{bmatrix} 10 & 1 & 1 & | & 12 \\ 0 & 9.8 & 0.8 & | & 10.6 \\ 0 & 0 & 9.653 & | & 9.653 \end{bmatrix} \quad R_3 - \dfrac{9}{49} R_2$$

$\Rightarrow \qquad 10x + y + z = 12$
$\qquad\qquad 9.8y + 0.8z = 10.6$
$\qquad\qquad\qquad 9.653z = 9.653$
$\therefore \qquad\qquad\qquad\quad z = 1$
$\qquad 9.8y + 0.8 \times 1 = 10.6$
$\Rightarrow \qquad\qquad\qquad\quad y = 1$
$\qquad\qquad 10x + 1 + 1 = 12$
$\Rightarrow \qquad\qquad\qquad\quad x = 1$

Hence, $\qquad x = 1, y = 1, z = 1$

2.13 GAUSS-JORDAN METHOD

This method is a modified form of Gauss elimination method. In this method, the coefficient matrix A of $AX = B$ is reduced to diagonal matrix or unit matrix by making all the elements above and below the principal diagonal of A as zero.

Here back substitution is not needed.

ILLUSTRATIVE EXAMPLES

Example 1. Solve the following system of equations by the Gauss-Jordan method:

$$10x + y + z = 12$$
$$2x + 10y + z = 13$$
$$x + y + 5z = 7$$

Solution: The given system can be written as

$$AX = B$$

i.e., $\begin{bmatrix} 10 & 1 & 1 \\ 2 & 10 & 1 \\ 1 & 1 & 5 \end{bmatrix} \begin{bmatrix} x \\ y \\ z \end{bmatrix} = \begin{bmatrix} 12 \\ 13 \\ 7 \end{bmatrix}$

$\therefore \qquad [A/B] = \begin{bmatrix} 10 & 1 & 1 & | & 12 \\ 2 & 10 & 1 & | & 13 \\ 1 & 1 & 5 & | & 7 \end{bmatrix}$

$\sim \begin{bmatrix} 1 & -8 & -44 & | & -51 \\ 2 & 10 & 1 & | & 13 \\ 1 & 1 & 5 & | & 7 \end{bmatrix} R_1 - 9 R_3$

$\sim \begin{bmatrix} 1 & -8 & -44 & | & -51 \\ 0 & 26 & 89 & | & 115 \\ 0 & 9 & 49 & | & 58 \end{bmatrix} R_2 - 2 R_1, R_3 - R_1,$

Solution of Algebraic and Transcendental Equations 55

$$\sim \begin{bmatrix} 1 & -8 & -44 & | & -51 \\ 0 & -1 & -58 & | & -59 \\ 0 & 9 & 49 & | & 58 \end{bmatrix} R_2 - 3R_3$$

$$\sim \begin{bmatrix} 1 & 0 & 420 & | & 421 \\ 0 & -1 & -58 & | & -59 \\ 0 & 0 & -473 & | & -473 \end{bmatrix} R_1 - 8R_2, R_3 + 9R_2$$

$$\sim \begin{bmatrix} 1 & 0 & 420 & | & 421 \\ 0 & 1 & 58 & | & 59 \\ 0 & 0 & 1 & | & 1 \end{bmatrix} -1 \times R_1, \frac{-1}{473} \times R_3$$

$$\sim \begin{bmatrix} 1 & 0 & 0 & | & 1 \\ 0 & 1 & 0 & | & 1 \\ 0 & 0 & 1 & | & 1 \end{bmatrix} R_1 - 420 R_3, R_2 - 58 R_3$$

Now $\qquad AX = B$

$$\Rightarrow \begin{bmatrix} 1 & 0 & 0 \\ 0 & 1 & 0 \\ 0 & 0 & 1 \end{bmatrix} \begin{bmatrix} x \\ y \\ z \end{bmatrix} = \begin{bmatrix} 1 \\ 1 \\ 1 \end{bmatrix}$$

i.e., $\qquad x = 1, y = 1, z = 1$

Example 2. Solve the following system of equations by the Gauss-Jordan method:

$$4x + y + 3z = 11$$
$$3x + 4y + 2z = 11$$
$$2x + 3y + z = 7$$

Solution: The given system can be written as
$$AX = B$$

i.e., $\begin{bmatrix} 4 & 1 & 3 \\ 3 & 4 & 2 \\ 2 & 3 & 1 \end{bmatrix} \begin{bmatrix} x \\ y \\ z \end{bmatrix} = \begin{bmatrix} 11 \\ 11 \\ 7 \end{bmatrix}$

$\therefore \qquad [A/B] = \begin{bmatrix} 4 & 1 & 3 & | & 11 \\ 3 & 4 & 2 & | & 11 \\ 2 & 3 & 1 & | & 7 \end{bmatrix}$

$$\sim \begin{bmatrix} 1 & -3 & 1 & | & 0 \\ 3 & 4 & 2 & | & 11 \\ 2 & 3 & 1 & | & 7 \end{bmatrix} R_1 - R_2$$

$$\sim \begin{bmatrix} 1 & -3 & 1 & | & 0 \\ 0 & 13 & -1 & | & 11 \\ 0 & 9 & -1 & | & 7 \end{bmatrix} \quad R_2 - 3R_1,\ R_3 - 2R_1$$

$$\sim \begin{bmatrix} 1 & -3 & 1 & | & 0 \\ 0 & 4 & 0 & | & 4 \\ 0 & 9 & -1 & | & 7 \end{bmatrix} \quad R_2 - R_3$$

$$\sim \begin{bmatrix} 1 & -3 & 1 & | & 0 \\ 0 & 1 & 0 & | & 1 \\ 0 & 9 & -1 & | & 7 \end{bmatrix} \quad \frac{1}{4} \times R_2$$

$$\sim \begin{bmatrix} 1 & -3 & 1 & | & 0 \\ 0 & 1 & 0 & | & 1 \\ 0 & 0 & -1 & | & -2 \end{bmatrix} \quad R_3 - 9 \times R_2$$

$$\sim \begin{bmatrix} 1 & -3 & 1 & | & 0 \\ 0 & 1 & 0 & | & 1 \\ 0 & 0 & 1 & | & 2 \end{bmatrix} \quad -1 \times R_3$$

$$\sim \begin{bmatrix} 1 & 0 & 1 & | & 3 \\ 0 & 1 & 0 & | & 1 \\ 0 & 0 & 1 & | & 2 \end{bmatrix} \quad R_1 + 3R_2$$

$$\sim \begin{bmatrix} 1 & 0 & 0 & | & 1 \\ 0 & 1 & 0 & | & 1 \\ 0 & 0 & 1 & | & 2 \end{bmatrix} \quad R_1 - R_3$$

Now $\qquad AX = B$

$$\Rightarrow \begin{bmatrix} 1 & 0 & 0 \\ 0 & 1 & 0 \\ 0 & 0 & 1 \end{bmatrix} \begin{bmatrix} x \\ y \\ z \end{bmatrix} = \begin{bmatrix} 1 \\ 1 \\ 2 \end{bmatrix}$$

i.e., $\qquad x = 1,\ y = 1,\ z = 2$

Example 3. Solve the following system of equations by Gauss-Jordan method:

$x + 2y + z = 8$

$2x + 3y + 4z = 20$

$4x + 3y + 2z = 16$

Solution: The given system can be written as

$$AX = B$$

i.e., $\begin{bmatrix} 1 & 2 & 1 \\ 2 & 3 & 4 \\ 4 & 3 & 2 \end{bmatrix} \begin{bmatrix} x \\ y \\ z \end{bmatrix} = \begin{bmatrix} 8 \\ 20 \\ 16 \end{bmatrix}$

$\therefore \quad [A/B] = \begin{bmatrix} 1 & 2 & 1 & | & 8 \\ 2 & 3 & 4 & | & 20 \\ 4 & 3 & 2 & | & 16 \end{bmatrix}$

$\sim \begin{bmatrix} 1 & 2 & 1 & | & 8 \\ 0 & -1 & 2 & | & 4 \\ 0 & -5 & -2 & | & -16 \end{bmatrix} \quad R_2 - 2R_1, R_3 - 4R_1$

$\sim \begin{bmatrix} 1 & 0 & 5 & | & 16 \\ 0 & -1 & 2 & | & 4 \\ 0 & 0 & -12 & | & -36 \end{bmatrix} \quad R_1 + 2R_2, R_3 - 5R_2$

$\sim \begin{bmatrix} 1 & 0 & 5 & | & 16 \\ 0 & 1 & -2 & | & -4 \\ 0 & 0 & 1 & | & 3 \end{bmatrix} \quad -1 \times R_2, \frac{-1}{12} \times R_3$

$\sim \begin{bmatrix} 1 & 0 & 0 & | & 1 \\ 0 & 1 & 0 & | & 2 \\ 0 & 0 & 1 & | & 3 \end{bmatrix} \quad R_1 - 5R_3, R_2 + 2R_3$

Now $\quad AX = B$

$\Rightarrow \begin{bmatrix} 1 & 0 & 0 \\ 0 & 1 & 0 \\ 0 & 0 & 1 \end{bmatrix} \begin{bmatrix} x \\ y \\ z \end{bmatrix} = \begin{bmatrix} 1 \\ 2 \\ 3 \end{bmatrix}$

i.e., $\quad x = 1, y = 2, z = 3$

Example 4. Solve the following system of equations by Gausss-Jordan method:

$2x - 6y + 8z = 24$
$5x + 4y - 3z = 2$
$3x + y + 2z = 16$

Solution: The given system can be written as
$$AX = B$$

i.e., $\begin{bmatrix} 2 & -6 & 8 \\ 5 & 4 & -3 \\ 3 & 1 & 2 \end{bmatrix} \begin{bmatrix} x \\ y \\ z \end{bmatrix} = \begin{bmatrix} 24 \\ 2 \\ 16 \end{bmatrix}$

$$\therefore \quad [A/B] = \begin{bmatrix} 2 & -6 & 8 & | & 24 \\ 5 & 4 & -3 & | & 2 \\ 3 & 1 & 2 & | & 16 \end{bmatrix}$$

$$\sim \begin{bmatrix} 2 & -6 & 8 & | & 24 \\ 5 & 4 & -3 & | & 2 \\ 1 & 7 & -6 & | & -8 \end{bmatrix} \quad R_3 - R_1$$

$$\sim \begin{bmatrix} 1 & 7 & -6 & | & -8 \\ 5 & 4 & -3 & | & 2 \\ 2 & -6 & 8 & | & 24 \end{bmatrix} \quad R_1 \leftrightarrow R_3$$

$$\sim \begin{bmatrix} 1 & 7 & -6 & | & -8 \\ 0 & -31 & 27 & | & 42 \\ 0 & -20 & 20 & | & 40 \end{bmatrix} \quad R_2 - 5R_1,\ R_3 - 2R_1$$

$$\sim \begin{bmatrix} 1 & 7 & -6 & | & -8 \\ 0 & -31 & 27 & | & 42 \\ 0 & -1 & 1 & | & 2 \end{bmatrix} \quad \frac{1}{20} \times R_3$$

$$\sim \begin{bmatrix} 1 & 0 & 1 & | & 6 \\ 0 & -1 & -3 & | & -18 \\ 0 & -1 & 1 & | & 2 \end{bmatrix} \quad R_1 + 7R_3,\ R_2 - 30R_3$$

$$\sim \begin{bmatrix} 1 & 0 & 1 & | & 6 \\ 0 & -1 & -3 & | & -18 \\ 0 & 0 & 4 & | & 20 \end{bmatrix} \quad R_3 - R_2$$

$$\sim \begin{bmatrix} 1 & 0 & 1 & | & 6 \\ 0 & -1 & -3 & | & -18 \\ 0 & 0 & 1 & | & 5 \end{bmatrix} \quad \frac{1}{4} \times R_3$$

$$\sim \begin{bmatrix} 1 & 0 & 0 & | & 1 \\ 0 & -1 & 0 & | & -3 \\ 0 & 0 & 1 & | & 5 \end{bmatrix} \quad R_1 - R_3,\ R_2 + 3R_3$$

$$\sim \begin{bmatrix} 1 & 0 & 0 & | & 1 \\ 0 & 1 & 0 & | & 3 \\ 0 & 0 & 1 & | & 5 \end{bmatrix} \quad -1 \times R_2$$

Now $AX = B$

$\Rightarrow \begin{bmatrix} 1 & 0 & 0 \\ 0 & 1 & 0 \\ 0 & 0 & 1 \end{bmatrix} \begin{bmatrix} x \\ y \\ z \end{bmatrix} = \begin{bmatrix} 1 \\ 3 \\ 5 \end{bmatrix}$

i.e., $x = 1, y = 3, z = 5$

EXERCISE 2.6

Solve the following system of equations by Gauss-Jordan method:
1. $10x + y + z = 12, x + 10y + z = 12, x + y + 10z = 12$
2. $x - y + z = 1, -3x + 2y - 3z = -63, 2x - 5y + 40z = 5$
3. $x + 2y + z = 3, 2x + 3y + 3z = 10, 3x - y + 2z = 13$
4. $6x - y + z = 13, x + y + z = 9, 10x + y - z = 19$
5. $10x + y + z = 12, x + 10y - z = 10, x - 2y + 10z = 9$
6. $4x - y - z = -7, x - 5y + z = -10, x + 2y + 6z = 9$

ANSWERS
---

1. 1,3,5
2. – 2,3,6
3. 2,–1,3
4. 2,3,4
5. 1,1,1
6. – 1,2,1

2.14 MATRIX-INVERSION METHOD

Consider the system of equations

$$a_{11}x + a_{12}y + a_{13}z = b_1$$
$$a_{21}x + a_{22}y + a_{23}z = b_2 \qquad \ldots(1)$$
$$a_{31}x + a_{32}y + a_{33}z = b_3$$

The above system of equations can be written in matrix form as

$$AX = B \qquad \ldots(2)$$

where,
$$A = \begin{bmatrix} a_{11} & a_{12} & a_{13} \\ a_{21} & a_{22} & a_{23} \\ a_{31} & a_{32} & a_{33} \end{bmatrix}$$

$$X = \begin{bmatrix} x \\ y \\ z \end{bmatrix} \text{ and } B = \begin{bmatrix} b_1 \\ b_2 \\ b_3 \end{bmatrix}$$

Let A be non-singular so that A^{-1} exists. Then premultiplying both sides of (2) by A^{-1}, we obtain

$$A^{-1}AX = A^{-1}B$$
$$IX = A^{-1}B$$

i.e., $X = A^{-1}B$

where I is an identity matrix.

ILLUSTRATIVE EXAMPLES

Example 1. Solve the following system of equations by the matrix-inversion method:

$$3x + y + 2z = 3$$
$$2x - 3y - z = -3$$
$$x + 2y + z = 4$$

Solution: Here determinant of A, i.e.,

$$|A| = \begin{vmatrix} 3 & 1 & 2 \\ 2 & -3 & -1 \\ 1 & 2 & 1 \end{vmatrix}$$

$$= 3(-3+2) - 1(2+1) + 2(4+3)$$
$$= -3 - 3 + 14 = 8$$

Also, cofactor matrix of $A = \begin{bmatrix} -1 & -3 & 7 \\ 3 & 1 & -5 \\ 5 & 7 & -11 \end{bmatrix}$

Adjoint of A, $\quad \text{adj } A = \begin{bmatrix} -1 & 3 & 5 \\ -3 & 1 & 7 \\ 7 & -5 & -11 \end{bmatrix}$

$\therefore \quad A^{-1} = \dfrac{\text{adj } A}{|A|} = \dfrac{1}{8}\begin{bmatrix} -1 & 3 & 5 \\ -3 & 1 & 7 \\ 7 & -5 & -11 \end{bmatrix}$

Thus, we have $X = A^{-1}B$

$$\begin{bmatrix} x \\ y \\ z \end{bmatrix} = \frac{1}{8}\begin{bmatrix} -1 & 3 & 5 \\ -3 & 1 & 7 \\ 7 & -5 & -11 \end{bmatrix}\begin{bmatrix} 3 \\ -3 \\ 4 \end{bmatrix}$$

$$\begin{bmatrix} x \\ y \\ z \end{bmatrix} = \frac{1}{8}\begin{bmatrix} 8 \\ 16 \\ -8 \end{bmatrix} = \begin{bmatrix} 1 \\ 2 \\ -1 \end{bmatrix}$$

i.e., $\quad x = 1, y = 2, z = -1$

Example 2. Solve the following system of equations by the matrix-Inversion mathod:

$$2x - y + z = 5$$
$$x + 3y - 2z = 7$$
$$x + 2y + 3z = 10$$

Solution: Here dederminant of A, i.e.,

$$|A| = \begin{vmatrix} 2 & -1 & 1 \\ 1 & 3 & -2 \\ 1 & 2 & 3 \end{vmatrix}$$

$$= 2(9+4) + 1(3+2) + 1(2-3)$$
$$= 26 + 5 - 1 = 30$$

Also, cofactor matrix of $A = \begin{bmatrix} 13 & -5 & -1 \\ 5 & 5 & -5 \\ -1 & 5 & 7 \end{bmatrix}$

Adjoint of A, $\quad \text{adj } A = \begin{bmatrix} 13 & 5 & -1 \\ -5 & 5 & 5 \\ -1 & -5 & 7 \end{bmatrix}$

$\therefore \quad A^{-1} = \dfrac{\text{adj } A}{|A|} = \dfrac{1}{30} \begin{bmatrix} 13 & 5 & -1 \\ -5 & 5 & 5 \\ -1 & -5 & 7 \end{bmatrix}$

Thus, we have $\quad X = A^{-1} B$

$$\begin{bmatrix} x \\ y \\ z \end{bmatrix} = \dfrac{1}{30} \begin{bmatrix} 13 & 5 & -1 \\ -5 & 5 & 5 \\ -1 & -5 & 7 \end{bmatrix} \begin{bmatrix} 5 \\ 7 \\ 10 \end{bmatrix}$$

$$\begin{bmatrix} x \\ y \\ z \end{bmatrix} = \dfrac{1}{30} \begin{bmatrix} 90 \\ 60 \\ 30 \end{bmatrix} = \begin{bmatrix} 3 \\ 2 \\ 1 \end{bmatrix}$$

i.e., $\quad x = 3, y = 2, z = 1$

Example 3. Solve the following system of equations by the matrix-inversion method:

$$x + y + z = 3$$
$$2x - y + 3z = 16$$
$$3x + y - z = -3$$

Solution: Here determinant of A, i.e.,

$$|A| = \begin{vmatrix} 1 & 1 & 1 \\ 2 & -1 & 3 \\ 3 & 1 & -1 \end{vmatrix}$$

$$= 1(1-3) - 1(-2-9) + 1(2+3)$$
$$= -2 + 11 + 5 = 14$$

Also, cofactor matrix of $A = \begin{bmatrix} -2 & 11 & 5 \\ 2 & -4 & 2 \\ 4 & -1 & -3 \end{bmatrix}$

Adjoint of A, $\quad \text{adj } A = \begin{bmatrix} -2 & 2 & 4 \\ 11 & -4 & -1 \\ 5 & 2 & -3 \end{bmatrix}$

$\therefore \quad A^{-1} = \dfrac{\text{adj } A}{|A|} = \dfrac{1}{14} \begin{bmatrix} -2 & 2 & 4 \\ 11 & -4 & -1 \\ 5 & 2 & -3 \end{bmatrix}$

Thus, we have $\quad X = A^{-1} B$

$$\begin{bmatrix} x \\ y \\ z \end{bmatrix} = \dfrac{1}{14} \begin{bmatrix} -2 & 2 & 4 \\ 11 & -4 & -1 \\ 5 & 2 & -3 \end{bmatrix} \begin{bmatrix} 3 \\ 16 \\ -3 \end{bmatrix}$$

$$\begin{bmatrix} x \\ y \\ z \end{bmatrix} = \dfrac{1}{14} \begin{bmatrix} 14 \\ -28 \\ 56 \end{bmatrix} = \begin{bmatrix} 1 \\ -2 \\ 4 \end{bmatrix}$$

i.e., $\quad x = 1, y = -2, z = 4$

EXERCISE 2.7

Solve the following system of equations by matrix-inversion method:
1. $3x + y - z = 3, 2x - 8y + z = -5, x - 2y + 9z = 8$
2. $x + 2y - z = 3, x - y + z = -1, 2x - 2y + 3z = 2$
3. $2x - 6y + 8z = 24, 3x + y + 2z = 16, 5x + 4y - 3z = 2$
4. $x - y + z = 1, -3x + 2y - 3z = -6, 2x - 5y + 4z = 5$

ANSWERS
---

1. 1,1,1 2. −1,4,4 3. 1,3,5 4. −2,3,6

2.15 METHOD OF TRIANGULARISATION OR METHOD OF FACTORIZATION

This method is based on the fact that any square matrix can be expressed as the product of a lower and an upper triangular matrix, provided all the minors of the matrix are non-singular.

Consider the system of equations

$$a_{11}x + a_{12}y + a_{13}z = b_1$$
$$a_{21}x + a_{22}y + a_{23}z = b_2 \qquad \ldots(1)$$
$$a_{31}x + a_{32}y + a_{33}z = b_3$$

These equations can be written in matrix form as
$$AX = B \qquad \ldots(2)$$

where,
$$A = \begin{bmatrix} a_{11} & a_{12} & a_{13} \\ a_{21} & a_{22} & a_{23} \\ a_{31} & a_{32} & a_{33} \end{bmatrix}$$

$$X = \begin{bmatrix} x \\ y \\ z \end{bmatrix} \text{ and } B = \begin{bmatrix} b_1 \\ b_2 \\ b_3 \end{bmatrix}$$

Now, let
$$A = LU \qquad \ldots(3)$$

where
$$L = \begin{bmatrix} 1 & 0 & 0 \\ l_{21} & 1 & 0 \\ l_{31} & l_{32} & 1 \end{bmatrix}$$

and
$$U = \begin{bmatrix} u_{11} & u_{12} & u_{13} \\ 0 & u_{22} & u_{23} \\ 0 & 0 & u_{33} \end{bmatrix}$$

Therefore, by (3)
$$LUX = B \qquad \ldots(4)$$

Assuming
$$UX = Y \qquad \ldots(5)$$

where
$$Y = \begin{bmatrix} y_1 \\ y_2 \\ y_3 \end{bmatrix}$$

Equation (4) takes the form $LY = B$

i.e.,
$$\begin{bmatrix} 1 & 0 & 0 \\ l_{21} & 1 & 0 \\ l_{31} & l_{32} & 1 \end{bmatrix} \begin{bmatrix} y_1 \\ y_2 \\ y_3 \end{bmatrix} = \begin{bmatrix} b_1 \\ b_2 \\ b_3 \end{bmatrix}$$

$\therefore \qquad y_1 = b_1$
$$l_{21}y_1 + y_2 = b_2$$
and $\qquad l_{31}y_1 + l_{32}y_2 + y_3 = b_3$

Solving these equations, we get the values of y_1, y_2 and y_3.

By (5),
$$\begin{bmatrix} u_{11} & u_{12} & u_{13} \\ 0 & u_{22} & u_{23} \\ 0 & 0 & u_{33} \end{bmatrix} \begin{bmatrix} x_1 \\ x_2 \\ x_3 \end{bmatrix} = \begin{bmatrix} y_1 \\ y_2 \\ y_3 \end{bmatrix}$$

This gives, $u_{11}x_1 + u_{12}x_2 + u_{13}x_3 = y_1$
$$u_{22}x_2 + u_{23}x_3 = y_2$$
$$u_{33}x_3 = y_3$$
Solving the above equations, we get the values of x_1, x_2 and x_3.

Here matrices L and U are obtained from the equation (3).

Note. This method is also known as Decomposition Method.

ILLUSTRATIVE EXAMPLES

Example 1. Solve the following system of equations by the method of factorization:
$$2x - 3y + 10z = 3$$
$$-x + 4y + 2z = 20$$
$$5x + 2y + z = -12$$

Solution: The given system can be written as $AX = B$

i.e., $\begin{bmatrix} 2 & -3 & 10 \\ -1 & 4 & 2 \\ 5 & 2 & 1 \end{bmatrix} \begin{bmatrix} x \\ y \\ z \end{bmatrix} = \begin{bmatrix} 3 \\ 20 \\ -12 \end{bmatrix}$

Let $LU = A$

i.e., $\begin{bmatrix} 1 & 0 & 0 \\ l_{21} & 1 & 0 \\ l_{31} & l_{32} & 1 \end{bmatrix} \begin{bmatrix} u_{11} & u_{12} & u_{13} \\ 0 & u_{22} & u_{23} \\ 0 & 0 & u_{33} \end{bmatrix} = \begin{bmatrix} 2 & -3 & 10 \\ -1 & 4 & 2 \\ 5 & 2 & 1 \end{bmatrix}$

$\Rightarrow \begin{bmatrix} u_{11} & u_{12} & u_{13} \\ l_{21}u_{11} & l_{21}u_{12} + u_{22} & l_{21}u_{13} + u_{23} \\ l_{31}u_{11} & l_{31}u_{11} + l_{32}u_{22} & l_{31}u_{13} + l_{32}u_{23} + u_{33} \end{bmatrix} = \begin{bmatrix} 2 & -3 & 10 \\ -1 & 4 & 2 \\ 5 & 2 & 1 \end{bmatrix}$

Hence, $u_{11} = 2, u_{12} = -3, u_{13} = 10$
$$l_{21}u_{11} = -1$$

$\Rightarrow$ $l_{21} = \dfrac{-1}{2}$

$l_{21}u_{12} + u_{22} = 4$

$\Rightarrow$ $\dfrac{-1}{2} \times -3 + u_{22} = 4$

$\Rightarrow$ $u_{22} = \dfrac{5}{2}$

$l_{21}u_{13} + u_{23} = 2$

$\Rightarrow$ $\dfrac{-1}{2} \times 10 + u_{23} = 2$

$\Rightarrow \qquad u_{23} = 7$

$\qquad\qquad l_{31} u_{11} = 5$

$\Rightarrow \qquad\qquad l_{31} = \dfrac{5}{2}$

$\qquad l_{31} u_{12} + l_{32} u_{22} = 2$

$\Rightarrow \qquad \dfrac{5}{2} \times -3 + l_{32} \times \dfrac{5}{2} = 2$

$\Rightarrow \qquad\qquad l_{32} = \dfrac{19}{5}$

$\qquad l_{31} u_{13} + l_{32} u_{23} + u_{33} = 1$

$\Rightarrow \qquad \dfrac{5}{2} \times 10 + \dfrac{19}{5} \times 7 + u_{33} = 1$

$\Rightarrow \qquad\qquad u_{33} = \dfrac{-253}{5}$

Now, we have

$$L = \begin{bmatrix} 1 & 0 & 0 \\ \dfrac{-1}{2} & 1 & 0 \\ \dfrac{5}{2} & \dfrac{19}{5} & 1 \end{bmatrix} \text{ and } U = \begin{bmatrix} 2 & -3 & 10 \\ 0 & \dfrac{5}{2} & 7 \\ 0 & 0 & \dfrac{-253}{5} \end{bmatrix}$$

Now $\qquad\qquad LUX = B$

Let $\qquad\qquad UX = Y$ then $LY = B$ where $Y = \begin{bmatrix} y_1 \\ y_2 \\ y_3 \end{bmatrix}$

From $\qquad\qquad LY = B$

$$\begin{bmatrix} 1 & 0 & 0 \\ \dfrac{-1}{2} & 1 & 0 \\ \dfrac{5}{2} & \dfrac{19}{5} & 1 \end{bmatrix} \begin{bmatrix} y_1 \\ y_2 \\ y_3 \end{bmatrix} = \begin{bmatrix} 3 \\ 20 \\ -12 \end{bmatrix}$$

$\Rightarrow y_1 = 3$; $\; y_1 \times -\dfrac{1}{2} + y_2 = 20$ and $y_1 \times \dfrac{5}{2} + y_2 \times \dfrac{19}{5} + y_3 = -12$

$\Rightarrow y_1 = 3;\; y_2 = \dfrac{43}{2}$ and $y_3 = -\dfrac{506}{5}$

Now $\qquad\qquad UX = Y$ gives

$$\begin{bmatrix} 2 & -3 & 10 \\ 0 & 5/2 & 7 \\ 0 & 0 & -\dfrac{253}{5} \end{bmatrix} \begin{bmatrix} x \\ y \\ z \end{bmatrix} = \begin{bmatrix} 3 \\ \dfrac{43}{2} \\ -\dfrac{506}{5} \end{bmatrix}$$

$\Rightarrow$ $\quad 2x - 3y + 10z = 3$

$\dfrac{5}{2}y + 7 \times 2 = \dfrac{43}{2}$

$\dfrac{-253}{5}z = \dfrac{-506}{5}$

$\therefore \quad z = 2$

$\dfrac{5}{2}y + 7 \times 2 = \dfrac{43}{2}$

$\Rightarrow \quad y = 3$

$2x - 3 \times 3 + 10 \times 2 = 3$

$\Rightarrow \quad x = -4$

Hence, the solution is $\quad x = -4, y = 3, z = 2$.

Example 2. Solve the following system of equations by the method of factorization

$$x + y + z = 1$$
$$4x + 3y - z = 6$$
$$3x + 5y + 3z = 4$$

Solution: The given system can be written as $AX = B$

i.e., $\begin{bmatrix} 1 & 1 & 1 \\ 4 & 3 & -1 \\ 3 & 5 & 3 \end{bmatrix} \begin{bmatrix} x \\ y \\ z \end{bmatrix} = \begin{bmatrix} 1 \\ 6 \\ 4 \end{bmatrix}$

Let $LU = A$

i.e., $\begin{bmatrix} 1 & 0 & 0 \\ l_{21} & 1 & 0 \\ l_{31} & l_{32} & 1 \end{bmatrix} \begin{bmatrix} u_{11} & u_{12} & u_{13} \\ 0 & u_{22} & u_{23} \\ 0 & 0 & u_{33} \end{bmatrix} = \begin{bmatrix} 1 & 1 & 1 \\ 4 & 3 & -1 \\ 3 & 5 & 3 \end{bmatrix}$

$\Rightarrow \begin{bmatrix} u_{11} & u_{12} & u_{13} \\ l_{21}u_{11} & l_{21}u_{12} + u_{22} & l_{21}u_{13} + u_{23} \\ l_{31}u_{11} & l_{31}u_{12} + l_{32}u_{22} & l_{31}u_{13} + l_{32}u_{23} + u_{33} \end{bmatrix} = \begin{bmatrix} 1 & 1 & 1 \\ 4 & 3 & -1 \\ 3 & 5 & 3 \end{bmatrix}$

Hence, $\quad u_{11} = 1, u_{12} = 1, u_{13} = 1$

$l_{21}u_{11} = 4$

$\Rightarrow \quad l_{21} = 4$

$l_{21}u_{12} + u_{22} = 1$

$\Rightarrow \quad 4 \times 1 + u_{22} = 3$

$\Rightarrow \quad u_{22} = -1$

$l_{21}u_{13} + u_{23} = -1$

$\Rightarrow \quad 4 \times 1 + u_{23} = -1$

$\Rightarrow \quad u_{23} = -5$

$l_{31}u_{11} = 3$

$$\Rightarrow \quad l_{31} = 3$$
$$l_{31}u_{12} + l_{32}u_{22} = 5$$
$$\Rightarrow \quad 3 \times 1 + l_{32} \times -1 = 5$$
$$\Rightarrow \quad l_{32} = -2$$
$$l_{31}u_{13} + l_{32}u_{23} + u_{33} = 3$$
$$\Rightarrow \quad 3 \times 1 + (-2) \times -5 + u_{33} = 3$$
$$\Rightarrow \quad u_{33} = -10$$

Now, we have

$$L = \begin{bmatrix} 1 & 0 & 0 \\ 4 & 1 & 0 \\ 3 & -2 & 1 \end{bmatrix} \text{ and } U = \begin{bmatrix} 1 & 1 & 1 \\ 0 & -1 & -5 \\ 0 & 0 & -10 \end{bmatrix}$$

Now $\quad LUX = B$

Let $\quad UX = Y$ then $LY = B$ where $Y = \begin{bmatrix} y_1 \\ y_2 \\ y_3 \end{bmatrix}$

From $\quad LY = B$

$$\begin{bmatrix} 1 & 0 & 0 \\ 4 & 1 & 0 \\ 3 & -2 & 1 \end{bmatrix} \begin{bmatrix} y_1 \\ y_2 \\ y_3 \end{bmatrix} = \begin{bmatrix} 1 \\ 6 \\ 4 \end{bmatrix}$$

$\Rightarrow \quad y_1 = 1 \,;\, 4y_1 + y_2 = 6$ and $3y_1 - 2y_2 + y_3 = 4$

$\Rightarrow \quad y_1 = 1 \,;\, y_2 = 2$ and $y_3 = 5$

Now $\quad UX = Y$ gives

$$\begin{bmatrix} 1 & 1 & 1 \\ 0 & -1 & -5 \\ 0 & 0 & -10 \end{bmatrix} \begin{bmatrix} x \\ y \\ z \end{bmatrix} = \begin{bmatrix} 1 \\ 2 \\ 5 \end{bmatrix}$$

$\Rightarrow \quad x + y + z = 1$
$\quad -y - 5z = 2$
$\quad -10z = 5$

$\therefore \quad z = \dfrac{-1}{2}$

$\quad -y - 5 \times \dfrac{-1}{2} = 2$

$\Rightarrow \quad y = \dfrac{1}{2}$

$\quad x + \dfrac{1}{2} - \dfrac{1}{2} = 1$

$\Rightarrow \quad x = 1$

Hence, the solution is $\quad x = 1, y = \dfrac{1}{2}, z = -\dfrac{1}{2}$.

Example 3. Solve the following system of equations by the method of factorization:
$$x + 2y + 3z = 14$$
$$2x + 3y + 4z = 20$$
$$3x + 4y + z = 14$$

Solution: The given system can be written as $AX = B$

i.e., $\begin{bmatrix} 1 & 2 & 3 \\ 2 & 3 & 4 \\ 3 & 4 & 1 \end{bmatrix} \begin{bmatrix} x \\ y \\ z \end{bmatrix} = \begin{bmatrix} 14 \\ 20 \\ 14 \end{bmatrix}$

Let $LU = A$

i.e., $\begin{bmatrix} 1 & 0 & 0 \\ l_{21} & 1 & 0 \\ l_{31} & l_{32} & 1 \end{bmatrix} \begin{bmatrix} u_{11} & u_{12} & u_{13} \\ 0 & u_{22} & u_{23} \\ 0 & 0 & u_{33} \end{bmatrix} = \begin{bmatrix} 1 & 2 & 3 \\ 2 & 3 & 4 \\ 3 & 4 & 1 \end{bmatrix}$

$\Rightarrow \begin{bmatrix} u_{11} & u_{12} & u_{13} \\ l_{21}u_{11} & l_{21}u_{12} + u_{22} & l_{21}u_{13} + u_{23} \\ l_{31}u_{11} & l_{31}u_{12} + l_{32}u_{22} & l_{31}u_{13} + l_{32}u_{23} + u_{33} \end{bmatrix} = \begin{bmatrix} 1 & 2 & 3 \\ 2 & 3 & 4 \\ 3 & 4 & 1 \end{bmatrix}$

Hence, $u_{11} = 1, u_{12} = 2, u_{13} = 3$
$$l_{21}u_{11} = 2$$
$\Rightarrow \qquad l_{21} = 2$
$$l_{21}u_{12} + u_{22} = 3$$
$\Rightarrow \qquad 2 \times 2 + u_{22} = 3$
$\Rightarrow \qquad u_{22} = -1$
$$l_{21}u_{13} + u_{23} = 4$$
$\Rightarrow \qquad 2 \times 3 + u_{23} = -4$
$\Rightarrow \qquad u_{23} = -2$
$$l_{31}u_{11} = 3$$
$\Rightarrow \qquad l_{31} = 3$
$$l_{31}u_{12} + l_{32}u_{22} = 4$$
$\qquad 3 \times 2 + l_{32} \times -1 = 4$
$\Rightarrow \qquad l_{32} = 2$
$$l_{31}u_{13} + l_{32}u_{23} + u_{33} = 1$$
$\Rightarrow \qquad 3 \times 3 + 2 \times -2 + u_{33} = 1$
$\Rightarrow \qquad u_{33} = -4$

Now, we have
$$L = \begin{bmatrix} 1 & 0 & 0 \\ 2 & 1 & 0 \\ 3 & 2 & 1 \end{bmatrix} \text{ and } U = \begin{bmatrix} 1 & 2 & 3 \\ 0 & -1 & -2 \\ 0 & 0 & -4 \end{bmatrix}$$

Now, $\qquad LUX = B$
Let $\qquad UX = Y$
then $\qquad LY = B$

where $\qquad Y = \begin{bmatrix} y_1 \\ y_2 \\ y_3 \end{bmatrix}$

From $\qquad LY = B$, we have $\begin{bmatrix} 1 & 0 & 0 \\ 2 & 1 & 0 \\ 3 & 2 & 1 \end{bmatrix} \begin{bmatrix} y_1 \\ y_2 \\ y_3 \end{bmatrix} = \begin{bmatrix} 14 \\ 20 \\ 14 \end{bmatrix}$

$\Rightarrow \qquad y_1 = 14\,;\, 2y_1 + y_2 = 20$ and $3y_1 + 2y_2 + y_3 = 14$
$\Rightarrow \qquad y_1 = 14\,;\, y_2 = -8$ and $y_3 = -12$
Now $\qquad UX = Y$ gives

$\begin{bmatrix} 1 & 2 & 3 \\ 0 & -1 & -2 \\ 0 & 0 & -4 \end{bmatrix} \begin{bmatrix} x \\ y \\ z \end{bmatrix} = \begin{bmatrix} 14 \\ -8 \\ -12 \end{bmatrix}$

$\Rightarrow \qquad x + 2y + 3z = 14$
$\qquad\qquad -y - 2z = -8$
$\qquad\qquad -4z = -12$
$\therefore \qquad z = 3$
$\qquad\qquad -y - 2 \times 3 = -8$
$\Rightarrow \qquad y = 2$
$\qquad\qquad x + 2 \times 2 + 3 \times 3 = 14$
$\Rightarrow \qquad x = 1$

Hence, the solution is $\qquad x = 1, y = 2, z = 3$.

Example 4. Solve the following system of equations by the method of factorization:

$$2x + y + 4z = 12$$
$$8x + 3y + 2z = 20$$
$$4x + 11y - z = 23$$

Solution: The given system can be written as $AX = B$

i.e., $\qquad \begin{bmatrix} 2 & 1 & 4 \\ 8 & 3 & 2 \\ 4 & 11 & -1 \end{bmatrix} \begin{bmatrix} x \\ y \\ z \end{bmatrix} = \begin{bmatrix} 12 \\ 20 \\ 23 \end{bmatrix}$

Let $\qquad LU = A$

70 *Numerical and Statistical Techniques*

i.e.,
$$\begin{bmatrix} 1 & 0 & 0 \\ l_{21} & 1 & 0 \\ l_{31} & l_{32} & 1 \end{bmatrix} \begin{bmatrix} u_{11} & u_{12} & u_{13} \\ 0 & u_{22} & u_{23} \\ 0 & 0 & u_{33} \end{bmatrix} = \begin{bmatrix} 2 & 1 & 4 \\ 8 & 3 & 2 \\ 4 & 11 & -1 \end{bmatrix}$$

$$\Rightarrow \begin{bmatrix} u_{11} & u_{12} & u_{13} \\ l_{21}u_{11} & l_{21}u_{12}+u_{22} & l_{21}u_{13}+u_{23} \\ l_{31}u_{11} & l_{31}u_{12}+l_{32}u_{22} & l_{31}u_{13}+l_{32}u_{23}+u_{33} \end{bmatrix} = \begin{bmatrix} 2 & 1 & 4 \\ 8 & 3 & 2 \\ 4 & 11 & -1 \end{bmatrix}$$

Hence, $u_{11} = 2$, $u_{12} = 1$, $u_{13} = 4$

$l_{21}u_{11} = 8 \Rightarrow l_{21} = 4$

$l_{21}u_{12} + u_{22} = 3 \Rightarrow 4 \times 1 + u_{22} = 3$

$\Rightarrow u_{22} = -1$

$l_{21}u_{13} + u_{23} = 2 \Rightarrow 4 \times 4 + u_{23} = 2$

$\Rightarrow u_{23} = -14$

$l_{31}u_{11} = 4 \Rightarrow l_{31} = 2$

$l_{31}u_{12} + l_{32} u_{22} = 11 \Rightarrow 2 \times 1 + l_{32} \times -1 = 11$

$\Rightarrow l_{32} = -9$

$l_{31}u_{13} + l_{32} u_{23} + u_{33} = -1 \Rightarrow 2 \times 4 - 9 \times -14 + u_{33} = -1$

$\Rightarrow u_{33} = -135$

Now, we have

$$L = \begin{bmatrix} 1 & 0 & 0 \\ 4 & 1 & 0 \\ 2 & -9 & 1 \end{bmatrix} \text{ and } U = \begin{bmatrix} 2 & 1 & 4 \\ 0 & -1 & -14 \\ 0 & 0 & -135 \end{bmatrix}$$

Now $\qquad LUX = B$

Let $\qquad UX = Y$ then $LY = B$ where $Y = \begin{bmatrix} y_1 \\ y_2 \\ y_3 \end{bmatrix}$

From $\qquad LY = B$, we have $\begin{bmatrix} 1 & 0 & 0 \\ 4 & 1 & 0 \\ 2 & -9 & 1 \end{bmatrix} \begin{bmatrix} y_1 \\ y_2 \\ y_3 \end{bmatrix} = \begin{bmatrix} 12 \\ 20 \\ 23 \end{bmatrix}$

$\Rightarrow \qquad y_1 = 12$; $4y_1 + y_2 = 20$ and $2y_1 - 9y_2 + y_3 = 23$

$\Rightarrow \qquad y_1 = 12$; $y_2 = -28$ and $y_3 = -253$

Now $\qquad UX = Y$ gives

$$\begin{bmatrix} 2 & 1 & 4 \\ 0 & -1 & -14 \\ 0 & 0 & -135 \end{bmatrix} \begin{bmatrix} x \\ y \\ z \end{bmatrix} = \begin{bmatrix} 12 \\ -28 \\ -253 \end{bmatrix}$$

$\Rightarrow \qquad 2x + y + 4z = 12$

$-y - 14z = -28$

$$-135z = -253$$
$$\therefore \quad z = 1.874$$
$$-y - 14 \times 1.874 = -28$$
$$\Rightarrow \quad y = 1.764$$
$$2x + 1.764 + 4 \times 1.874 = 12$$
$$\Rightarrow \quad x = 1.37$$

Hence, the solution is $\quad x = 1.37, y = 1.764, z = 1.874$.

EXERCISE 2.8

Solve the following system of equations by method of factorization:
1. $x + 3y + 8z = 4$, $x + 4y + 3z = -2$, $x + 3y + 4z = 1$
2. $10x + y + z = 12$, $2x + 10y + z = 13$, $2x + 2y + 10z = 14$
3. $x + 3y + 6z = 2$, $x - 4y + 2z = 7$, $3x - y + 4z = 9$
4. $5x - 2y + z = 4$, $7x + y - 5z = 8$, $3x + 7y + 4z = 10$
5. $x + y - z = 1$, $3x + y + z = 1$, $4x + 3y + 2z = -1$
6. $3x + y + 2z = 16$, $2x - 6y + 8z = 24$, $5x + 4y - 3z = 2$

ANSWERS
---

1. $\dfrac{-29}{4}, \dfrac{7}{4}, \dfrac{3}{4}$ 2. $1, 1, 1$ 3. $2, -1, \dfrac{1}{2}$

4. $\dfrac{366}{327}, \dfrac{284}{327}, \dfrac{46}{327}$ 5. $1, -1, -1$ 6. $1, 3, 5$

2.16 GAUSS-JACOBI METHOD OR JACOBI METHOD OF ITERATION

Consider the system of equations

$$\left.\begin{array}{r}a_1 + b_1y + c_1z = d_1 \\ a_2x + b_2y + c_2z = d_2 \\ a_3x + b_3y + c_3z = d_3\end{array}\right\} \quad ...(1)$$

Let $|a_1| > |b_1| + |c_1|$; $|b_2| > |a_2| + |c_2|$; $|c_3| > |a_3| + |b_3|$
Then, this iterative method can be used on the system of equations (1).
Solving for x, y and z, respectively, we get

$$\left.\begin{array}{r}x = \dfrac{1}{a_1}(d_1 - b_1y - c_1z) \\ y = \dfrac{1}{b_2}(d_2 - a_2x - c_2z) \\ z = \dfrac{1}{c_3}(d_3 - a_3x - b_3y)\end{array}\right\} \quad ...(2)$$

72 Numerical and Statistical Techniques

Let $x^{(0)}$, $y^{(0)}$ and $z^{(0)}$ are the initial approximations of the unknowns x, y and z, then from (2), we have

$$\left. \begin{array}{l} x^{(1)} = \dfrac{1}{a_1}\left(d_1 - b_1 y^{(0)} - c_1 z^{(0)}\right) \\[4pt] y^{(1)} = \dfrac{1}{b_2}\left(d_2 - a_2 x^{(0)} - c_2 z^{(0)}\right) \\[4pt] z^{(1)} = \dfrac{1}{c_3}\left(d_3 - a_3 x^{(0)} - b_3 y^{(0)}\right) \end{array} \right\} \quad \ldots(3)$$

Again, using $x^{(1)}$, $y^{(1)}$ and $z^{(1)}$ as first approximations in (3), we get the second approximations as

$$\left. \begin{array}{l} x^{(2)} = \dfrac{1}{a_1}\left(d_1 - b_1 y^{(1)} - c_1 z^{(1)}\right) \\[4pt] y^{(2)} = \dfrac{1}{b_2}\left(d_2 - a_2 x^{(1)} - c_2 z^{(1)}\right) \\[4pt] z^{(2)} = \dfrac{1}{c_3}\left(d_3 - a_3 x^{(1)} - b_3 y^{(1)}\right) \end{array} \right\} \quad \ldots(4)$$

Now, this process is continued till the convergence is assured *i.e.*, the difference between any two consecutive values is negligible.

Note. In the absence of the initial values, we take (0,0,0) as the initial approximations.

--- **ILLUSTRATIVE EXAMPLES** ---

Example 1. Solve the following system of equations using gauss-jacobi method:
$$10x - 5y - 2z = 3$$
$$4x - 10y + 3z = -3$$
$$x + 6y + 10z = -3$$

Solution: Here we see that
$$|10| > |-5| + |-2| \ ; \ |-10| > |4| + |3| \ ; \ |10| > |1| + |6|$$

Now the given equations can be written as

$$x = \frac{1}{10}(3 + 5y + 2z)$$

$$y = \frac{1}{10}(3 + 4x + 3z) \quad \ldots(1)$$

$$z = \frac{1}{10}(-3 - x - 6z)$$

First Iteration: Let the initial values be (0,0,0), then from (1), we get

$$x^{(1)} = \frac{1}{10}[3 + 5(0) + 2(0)] = 0.3$$

$$y^{(1)} = \frac{1}{10}[3 + 4(0) + 3(0)] = 0.3$$

$$z^{(1)} = \frac{1}{10}[-3 - 0 - 6(0)] = -0.3$$

Second iteration: Taking $x^{(1)}$, $y^{(1)}$ and $z^{(1)}$ as the first approximations, we get

$$x^{(2)} = \frac{1}{10}[3 + 5(0.3) + 2(-0.3)] = 0.39$$

$$y^{(2)} = \frac{1}{10}[3 + 4(0.3) + 3(-0.3)] = 0.33$$

$$z^{(2)} = \frac{1}{10}[-3 - (0.3) - 6(0.3)] = -0.51$$

Third iteration: Taking $x^{(2)}$, $y^{(2)}$ and $z^{(2)}$ as the second approximations, we get

$$x^{(3)} = \frac{1}{10}[3 + 5(0.33) + 2(-0.51)] = 0.363$$

$$y^{(3)} = \frac{1}{10}[3 + 4(0.39) + 3(-0.51)] = 0.303$$

$$z^{(3)} = \frac{1}{10}[-3 - (0.39) - 6(0.33)] = -0.537$$

Fourth iteration: Taking $x^{(3)}$, $y^{(3)}$ and $z^{(3)}$ as the third approximations, we get

$$x^{(4)} = \frac{1}{10}[3 + 5(0.303) + 2(-0.537)] = 0.3441$$

$$y^{(4)} = \frac{1}{10}[3 + 4(0.363) + 3(-0.537)] = 0.2841$$

$$z^{(4)} = \frac{1}{10}[-3 - (0.363) - 6(0.303)] = -0.5181$$

Fifth iteration: Taking $x^{(4)}$, $y^{(4)}$ and $z^{(4)}$ as the fourth approximations, we get

$$x^{(5)} = \frac{1}{10}[3 + 5(0.2841) + 2(-0.5181)] = 0.33843$$

$$y^{(5)} = \frac{1}{10}[3 + 4(0.3441) + 3(-0.5181)] = 0.2822$$

$$z^{(5)} = \frac{1}{10}[-3 - (0.3441) - 6(0.2841)] = -0.50487$$

Sixth iteration: Taking $x^{(5)}$, $y^{(5)}$ and $z^{(5)}$ as the fifth approximtions, we get

$$x^{(6)} = \frac{1}{10}\,[3 + 5(0.2822) + 2(-0.50487)] = 0.340126$$

$$y^{(6)} = \frac{1}{10}\,[3 + 4(0.33843) + 3(-0.50487)] = 0.283911$$

$$z^{(6)} = \frac{1}{10}\,[-3 - (0.33843) - 6(0.2822)] = -0.503163$$

Seventh iteration: Taking $x^{(6)}$, $y^{(6)}$ and $z^{(6)}$ as the sixth approximations, we get

$$x^{(7)} = \frac{1}{10}\,[3 + 5(0.283911) + 2(-0.503163)]$$
$$= 0.3413229$$

$$y^{(7)} = \frac{1}{10}\,[3 + 4(0.340126) + 3(-0.503163)]$$
$$= 0.2851015$$

$$z^{(7)} = \frac{1}{10}\,[-3 - (0.340126) - 6(0.283911)]$$
$$= -0.5043592$$

Eighth iteration: Taking $x^{(7)}$, $y^{(7)}$ and $z^{(7)}$ as the seventh approximations, we get

$$x^{(8)} = \frac{1}{10}\,[3 + 5(0.2851015) + 2(-0.5043592)]$$
$$= 0.34167891$$

$$y^{(8)} = \frac{1}{10}\,[3 + 4(0.3413229) + 3(-0.5043592)]$$
$$= 0.2852214$$

$$z^{(8)} = \frac{1}{10}\,[-3 - (0.3413229) - 6(0.2851015)]$$
$$= -0.50519319$$

Ninths iteration: Taking $x^{(8)}$, $y^{(8)}$ and $z^{(8)}$ as the eighth approximations, we get

$$x^{(9)} = \frac{1}{10}\,[3 + 5(0.2852214) + 2(-0.50519319)]$$
$$= 0.34157206$$

$$y^{(9)} = \frac{1}{10}\,[3 + 4(0.33167891) + 3(-0.50519319)]$$
$$= 0.285113607$$

Solution of Algebraic and Transcendental Equations 75

$$z^{(9)} = \frac{1}{10}[-3 - (0.34167891) - 6(0.2852214)]$$
$$= -0.505300731$$

Hence, correct to three decimal places, the solution of the given system of equations
is $x = 0.342$, $y = 0.285$, $z = -0.505$.

Example 2. Solve the following system of equations using Gauss-Jacobi method:
$$5x - y + z = 10$$
$$2x + 4y = 12$$
$$x + y + 5z = -1$$

Solution: Here we see that
$|5| > |-1| + |1|$; $|4| > |2| + |0|$; $|5| > |1| + |1|$
Given equations can be written as

$$x = \frac{1}{5}(10 + y - z)$$
$$y = \frac{1}{4}(12 - 2x) \qquad \qquad ...(1)$$
$$z = \frac{1}{5}(-1 - x - y)$$

First iteration: Let the initial values be (0,0,0), then from (1), we get

$$x^{(1)} = \frac{1}{5}[10 + (0) - (0)] = 2$$
$$y^{(1)} = \frac{1}{4}[12 - 2(0)] = 3$$
$$z^{(1)} = \frac{1}{5}[-1 - (0) - (0)] = -0.2$$

Second iteration: Taking $x^{(1)}$, $y^{(1)}$ and $z^{(1)}$ as the first approximations, we get

$$x^{(2)} = \frac{1}{5}[10 + (2) - (-0.2)] = 2.44$$
$$y^{(2)} = \frac{1}{4}[12 - 2(2)] = 2$$
$$z^{(2)} = \frac{1}{5}[-1 - (2) - (3)] = -1.2$$

Third iteration: Taking $x^{(2)}$, $y^{(2)}$ and $z^{(2)}$ as the second approximations, we get

$$x^{(3)} = \frac{1}{5}[10 + (2) - (-1.2)] = 2.64$$

$$y^{(3)} = \frac{1}{4}[12 - 2(2.44)] = 1.78$$

$$z^{(3)} = \frac{1}{5}[-1 - (2.44) - (2)] = -1.088$$

Fourth iteration: Taking $x^{(3)}$, $y^{(3)}$ and $z^{(3)}$ as the third approximations, we get

$$x^{(4)} = \frac{1}{5}[10 + (1.78) - (-1.088)] = 2.5736$$

$$y^{(4)} = \frac{1}{4}[12 - 2(2.64)] = 1.68$$

$$z^{(4)} = \frac{1}{5}[-1 - (2.64) - (1.78)] = -1.084$$

Fifth iteration: Taking $x^{(4)}$, $y^{(4)}$ and $z^{(4)}$ as the fourth approximations, we get

$$x^{(5)} = \frac{1}{5}[10 + (1.64) - (-1.084)] = 2.5528$$

$$y^{(5)} = \frac{1}{4}[12 - 2(2.5736)] = 1.7132$$

$$z^{(5)} = \frac{1}{5}[-1 - (2.5736) - (1.68)] = -1.05072$$

Sixth iteration: Taking $x^{(5)}$, $y^{(5)}$ and $z^{(5)}$ as the fifth approximations, we get

$$x^{(6)} = \frac{1}{5}[10 + (1.7132) - (-1.05072)] = 2.552784$$

$$y^{(6)} = \frac{1}{4}[12 - 2(2.5528)] = 1.7236$$

$$z^{(6)} = \frac{1}{5}[-1 - (2.5528) - (1.7132)] = -1.0532$$

Seventh iteration: Taking $x^{(6)}$ $y^{(6)}$ and $z^{(6)}$ as the sixth approximations, we get

$$x^{(7)} = \frac{1}{5}[10 + (1.7236) - (-1.0532)] = 2.55536$$

$$y^{(7)} = \frac{1}{4}[12 - 2(2.552784)] = 1.723608$$

$$z^{(7)} = \frac{1}{5}[-1 - (2.552784) - (1.7236)] = -1.0552768$$

Eighth iteraion: Taking $x^{(7)}$, $y^{(7)}$ and $z^{(7)}$ as the seventh approximations, we get

$$x^{(8)} = \frac{1}{5}[10 + (1.723608) - (-1.0552768)] = 2.555777$$

$$y^{(8)} = \frac{1}{4}[12 - 2(2.55536)] = 1.72232$$

$$z^{(8)} = \frac{1}{5}[-1 - (2.55536) - (1.723608)] = -1.0557936$$

Hence, approximate solution of the given system of equations upto 3 decimal places is

$$x = 2.555, \ y = 1.723, \ z = -1.055.$$

EXERCISE 2.9

Solve the following system of equations by gauss-jacobi method:
1. $8x - 3y + 2z = 20, \ 4x + 11y - z = 33, \ 6x + 3y + 12z = 35$
2. $x + y + 54z = 110, \ 27x + 6y - z = 85, \ 6x + 15y + 2z = 72$
3. $5x - 2y + z = -4, \ x + 6y - 2z = -1, \ 3x + y + 5z = 13$
4. $30x - 2y + 3z = 75, \ 2x + 2y + 18z = 30, \ x + 17y - 2z = 48$

ANSWERS

1. 3.0167, 1.9857, 0.9116
2. 2.4252, 3.5727, 1.9259
3. −1.001, 0.999, 3
4. 2.5796, 2.7976, 1.0693

2.17 GAUSS-SEIDEL METHOD

Let us consider a system of equations in three unknowns x, y, z

$$a_1 x + b_1 y + c_1 z = d_1$$
$$a_2 x + b_2 y + c_2 z = d_2$$
$$a_3 x + b_3 y + c_3 z = d_3$$

Rewrite the above equations as

$$x = \frac{1}{a_1}[d_1 - b_1 y - c_1 z] \qquad \ldots(1)$$

$$y = \frac{1}{b_2}[d_2 - a_2 x - c_2 z] \qquad \ldots(2)$$

$$z = \frac{1}{c_3}[d_3 - a_3 x - b_3 y] \qquad \ldots(3)$$

Now this method can be used to the above equations if

$$|a_1| > |b_1| + |c_1|$$

$$|b_2| > |a_2| + |c_2|$$
$$|c_3| > |a_3| + |b_3|$$

Now taking $y^{(0)}$, $z^{(0)}$ as initial values for y, z; we get $x^{(1)}$ from equation (1)

$$x^{(1)} = \frac{1}{a_1}\left[d_1 - b_1 y^{(0)} - c_1 z^{(0)}\right]$$

Now using $x^{(1)}$, $z^{(0)}$ as initial values for x, z; we get $y^{(1)}$ from equation (2)

$$z^{(1)} = \frac{1}{b_2}\left[d_2 - a_2 x^{(1)} - c_2 z^{(0)}\right]$$

Now using $x^{(1)}$, $y^{(1)}$ as initial values for x, y; we get $z^{(1)}$ from equation (3)

$$z^{(1)} = \frac{1}{c_3}\left[d_3 - a_3 x^{(1)} - b_3 y^{(1)}\right]$$

Now, repeat this process taking $x^{(1)}$, $y^{(1)}$, $z^{(1)}$ as initial values.

This whole process is repeated till we get the required solution.

ILLUSTRATIVE EXAMPLES

Example 1. Solve the following system of equations by Gauss-Seidel method:
$$27x + 6y - z = 85$$
$$6x + 15y + 2z = 72$$
$$x + y + 54z = 110$$

Solution. Here, $|27| > |6| + |1|$
$|15| > |6| + |2|$
$|54| > |1| + |1|$

We have seen that the diagonal elements are dominant. Hence the Gauss-Seidel method can be applied.

The above equations can be written as

$$x = \frac{1}{27}[85 - 6y + z] \quad \ldots(1)$$

$$y = \frac{1}{15}[72 - 6x - 2z] \quad \ldots(2)$$

$$z = \frac{1}{54}[110 - x - y] \quad \ldots(3)$$

Let the initial values be $y = 0$, $z = 0$

First iteration:

$$x^{(1)} = \frac{1}{27}[85 - 6(0) + z(0)] = 3.148$$

$$y^{(1)} = \frac{1}{15}[72 - 6(3.148) - 2(0)] = 3.5408$$

$$z^{(1)} = \frac{1}{54}[110 - 3.148 - 3.5408] = 1.9132$$

Second iteration:

$$x^{(2)} = \frac{1}{27}[85 - 6(3.5408) + (1.9132)] = 2.4322$$

$$y^{(2)} = \frac{1}{15}[72 - 6(2.4322) - 2(1.9132)] = 3.5720$$

$$z^{(2)} = \frac{1}{54}[110 - 2.4322 - 3.572] = 1.9259$$

Third iteration:

$$x^{(3)} = \frac{1}{27}[85 - 6(3.572) + (1.9259)] = 2.4257$$

$$y^{(3)} = \frac{1}{15}[72 - 6(1.9259) - 2(2.4257)] = 3.7062$$

$$z^{(3)} = \frac{1}{54}[110 - 2.4257 - 3.7062] = 1.9235$$

Fourth iteration:

$$x^{(4)} = \frac{1}{27}[85 - 6(3.7062) + (1.9235)] = 2.3958$$

$$y^{(4)} = \frac{1}{15}[72 - 6(1.9235) - 2(2.3958)] = 3.7112$$

$$z^{(4)} = \frac{1}{54}[110 - 2.3958 - 3.7112] = 1.9239$$

Fifth iteration:

$$x^{(5)} = \frac{1}{27}[85 - 6(3.7112) + (1.9239)] = 2.3947$$

$$y^{(5)} = \frac{1}{15}[72 - 6(1.9239) - 2(2.3947)] = 3.7112$$

$$z^{(5)} = \frac{1}{54}[110 - 2.3947 - 3.7112] = 1.9239$$

Hence, the required solution is $x = 2.395$, $y = 3.7112$, $z = 1.9239$.

Example 2. Solve the following system of equations by Gauss-Seidel method:

$$2x + 5y - 2z = 3$$
$$x + y - 3z = -16$$
$$8x - y + z = 18$$

Solution. Rewrite the above equations as

$$8x - y + z = 18$$
$$2x + 5y - 2z = 3$$
$$x + y - 3z = -16$$

Here,
$$|8| > |-1| + |1|$$
$$|5| > |2| + |-2|$$
$$|-3| > |1| + |1|$$

∴ The above equations can be written as

$$x = \frac{1}{8}[18 + y - z] \qquad \ldots(1)$$

$$y = \frac{1}{5}[3 - 2x + 2z] \qquad \ldots(2)$$

$$z = \frac{1}{3}[16 + x + y] \qquad \ldots(3)$$

Let the initial values be $y = 0, z = 0$

First iteration:

$$x^{(1)} = \frac{1}{8}[18 + 0 - 0] = 2.25$$

$$y^{(1)} = \frac{1}{5}[3 - 2(2.25) + 2(0)] = -0.3$$

$$z^{(1)} = \frac{1}{3}[16 + 2.25 - 0.3] = 5.983$$

Second iteration:

$$x^{(2)} = \frac{1}{8}[18 - 0.3 - 5.983] = 1.4646$$

$$y^{(2)} = \frac{1}{5}[3 - 2(1.4646) + 2(5.983)] = 2.407$$

$$z^{(2)} = \frac{1}{3}[16 + 1.4646 + 2.407] = 6.6239$$

Third iteration:

$$x^{(3)} = \frac{1}{8}[18 + 2.407 - 6.6239] = 1.7228$$

$$y^{(3)} = \frac{1}{5}[3 - 2(1.7228) + 2(6.6239)] = 2.5604$$

$$z^{(3)} = \frac{1}{3}[16 + 1.7228 + 2.5604] = 6.7611$$

Fourth iteration:

$$x^{(4)} = \frac{1}{8}[18 + 2.5604 - 6.7611] = 1.7249$$

$$y^{(4)} = \frac{1}{5}[3-2(1.7249)+2(6.7611)] = 2.6144$$

$$z^{(4)} = \frac{1}{3}[16+1.7249+2.6144] = 6.7797$$

Fifth iteration:

$$x^{(5)} = \frac{1}{8}[18+2.6144-6.7797] = 1.7293$$

$$y^{(5)} = \frac{1}{5}[3-2(1.7293)+2(6.7797)] = 2.620$$

$$z^{(5)} = \frac{1}{3}[16+1.7293+2.620] = 6.783$$

Sixth iteration:

$$x^{(6)} = \frac{1}{8}[18+2.62-6.783] = 1.7296$$

$$y^{(6)} = \frac{1}{5}[3-2(1.7296)+2(6.783)] = 2.621$$

$$z^{(6)} = \frac{1}{3}[16+1.7296+2.621] = 6.7835$$

Hence, the required solution is $x = 1.729$, $y = 2.62$, $z = 6.7835$

Example 3. Solve the following system of equations by Gauss-Seidel method

$10x + y + 2z = 44$
$2x + 10y + z = 51$
$x + 2y + 10z = 61$ (I.U. 2008-09)

Solution. Here, $|10| > |1| + |2|$
$|10| > |2| + |1|$
$|10| > |1| + |2|$

∴ The above equations can be written as

$$x = \frac{1}{10}[44-y-2z] \qquad ...(1)$$

$$y = \frac{1}{10}[51-2x-z] \qquad ...(2)$$

$$z = \frac{1}{10}[61-x-2y] \qquad ...(3)$$

Let the initial values be $y = 0$, $z = 0$
First iteration:

$$x^{(1)} = \frac{1}{10}\left[44 - 0 - 2(0)\right] = 4.4$$

$$y^{(1)} = \frac{1}{10}\left[51 - 2(4.4) - 0\right] = 4.22$$

$$z^{(1)} = \frac{1}{10}\left[61 - 4.4 - 2(4.22)\right] = 4.816$$

Second iteration:

$$x^{(2)} = \frac{1}{10}\left[44 - 4.22 - 2(4.816)\right] = 3.0148$$

$$y^{(2)} = \frac{1}{10}\left[51 - 2(3.0148) - 4.816\right] = 4.0154$$

$$z^{(2)} = \frac{1}{10}\left[61 - 3.0148 - 2(4.0154)\right] = 4.995$$

Third iteration:

$$x^{(3)} = \frac{1}{10}\left[44 - 4.0154 - 2(4.995)\right] = 2.999$$

$$y^{(3)} = \frac{1}{10}\left[51 - 2(2.999) - 4.995\right] = 4.0007$$

$$z^{(3)} = \frac{1}{10}\left[61 - 2.999 - 2(4.0007)\right] = 4.9999$$

Fourth iteration:

$$x^{(4)} = \frac{1}{10}\left[44 - 4.0007 - 2(4.9999)\right] = 2.9999$$

$$y^{(4)} = \frac{1}{10}\left[51 - 2(2.9999) - 4.9999\right] = 4.0000$$

$$z^{(4)} = \frac{1}{10}\left[61 - 2.9999 - 2(4.0000)\right] = 5.0000$$

Fifth iteration:

$$x^{(5)} = \frac{1}{10}\left[44 - 4.0000 - 2(5.0)\right] = 3.0000$$

$$y^{(5)} = \frac{1}{10}\left[51 - 2(3.0) - 5.0\right] = 4.0000$$

$$z^{(5)} = \frac{1}{10}\left[61 - 3.0 - 2(4.0)\right] = 5.0000$$

Hence, the required solution is $x = 3, y = 4, z = 5$

EXERCISE 2.10

Solve the following system of equations by Gauss-Seidel Method:
1. $8x + y + z = 8$, $2x + 4y + z = 4$, $x + 3y + 5z = 5$
2. $8x - 3y + 2z = 20$, $4x + 11y - z = 33$, $6x + 3y + 12z = 35$
3. $2x - y + z = 5$, $x + 3y - 2z = 7$, $x + 2y + 3z = 10$
4. $5x - y + z = 10$, $2x + 4y = 12$, $x + y + 15z + 1 = 0$
5. $x + 3y + 6z = 2$, $x - 4y + 2z = 7$, $3x - y + 4z = 9$

ANSWERS

1. 0.876, 0.419, 0.574
2. 3.0168, 1.9859, 0.9118
3. 3, 2, 1
4. 2.556, 1.722, −1.055
5. 2, −1, 0.5

2.18 LIN-BAIRSTOW'S METHOD

This method is applied to find the complex roots of an algebraic equation with real coefficients. The complex roots of this type of equation occur in pairs $\alpha \pm i\beta$. Each such pair corresponds to a quadratic factor

$$\{x - (\alpha + i\beta)\}\{x - (\alpha - i\beta)\} = x^2 - 2\alpha x + \alpha^2 + \beta^2$$
$$= x^2 + px + q$$

where $p = -2\alpha$ and $q = \alpha^2 + \beta^2$ are real.

Let, $\quad f(x) = x^n + a_1 x^{n-1} + a_2 x^{n-2} + \ldots + a_{n-1} x + a_n$

On dividing $f(x)$ by $x^2 + px + q$, we get a quotient

$$Q_{n-2} = x^{n-2} + b_1 x^{n-3} + b_2 x^{n-4} + \ldots + b_{n-2}]$$

and a remainder $R_n = Rx + S$

Thus, $\quad f(x) = (x^2 + px + q)(x^{n-2} + b_1 x^{n-3} + \ldots + b_{n-2}) + Rx + S$

...(1)

If $(x^2 + px + q)$ divides $f(x)$ completely, then the remainder $R_n = Rx + S = 0$ i.e., $R = 0$, $S = 0$. Obviously R and S both depend upon p and q.

So our problem is to find p and q such that

$$R(p, q) = 0, \quad S(p, q) = 0 \qquad \ldots(2)$$

Let $p + \Delta p$, $q + \Delta q$ be the actual values of p and q which satisfy (2), then

$$R(p + \Delta p, q + \Delta q) = 0, \quad S(p + \Delta p, q + \Delta q) = 0$$

To find the corrections Δp, Δq we use the following equations

$$c_{n-2} \Delta p + c_{n-3} \Delta q = b_{n-1}$$
$$(c_{n-1} - b_{n-1}) \Delta p + c_{n-2} \Delta q = b_n$$

After finding the values of b_i's and c_i's by synthetic division method, we obtain approximate values of Δp and Δq say Δp_0 and Δq_0.

If p_0, q_0 be the initial approximations then their improved values are
$$p_1 = p_0 + \Delta p_0, \; q_1 = q_0 + \Delta q_0.$$

Now, taking p_1 and q_1 as the initial values and repeating the process, we can get better value of p and q.

Note:

1. Synthetic division method can be summarized as follows:

$a_0(=1)$	a_1	a_2	...	a_{n-1}	a_n	
	$-pb_0$	$-pb_1$	...	$-pb_{n-2}$	$-pb_{n-1}$	$-p$
		$-qb_0$	...	$-qb_{n-3}$	$-qb_{n-2}$	$-q$
$b_0(=1)$	b_1	b_2	...	b_{n-1}	bn	
	$-pc_0$	$-pc_1$	...	$-pc_{n-2}$		$-p$
		$-qc_0$	...	$-qc_{n-3}$		$-q$
$c_0(=1)$	c_1	c_2	...	c_{n-1}		

2. Generally, the values of p_0 and q_0 are given, otherwise we take the values of p and q in such a way that both R and S becomes zero.

3. Bairstow's method works well if the starting values of p and q are close to the correct values.

4. Process is repeated until approximate error comes out below to the pre specified tolerance.

$$|\varepsilon_p| = \left|\frac{\Delta p_i}{p_{i+1}}\right| \times 100\%$$

and

$$|\varepsilon_q| = \left|\frac{\Delta q_i}{q_{i+1}}\right| \times 100\%$$

ILLUSTRATIVE EXAMPLES

Example 1. Find the quadratic factor of $x^4 + 5x^3 + 3x^2 - 5x - 9 = 0$ starting with $p_0 = 3$, $q_0 = -5$ by using Bairstow's method.

Solution: We have

1	5	3	−5	−9	
	−3	−6	−6	3	−3
		5	10	10	5
1	2	2	−1	4	
	−3	3	−30		−3
		5	−5		5
1	−1	10	−36		

Here, $b_n = 4$, $b_{n-1} = -1$, $c_{n-1} = -36$, $c_{n-2} = 10$ and $c_{n-3} = -1$
$$c_{n-1} - b_{n-1} = -36 + 1 = -35$$
Corrections Δp_0 and Δq_0 can be obtained as follows
$$c_{n-2} \Delta p_0 + c_{n-3}\Delta q_0 = b_{n-1} \Rightarrow 10 \Delta p_0 - \Delta q_0 = -1$$
and $\quad (c_{n-1} - b_{n-1}) \Delta p_0 + c_{n-2} \Delta q_0 = b_n \Rightarrow -35\Delta p_0 + 10 \Delta q_0 = 4$
On solving these equations for Δp_0 and Δq_0, we have
$$\Delta p_0 = -0.09, \Delta q_0 = 0.08$$
Thus, the first approximation of p and q are given by
$$p_1 = p_0 + \Delta p_0 = 3 - 0.09 = 2.91$$
$$q_1 = q_0 + \Delta q_0 = -5 + 0.08 = -4.92$$

$$|\varepsilon_p| = \left|\frac{\Delta p_0}{p_1}\right| \times 100\% = \left|\frac{-0.09}{2.91}\right| \times 100\% = 3.0927\%$$

and $\quad |\varepsilon_q| = \left|\frac{\Delta q_0}{q_1}\right| \times 100\% = \left|\frac{0.08}{-4.92}\right| \times 100\% = 1.6260\%$

Repeating the same process *i.e.*, dividing the $f(x)$ by $x^2 + 2.91x - 4.92$, we get

1	5	3	−5	−9	
	−2.91	−6.08	−5.35	0.20	−2.91
		4.92	10.28	9.05	4.92
1	2.09	1.84	−0.07	0.25	
	−2.91	2.37	−26.57		−2.91
		4.92	−4.03		4.92
1	−0.82	9.13	−30.67		

At this step, the corrections Δp_1 and Δq_1 can be obtained as follows
$$9.13 \Delta p_1 - 0.82 \Delta q_1 = -0.07$$
and $-30.60 \Delta p_1 + 9.13 \Delta q_1 = 0.25$
On solving these equations for Δp_1 and Δq_1, we have
$$\Delta p_1 = -0.00745, \Delta q_1 = 0.00241$$
Thus, the second approximation of p and q are given by
$$p_2 = p_1 + \Delta p_1 = 2.91 - 0.00745 = 2.90255$$
$$q_2 = q_1 + \Delta q_1 = -4.92 + 0.00241 = -4.91759$$

$$|\varepsilon_p| = \left|\frac{\Delta p_1}{p_2}\right| \times 100\% = \left|\frac{-0.00745}{2.90255}\right| \times 100\% = 0.2566\%$$

and
$$|\varepsilon_q| = \left|\frac{\Delta q_1}{q_2}\right| \times 100\% = \left|\frac{0.00241}{-4.91759}\right| \times 100\% = 0.4901\%$$

Thus, a quadratic factor is
$$x^2 + 2.90255x - 4.91759$$

Dividing the given equation by this factor, we can obtain the other quadratic factor.

EXERCISE 2.11

1. Find the quadratic factor of the polynomial $x^4 - 5x^3 + 20x^2 - 40x + 60 = 0$ with $p_0 = -4$, $q_0 = 8$ by using Bairstow's method.
2. Find the quadratic factor of $x^4 - 3x^3 + 20x^2 + 44x + 54 = 0$ close to $x^2 + 2x + 2$ with $p_0 = 2$, $q_0 = 2$ by using Bairstow's method.

ANSWERS

1. $x^2 - 1.17x + 8.2125$ 2. $x^2 + 1.9412x + 1.953$

Chapter 3

Finite Differences

The calculus of finite differences deals with the changes that take place in the value of the function (dependent variable) due to finite changes. Assume that we have a table of values (x_i, y_i), $i = 0, 1, 2, ..., n$ of any function $y = f(x)$, the value of x being equally spaced with interval h i.e., $x_i = x_0 + ih$, $i = 0,1, 2, ..,n$. The independent variable x is called the argument and the corresponding dependent value y is called entry.

3.1 FORWARD DIFFERENCES

If $y_0, y_1, y_2, ..., y_n$ denote a set of values of y, then $y_1 - y_0, y_2 - y_1 .... , y_n - y_{n-1}$ are called the differences of y and these differences are denoted by $\Delta y_0, \Delta y_1, ..., \Delta y_{n-1}$ respectively i.e.,

$$\Delta y_0 = y_1 - y_0$$
$$\Delta y_1 = y_2 - y_1$$

In general $\Delta y_{n-1} = y_n - y_{n-1}$

Here Δ is called the forward difference operator and $\Delta y_0, \Delta y_1 .....$ are called first forward differences. The differences of the first forward differences are called second forward differences and are denoted by $\Delta^2 y_0, \Delta^2 y_1,.....$. Similarly, we can define third forward differences fourth forward differences etc. Thus $\Delta^2 y_0 = \Delta(\Delta y_0) = \Delta(y_1 - y_0)$

$$= \Delta y_1 - \Delta y_0$$
$$= (y_2 - y_1) - (y_1 - y_0)$$

$\Rightarrow \qquad \Delta^2 y_0 = y_2 - 2y_1 + y_0$

$$\Delta^2 y_1 = \Delta(\Delta y_1)$$
$$= \Delta(y_2 - y_1)$$
$$= \Delta y_2 - \Delta y_1$$
$$= (y_3 - y_2) - (y_2 - y_1)$$

$\Rightarrow \qquad \Delta^2 y_1 = y_3 - 2y_2 + y_1$

In general $\quad \Delta^2 y_r = \Delta(\Delta y_r) = \Delta(y_{r+1} - y_r)$
$\quad\quad\quad\quad\quad = \Delta y_{r+1} - \Delta y_r = y_{r+2} - 2y_{r+1} + y_r$

Similarly, $\quad \Delta^3 y_0 = \Delta(\Delta^2 y_0) = \Delta(y_2 - 2y_1 + y_0)$
$\quad\quad\quad\quad\quad = \Delta y_2 - 2 \Delta y_1 + \Delta y_0$
$\quad\quad\quad\quad\quad = (y_3 - y_2) - 2(y_2 - y_1) + (y_1 - y_0)$
$\quad\quad\quad\quad\quad = y_3 - 3y_2 + 3y_1 - y_0$

and $\quad \Delta^4 y_0 = \Delta(\Delta^3 y_0) = \Delta(y_3 - 3y_2 + 3y_1 - y_0)$
$\quad\quad\quad\quad\quad = y_4 - 4y_3 + 6y_2 - 4y_1 + y_0$

In general, n^{th} forward differences are defined as,

$$\boxed{\Delta^n y_r = \Delta^{n-1} y_{r+1} - \Delta^{n-1} y_r}$$

Usually, the arguments are taken as

$$x_0, x_0 + h, x_0 + 2h, \ldots$$

so that $x_1 - x_0 = x_2 - x_1 = x_3 - x_2 = \ldots = h$, where h is the interval of differencing.

Therefore, $\Delta f(x) = f(x + h) - f(x)$
$\Delta^2 f(x) = \Delta(\Delta f(x)) = \Delta(f(x + h) - f(x))$
$\quad\quad\quad = \Delta f(x + h) - \Delta f(x)$
$\quad\quad\quad = [f(x + 2h) - f(x + h)] - [f(x + h) - f(x)]$
$\Delta^2 f(x) = f(x + 2h) - 2f(x + h) + f(x)$
$\Delta^3 f(x) = \Delta^2 [\Delta f(x)] = \Delta [\Delta^2 f(x)]$
$\quad\quad\quad = \Delta [f(x + 2h) - 2f(x + h) + f(x)]$
$\quad\quad\quad = [f(x + 3h) - f(x + 2h)] - 2[f(x + 2h) - f(x + h)]$
$\quad\quad\quad\quad + [f(x + h) - f(x)]$
$\quad\quad\quad = f(x + 3h) - 3f(x + 2h) + 3f(x + h) - f(x)$ and so on.

It is clear that any higher order difference can easily be expressed in terms of the ordinates, since the coefficients occuring on the right hand side are the binomial coefficients*.

* $\quad\quad \Delta^n y_0 = y_n - {}^n C_1 y_{n-1} + {}^n C_2 y_{n-2} + \ldots + (-1)^n y_0.$

The following table shows how the forward differences of all orders can be formed:

Forward Difference Table

x	y	Δy	$\Delta^2 y$	$\Delta^3 y$	$\Delta^4 y$	$\Delta^5 y$
x_0	y_0					
$x_1 = x_0 + h$	y_1	Δy_0				
$x_2 = x_0 + 2h$	y_2	Δy_1	$\Delta^2 y_0$			
$x_3 = x_0 + 3h$	y_3	Δy_2	$\Delta^2 y_1$	$\Delta^3 y_0$		
$x_4 = x_0 + 4h$	y_4	Δy_3	$\Delta^2 y_2$	$\Delta^3 y_1$	$\Delta^4 y_0$	
$x_5 = x_0 + 5h$	y_5	Δy_4	$\Delta^2 y_3$	$\Delta^3 y_2$	$\Delta^4 y_1$	$\Delta^5 y_0$

Remark 1. The first entry y_0 is called leading term and $\Delta y_0, \Delta^2 y_0, \Delta^3 y_0, \ldots,$ are called leading differences.

2. The difference $\Delta y_0 = y_1 - y_0$ is written in the next column in between y_1 and y_0.
3. The forward differences is linear in nature *i.e.*, $\Delta[a f(x) + b g(x)] = a \Delta f(x) + b \Delta g(x)$; a and b being constants.

3.2 BACKWARD DIFFERENCES

The differences $y_1 - y_0, y_2 - y_1, \ldots, y_n - y_{n-1}$ are called first backward differences and is denoted by $\nabla y_1, \nabla y_2, \ldots, \nabla y_n$ respectively. That is,

$$\nabla y_1 = y_1 - y_0$$
$$\nabla y_2 = y_2 - y_1$$
$$\nabla y_n = y_n - y_{n-1}$$

where ∇ is called the backward difference operator.

Similarly, we can define backward differences of higher orders as

$$\nabla^2 y_2 = \nabla y_2 - \nabla y_1 = (y_2 - y_1) - (y_1 - y_0) = y_2 - 2y_1 + y_0$$
$$\nabla^3 y_3 = \nabla^2 y_3 - \nabla^2 y_2 = y_3 - 3y_2 + 3y_1 - y_0 \text{ etc.}$$

In general, $\nabla^n y_k = \nabla^{n-1} y_k - \nabla^{n-1} y_{k-1}$

Also, $\nabla f(x) = f(x) - f(x - h)$
$$\nabla^2 f(x) = \nabla[f(x) - f(x-h)] = \nabla f(x) - \nabla f(x-h)$$
$$= [f(x) - f(x-h)] - [f(x-h) - f(x-2h)]$$
$$= f(x) - 2f(x-h) + f(x-2h) \text{ and so on.}$$

Backward difference table

x	y	∇y	$\nabla^2 y$	$\nabla^3 y$	$\nabla^4 y$	$\nabla^5 y$
x_0	y_0					
$x_1 = x_0 + h$	y_1	∇y_1				
$x_2 = x_0 + 2h$	y_2	∇y_2	$\nabla^2 y_2$			
$x_3 = x_0 + 3h$	y_3	∇y_3	$\nabla^2 y_3$	$\nabla^3 y_3$		
$x_4 = x_0 + 4h$	y_4	∇y_4	$\nabla^2 y_4$	$\nabla^3 y_4$	$\nabla^4 y_4$	
$x_5 = x_0 + 5h$	y_5	∇y_5	$\nabla^2 y_5$	$\nabla^3 y_5$	$\nabla^4 y_5$	$\nabla^5 y_5$

Remark 1. In backward difference table y_5 is the leading term and ∇y_5, $\nabla^2 y_5$, ..., are called leading differences.

2. Backward difference is also linear in nature.

3.3 CENTRAL DIFFERENCES

The central difference operator δ is defined by

$$y_1 - y_0 = \delta y_{1/2}, \ y_2 - y_1 = \delta y_{3/2}, \ \ldots, \ y_n - y_{n-1} = \delta y_{n-1/2}$$

Also, $\delta f(x) = f(x + h/2) - f(x - h/2)$

or $\delta y_x = y_{x+h/2} - y_{x-h/2}$

Similarly, higher order central differences are defined as:

$\delta y_{3/2} - \delta y_{1/2} = \delta^2 y_1$, $\delta y_{5/2} - \delta y_{3/2} = \delta^2 y_2$ and so on.

Central difference table

x	y	δy	$\delta^2 y$	$\delta^3 y$	$\delta^4 y$	$\delta^5 y$
x_0	y_0					
x_1	y_1	$\delta y_{1/2}$				
x_2	y_2	$\delta y_{3/2}$	$\delta^2 y_1$			
x_3	y_3	$\delta y_{5/2}$	$\delta^2 y_2$	$\delta^3 y_{3/2}$		
x_4	y_4	$\delta y_{7/2}$	$\delta^2 y_3$	$\delta^3 y_{5/2}$	$\delta^4 y_2$	
x_5	y_5	$\delta y_{9/2}$	$\delta^2 y_4$	$\delta^3 y_{7/2}$	$\delta^4 y_3$	$\delta^5 y_{5/2}$

It is clear from the three tables that in a definite numerical case, the same numbers occur in the same positions whether we use forward, backward or central differences.

Thus, we obtain

$$\Delta y_0 = \nabla y_1 = \delta y_{1/2}$$

3.4 DIFFERENT TYPES OF OPERATORS

1. **Averaging operator (μ):** The averaging operator μ is defined by the equation

$$\mu y_x = \frac{1}{2}[y_{x+h/2} + y_{x-h/2}]$$

or

$$\mu f(x) = \frac{1}{2}[f(x+h/2) + f(x-h/2)]$$

where h is called interval of differencing.

2. **Shift operator E:** A shift operator E is defined by the equation

$$E y_x = y_{x+h}$$

or $\quad E f(x) = f(x+h)$

In particular $\quad E y_1 = y_2$, $E y_2 = y_3$ etc.

which shows that the effect of E is to shift the functional value y_x to the next higher value y_{x+h}. A second operation with E gives

$$E^2 y_x = E(E y_x) = E(y_{x+h}) = y_{x+2h}$$

$$E^3 y_x = y_{x+3h}$$

In general $\quad E^n y_x = y_{x+nh}$ and $E^n f(x) = f(x+nh)$

The inverse shift operator E^{-1} is defined by,

$$E^{-1} f(x) = f(x-h)$$

or $\quad E^{-1} y_x = y_{x-h}$

similarly $\quad E^{-r} y_x = y_{x-rh}$ or $E^{-r} f(x) = f(x-rh)$.

3. **Central difference operator (δ):** The central difference operator δ is defined by
$$\delta y_x = (y_{x+h/2} - y_{x-h/2})$$
or $\quad\quad \delta f(x) = f(x+h/2) - f(x-h/2)]$

4. **Differential operator D:** The differential operator D is defined by
$$Df(x) = \frac{d}{dx}[f(x)]$$

$$D^2 f(x) = \frac{d^2}{dx^2}[f(x)]$$

similarly $\quad D^n f(x) = \dfrac{d^n}{dx^n}[f(x)]$

3.5 RELATION BETWEEN OPERATORS

1. **Relation between Δ and E**
$$\Delta = E - 1 \text{ or } E = 1 + \Delta$$
Proof: We know that
$$\Delta y_0 = y_1 - y_0 = E y_0 - y_0 = (E - 1) y_0$$
$\Rightarrow \quad\quad \Delta = E - 1$
or $\quad\quad \boxed{E = 1 + \Delta}$

Note: Here 1 is not the numeral 1 but it the unit operator 1 which means $1\, f(x) = f(x)$.

2. **Relation between E and ∇**
$$\nabla = 1 - E^{-1} \text{ or } E = (1 - \nabla)^{-1}$$
Proof: We know that
$$\nabla y_x = y_x - y_{x-h} = y_x - E^{-1} y_x$$
$$\nabla y_x = (1 - E^{-1}) y_x$$
$\Rightarrow \quad\quad \nabla = 1 - E^{-1}$
$\Rightarrow \quad\quad E^{-1} = 1 - \nabla$
$\Rightarrow \quad\quad E = (1 - \nabla)^{-1}$ since $(E^{-1})^{-1} = E$

3. **Relation between E and δ**
$$\delta = E^{1/2} - E^{-1/2}$$
Proof: We know that
$$\delta y_x = (y_{x+h/2} - y_{x-h/2})$$
$$= (E^{1/2} y_x - E^{-1/2} y_x)$$
$$= (E^{1/2} - E^{-1/2}) y_x$$
$\therefore \quad\quad \delta = E^{1/2} - E^{-1/2}$

4. Relation between E and μ:
Proof. We know that

$$\mu y_x = \frac{1}{2}(y_{x+h/2} + y_{x-h/2})$$

$$\mu y_x = \frac{1}{2}(E^{1/2} + E^{-1/2})y_x$$

$$\mu = \frac{1}{2}(E^{1/2} + E^{-1/2})$$

5. $\Delta = E\nabla = \nabla E = \delta E^{1/2}$
Proof. $E(\nabla y_x) = E(y_x - y_{x-h}) = y_{x+h} - y_x = \Delta y_x$
$\Rightarrow \quad E\nabla = \Delta$
$\nabla(Ey_x) = \nabla y_{x+h} = y_{x+h} - y_x = \Delta y_x$
$\Rightarrow \quad \nabla E = \Delta$
$\delta E^{1/2} y_x = \delta y_{x+h/2} = y_{x+h} - y_x = \Delta y_x$
$\Rightarrow \quad \delta E^{1/2} = \Delta$

6. Relation between μ and δ.

$$\mu^2 = 1 + \frac{1}{4}\delta^2$$

or $\quad \mu = \sqrt{1 + \frac{1}{4}\delta^2}$

Proof. We know that

$$\mu = \frac{1}{2}(E^{1/2} + E^{-1/2})$$

$\Rightarrow \quad \mu^2 = \frac{1}{4}(E^{1/2} + E^{-1/2})^2 = \frac{1}{4}(E + E^{-1} + 2)$

$\quad \mu^2 = \frac{1}{4}[(E^{1/2} - E^{-1/2})^2 + 4] = \frac{1}{4}(\delta^2 + 4)$

$\Rightarrow \quad \mu = \sqrt{1 + \frac{1}{4}\delta^2}$

7. $E = e^{hD}$
Proof: We know that
$Ef(x) = f(x+h)$

$= f(x) + hf'(x) + \frac{h^2}{2!}f''(x) + \dots$ (By Taylor's series)

$$= f(x) + hDf(x) + \frac{h^2}{2!}D^2f(x) + \ldots$$

$$= [1 + hD + \frac{(hD)^2}{2!} + \ldots]f(x)$$

$$Ef(x) = e^{hD}f(x)$$
$$\Rightarrow \quad E = e^{hD} \quad \text{or} \quad \Delta = e^{hD} - 1$$

3.6 DIFFERENCES OF A POLYNOMIAL

Theorem. The nth differences of a polynomial of the nth degree are constants. That is, if
$$f(x) = a_0 x^n + a_1 x^{n-1} + a_2 x^{n-2} + \ldots + a_n$$
Then $\quad \Delta^n f(x) = a_0 n! \, h^n$
where h is the interval of differencing.

Proof: Let $f(x) = a_0 x^n + a_1 x^{n-1} + a_2 x^{n-2} + \ldots + a_n$
$\Delta f(x) = f(x+h) - f(x)$
$\Delta f(x) = a_0 [(x+h)^n - x^n] + a_1 [(x+h)^{n-1} - x^{n-1}] + \ldots + a_n$
$\Delta f(x) = a_0 nh \, x^{n-1} + a'_1 x^{n-2} + a'_2 x^{n-3} + \ldots + k'x + l' \ldots(1)$

where $a_1', a_2', \ldots, l'$ are new constant coefficients.

$\Delta f(x) = a_0 nh \, x^{n-1}$ + terms involving powers of x less than $(n-1)$

That is, $\Delta f(x) = $ a polynomial of degree $(n-1)$.

$\Delta^2 f(x) = a_0 nh [(x+h)^{n-1} - x^{n-1}]$ + terms involving lesser degree
$= a_0 n(n-1) h^2 x^{n-2}$ + terms involving degree less than $(n-2)$

i.e., second difference of a polynomial of degree n is a polynomial of degree x^{n-2}.

Proceeding like this
$$\Delta^n f(x) = a_0 n! \, h^n x^0$$
$$\Delta^n f(x) = a_0 n! \, h^n$$

Note 1. The converse of the above theorem is also true. That is, if the nth differences of a tabulated function are constants, then the function is a polynomial of degree n.

    2. The $(n+1)^{\text{th}}$ and higher differences of a polynomial of degree n are zeros.

ILLUSTRATIVE EXAMPLES

Example 1. Prove that
  (i) $\Delta [f(x) g(x)] = f(x+h) \Delta g(x) + g(x) \Delta f(x)$

(ii) $\Delta\left[\dfrac{f(x)}{g(x)}\right] = \dfrac{g(x)\,\Delta f(x) - f(x)\,\Delta g(x)}{g(x+h)\,g(x)}$

Proof : (i) $\Delta[f(x)\,g(x)] = f(x+h)\,g(x+h) - f(x)\,g(x)$
$= [f(x+h)\,g(x+h) - f(x+h)\,g(x)]$
$\quad + [f(x+h)\,g(x) - f(x)\,g(x)]$
$= f(x+h)\,[g(x+h) - g(x)] + g(x)\,[f(x+h) - f(x)]$
$= f(x+h)\,\Delta g(x) + g(x)\,\Delta f(x)$

(ii) $\Delta\left[\dfrac{f(x)}{g(x)}\right] = \dfrac{f(x+h)}{g(x+h)} - \dfrac{f(x)}{g(x)}$

$= \dfrac{f(x+h)\,g(x) - g(x+h)\,f(x)}{g(x+h)\,g(x)}$

$= \dfrac{f(x+h)\,g(x) - f(x)\,g(x) + f(x)\,g(x) - g(x+h)\,f(x)}{g(x+h)\,g(x)}$

$= \dfrac{g(x)\,[f(x+h) - f(x)] - f(x)\,[g(x+h) - g(x)]}{g(x+h)\,g(x)}$

$= \dfrac{g(x)\,\Delta f(x) - f(x)\,\Delta g(x)}{g(x+h)\,g(x)}$

Example 2. Evaluate (i) $\Delta^n(e^{ax+b})$ (ii) $\Delta^n[\sin(ax+b)]$ (iii) $\Delta^n[\cos(ax+b)]$ (iv) $\Delta[\log(ax+b)]$ (v) $\Delta \log f(x)$

Solution. (i) $\Delta(e^{ax+b}) = e^{a(x+h)+b} - e^{ax+b} = e^{ax+b}(e^{ah} - 1)$
$\Delta^2(e^{ax+b}) = (e^{ah} - 1)\,\Delta(e^{ax+b}) = (e^{ah} - 1)^2\,e^{ax+b}$

Proceeding like this, we get
$\Delta^n(e^{ax+b}) = e^{ax+b}(e^{ah} - 1)^n$.

(ii) $\Delta[\sin(ax+b)] = \sin[a(x+h)+b] - \sin(ax+b)$

$= 2\cos\left(ax+b+\dfrac{ah}{2}\right)\sin\dfrac{ah}{2}$

$= 2\sin\dfrac{ah}{2} \cdot \sin\left(\dfrac{\pi}{2}+ax+b+\dfrac{ah}{2}\right)$

$= 2\sin\dfrac{ah}{2} \cdot \sin\left(ax+b+\dfrac{\pi+ah}{2}\right)$

$\Delta^2\sin(ax+b) = 2\sin\dfrac{ah}{2}\,\Delta\left[\sin\left(ax+b+\dfrac{\pi+ah}{2}\right)\right]$

$\Delta^2\sin(ax+b) = \left(2\sin\dfrac{ah}{2}\right)^2 \cdot \sin\left[ax+b+2\left(\dfrac{\pi+ah}{2}\right)\right]$

Proceeding like this, we get

$$\Delta^n \sin(ax+b) = \left(2\sin\frac{ah}{2}\right)^n \cdot \sin\left[ax+b+\frac{n(\pi+ah)}{2}\right]$$

(iii) $\Delta[\cos(ax+b)] = \cos(ax+ah+b) - \cos(ax+b)$

$$= -2\sin\left(ax+b+\frac{ah}{2}\right)\sin\frac{ah}{2}$$

$$= 2\sin\frac{ah}{2}\cos\left(\frac{\pi}{2}+ax+b+\frac{ah}{2}\right)$$

$$= 2\sin\frac{ah}{2}\cos\left(ax+b+\frac{\pi+ah}{2}\right)$$

$$\Delta^2[\cos(ax+b)] = \left(2\sin\frac{ah}{2}\right)^2 \cdot \cos\left(ax+b+\frac{2(\pi+ah)}{2}\right)$$

Proceeding like this, we get

$$\Delta^n[\cos(ax+b)] = \left(2\sin\frac{ah}{2}\right)^n \cdot \cos\left[ax+b+\frac{n(\pi+ah)}{2}\right]$$

(iv) $\Delta[\log(ax+b)] = \log(ax+ah+b) - \log(ax+b)$

$$= \log\left[\frac{ax+ah+b}{ax+b}\right]$$

$$= \log\left[1+\frac{ah}{ax+b}\right]$$

$$= \log\left[1+\frac{\Delta(ax+b)}{ax+b}\right]$$

(v) $\quad \Delta \log f(x) = \log f(x+h) - \log f(x)$

$$= \log\left[\frac{f(x+h)}{f(x)}\right] = \log\left[\frac{E f(x)}{f(x)}\right]$$

$$= \log\left[\frac{f(x)+\Delta f(x)}{f(x)}\right]$$

$$= \log\left[1+\frac{\Delta f(x)}{f(x)}\right]$$

Example 3. Prove that (i) $1 + \mu^2 \delta^2 = \left(1+\frac{1}{2}\delta^2\right)^2$ (ii) $E^{1/2} = \mu + \frac{1}{2}\delta$

(iii) $E^{-1/2} = \mu - \frac{1}{2}\delta$ (iv) $\mu\delta = \frac{1}{2}\Delta E^{-1} + \frac{1}{2}\Delta$.

Solution. (i) $1 + \mu^2 \delta^2 = \left(1+\frac{1}{2}\delta^2\right)^2$

$$1 + \mu^2 \delta^2 = 1 + \left(\frac{E^{1/2} + E^{-1/2}}{2}\right)^2 (E^{1/2} - E^{-1/2})^2$$

$$= 1 + \left(\frac{E - E^{-1}}{2}\right)^2$$

$$= \frac{4 + (E - E^{-1})^2}{4} = \left(\frac{E + E^{-1}}{2}\right)^2 \quad \ldots(1)$$

$$\left(1 + \frac{1}{2}\delta^2\right)^2 = \left[1 + \frac{1}{2}(E^{1/2} - E^{-1/2})^2\right]^2$$

$$= \left[1 + \frac{1}{2}(E + E^{-1} - 2)\right]^2$$

$$\left(1 + \frac{1}{2}\delta^2\right)^2 = \left[\frac{E + E^{-1}}{2}\right]^2 \quad \ldots(2)$$

Form (1) and (2), we get

$$1 + \frac{1}{2}\mu^2\delta^2 = \left(1 + \frac{1}{2}\delta^2\right)^2$$

(ii)
$$E^{\frac{1}{2}} = \mu + \frac{1}{2}\delta$$

$$\mu + \frac{1}{2}\delta = \frac{E^{\frac{1}{2}} + E^{-1/2}}{2} + \frac{E^{\frac{1}{2}} - E^{-1/2}}{2} = E^{\frac{1}{2}}$$

(iii)
$$E^{-1/2} = \mu - \frac{1}{2}\delta$$

$$\mu - \frac{1}{2}\delta = \frac{E^{\frac{1}{2}} + E^{-1/2}}{2} - \frac{1}{2}(E^{\frac{1}{2}} - E^{-1/2}) = E^{-1/2}$$

(iv)
$$\mu\delta = \frac{1}{2}\Delta E^{-1} + \frac{1}{2}\Delta$$

$$\frac{1}{2}\Delta E^{-1} + \frac{1}{2}\Delta = \frac{1}{2}\Delta(E^{-1} + 1) = \frac{1}{2}(E - 1)(E^{-1} + 1)$$

$$= \frac{1}{2}(E - E^{-1}) = \mu\delta$$

Example 4. Prove that (i) $\frac{1}{2}\delta^2 + \delta\sqrt{1+\frac{\delta^2}{4}} = \Delta$ (ii) $\nabla\Delta = \Delta - \nabla = \delta^2$

(iii) $\mu\delta = \frac{1}{2}(\Delta + \nabla)$ (iv) $(1 + \Delta)(1 - \nabla) = 1$.

Proof: (i) $\frac{1}{2}\delta^2 + \delta\sqrt{1+\frac{\delta^2}{4}}$

$$= \frac{1}{2}(E^{\frac{1}{2}} - E^{-1/2})^2 + (E^{1/2} - E^{-1/2})\sqrt{1+\frac{1}{4}(E^{1/2} - E^{-1/2})^2}$$

$$\frac{1}{2} + (E + E^{-1} - 2) + (E^{1/2} - E^{-1/2})\left(\frac{(E^{1/2} + E^{-1/2})}{2}\right)$$

$$= \frac{1}{2}(2E - 2) = E - 1 = \Delta.$$

(ii) $\nabla\Delta = (1 - E^{-1})(E - 1) = E + E^{-1} - 2 = (E^{1/2} - E^{-1/2})^2 = \delta^2$
$\Delta - \nabla = (E - 1) - (1 - E^{-1}) = E + E^{-1} - 2 = \delta^2$

(iii) $\mu\delta = \frac{1}{2}(\Delta + \nabla)$

$\frac{1}{2}(\Delta + \nabla) = \frac{1}{2}[E - 1 + 1 - E^{-1}] = \frac{1}{2}(E - E^{-1}) = \mu\delta$

(iv) $(1 + \Delta)(1 - \nabla) = E \cdot E^{-1} = 1$.

Example 5. Prove that
(i) $(E^{1/2} + E^{-1/2})(1 + \Delta)^{1/2} = \Delta + 2$
(ii) $\Delta^3 y_2 = \nabla^3 y_5$

Solution. (i) $(E^{1/2} + E^{-1/2})(1 + \Delta)^{1/2} = (E^{1/2} + E^{-1/2})E^{1/2}$
$= E + 1 = 1 + \Delta + 1 = \Delta + 2$

(ii) $\Delta^3 y_2 = (E - 1)^3 y_2$
$= (E^3 - 3E^2 + 3E - 1)y_2$
$= y_5 - 3y_4 + 3y_3 - y_2$
$\nabla^3 y_5 = (1 - E^{-1})^3 y_5$
$= (1 - 3E^{-1} + 3E^{-2} - E^{-3}) y_5 = y_5 - 3y_4 + 3y_3 - y_2.$

Example 6. Construct the forward difference table, given that

x :	0	1	2	3	4	5	6
y :	1	1	9	31	73	141	241

Solution. Forward difference table is as follows:

x	y	Δy	$\Delta^2 y$	$\Delta^3 y$	$\Delta^4 y$
0	1				
1	1	0			
2	9	8	8		
3	31	22	14	6	0
4	73	42	20	6	0
5	141	68	26	6	0
6	241	100	32	6	

Example 7. Construct the backward difference table, given that

x : 20   25   30   35   40   45
y : 354   332   291   260   231   204

Solution. Backward difference table is as follows:

x	y	∇y	$\nabla^2 y$	$\nabla^3 y$	$\nabla^4 y$	$\nabla^5 y$
20	354					
25	332	−22				
30	291	−41	−19			
35	260	−31	10	29		
40	231	−29	2	−8	−37	
45	204	−27	2	0	8	45

Note. To find y_k in terms of $y_0, \Delta y_0, \Delta^2 y_0, \ldots$

$$y_k = y_0 + k_{C_1} \Delta y_0 + k_{C_2} \Delta^2 y_0 + \ldots + \Delta^k y_0$$

(k is + ve integer)

Example 8. Find the first term of the series whose second and subsequent terms are 8, 3, 0, − 1, 0,......

Solution: Let y_0 be the first term.

∴ $y_1 = 8, y_2 = 3, y_3 = 0, y_4 = −1$ etc

The difference table is

x	y	Δy	$\Delta^2 y$	$\Delta^3 y$	$\Delta^4 y$
0	y_0				
1	8 (y_1)	$8 - y_0$ (Δy_0)			
2	3 (y_2)	−5 (Δy_1)	$-13 + y_0$		
3	0	−3	2	$15 - y_0$	$-15 + y_0$
4	−1	−1	2	0	0
5	0	1	2		

From the table, we have
$$\Delta y_1 = -5, \nabla^2 y_1 = 2, \Delta^3 y_1 = 0, \Delta^4 y_1 = 0$$

$$\therefore \quad y_0 = E^{-1} y_1 = (1 + \Delta)^{-1} y_1$$
$$= (1 - \Delta + \Delta^2 - \Delta^3 + \Delta^4 - \ldots) y_1$$
$$= y_1 - \Delta y_1 + \Delta^2 y_1 - \Delta^3 y_1 + \Delta^4 y_1 - \ldots$$
$$= 8 - (-5) + 2 - 0 + 0$$
$$y_0 = 15$$

Example 9. Given the set of values

| x : | 10 | 15 | 20 | 25 | 30 | 35 |
| y : | 20 | 22 | 23 | 24 | 25 | 27 |

form the difference table and find the values of $\Delta^2 y_{10}$, Δy_{20}, $\Delta^3 y_{15}$ and $\Delta^5 y_{10}$.

Solution Forward difference table is :

x	y	Δy	Δ²y	Δ³y	Δ⁴y	Δ⁵y
10	19.97					
		1.54				
15	21.51		−0.58			
		0.96		0.67		
20	22.47		0.09		−0.68	
		1.05		−0.01		0.72
25	23.52		0.08		0.04	
		1.13		0.03		
30	24.65		0.11			
		1.24				
35	25.89					

From the table,
$$\Delta^2 y_{10} = -0.58, \Delta y_{20} = 1.05$$
$$\Delta^3 y_{15} = -0.01$$
and $$\Delta^5 y_{10} = 0.72$$

Example 10. Construct a backward difference table, given

$$\sin 30° = 0.5000$$
$$\sin 35° = 0.5736$$
$$\sin 40° = 0.6428$$
and $$\sin 45° = 0.7071.$$

Assuming that the third difference are constants, find sin 50°.

Solution. Backward difference table is :

x	y = sin x	∇y	∇²y	∇³y
30°	0.5000			
		0.0736		
35°	0.5736		−0.0044	
		0.0692		−0.0005
40°	0.6428		−0.0049	
		0.0643		
45°	0.7071			

since third difference are constants

$$\therefore \quad \nabla^3 y_{50} = -0.0005$$
$$\Rightarrow \quad \nabla^2 y_{50} - \nabla^2 y_{45} = -0.0005$$
$$\Rightarrow \quad \Delta^2 y_{50} - (-0.0049) = -0.0005$$
$$\Rightarrow \quad \Delta^2 y_{50} = -0.0054$$

Again, $\quad \nabla y_{50} - \nabla y_{45} = -0.0054$

$$\Rightarrow \quad \nabla y_{50} = -0.0054 + \nabla y_{45}$$
$$= -0.0054 + 0.0643$$
$$\nabla y_{50} = 0.0589$$

Again, $\quad y_{50} - y_{45} = 0.0589$

$$\Rightarrow \quad y_{50} = y_{45} + 0.0589$$
$$\Rightarrow \quad y_{50} = 0.7071 + 0.0589$$
$$\Rightarrow \quad y_{50} = 0.7660$$
$$\therefore \quad \sin 50° = 0.7660$$

Example 11. If the third differences are constants, find u_6 given $u_0 = 9$, $u_1 = 18$, $u_2 = 20$ and $u_3 = 24$

Solution. Difference table is

x	y	Δy	$\Delta^2 y$	$\Delta^3 y$
0	9			
		9		
1	18		-7	
		2		9
2	20		2	
		4		$u_4 - 30$
3	24		$u_4 - 28$	
		$u_4 - 24$		$u_5 - 3u_4 + 52$
4	u_4		$u_5 - 2u_4 + 24$	
		$u_5 - u_4$		$u_6 - 3u_5 + 3u_4 - 24$
5	u_5		$u_6 - 2u_5 + u_4$	
		$u_6 - u_5$		
6	u_6			

since third difference is consatnt

$$\therefore \quad u_4 - 30 = 9$$
$$u_4 = 39$$

Now $\quad u_5 - 3u_4 + 52 = 9$

$$u_5 - 117 + 52 = 9$$
$$\Rightarrow \quad u_5 = 74$$

Again, $\quad u_6 - 3u_5 + 3u_4 - 24 = 9$

$$\Rightarrow \quad u_6 - 222 + 117 - 24 = 9$$
$$\Rightarrow \quad u_6 = 138$$

Example 12. Prove that $u_4 = u_3 + \Delta u_2 + \Delta^2 u_1 + \Delta^3 u_1$, by the method of separation of symbols.

Solution. $\quad$ R.H.S $= u_3 + \Delta u_2 + \Delta^2 u_1 + \Delta^3 u_1$
$$= E^2 u_1 + \Delta E u_1 + \Delta^2 (1 + \Delta) u_1$$
$$= (E^2 + \Delta E + \Delta^2 E) u_1$$

$$= E(E + \Delta + \Delta^2)u_1$$
$$= E(E + \Delta E)u_1$$
$$= E^2(1 + \Delta)u_1$$
$$= E^2 \cdot Eu_1 = E^3 u_1 = u_4$$
$$= \text{L.H.S.}$$

Example 13. Prove the following by the method of separation of symbols.

$$u_0 - u_1 + u_2 - u_3 + \ldots = \frac{u_0}{2} - \frac{\Delta u_0}{4} + \frac{\Delta^2 u_0}{8} - \frac{\Delta^3 u_0}{16}$$

Solution. L.H.S. $= u_0 - u_1 + u_2 - u_3 + \ldots$
$$= u_0 - Eu_0 + E^2 u_0 - E^3 u_0$$
$$= (1 - E + E^2 - E^3 + \ldots)u_0$$
$$= (1 + E)^{-1} u_0$$
$$= (1 + 1 + \Delta)^{-1} u_0 = (2 + \Delta)^{-1} u_0$$
$$= \frac{1}{2}\left(1 + \frac{\Delta}{2}\right)^{-1} u_0$$
$$= \frac{1}{2}\left(1 - \frac{\Delta}{2} + \frac{\Delta^2}{4} - \frac{\Delta^3}{8} + \ldots\right) u_0$$
$$= \frac{u_0}{2} - \frac{\Delta u_0}{4} + \frac{\Delta^2 u_0}{8} - \frac{\Delta^3 u_0}{16} + \ldots$$
$$= \text{R.H.S.}$$

Example 14. Calculate $\Delta^4 y_6$ if $y_6 = 2$, $y_7 = -7$, $y_8 = 9$, $y_9 = 11$, $y_{10} = 19$.

Solution.
$$\Delta^4 y_6 = (E - 1)^4 y_6$$
$$= (E^4 - 4E^3 + 6E^2 - 4E + 1)y_6$$
$$= E^4 y_6 - 4E^3 y_6 + 6E^2 y_6 - 4E y_6 + y_6$$
$$= y_{10} - 4y_9 + 6y_8 - 4y_7 + y_6$$
$$= 19 - 44 + 54 + 28 + 2 = 59$$

Example 15. Find the 6th term of the sequence, 7, 31, 72, 131, 225.
Solution. Difference table is

x	y	Δy	$\Delta^2 y$	$\Delta^3 y$	$\Delta^4 y$
0	7				
		24			
1	31		17		
		41		1	
2	72		18		16
		59		17	
3	131		35		
		94			
4	225				

102 Numerical and Statistical Techniques

$$6^{th} \text{ term} = y_5 = E^5 y_0 = (1 + \Delta)^5 y_0$$
$$= (1 + 5_{C_1}\Delta + 5_{C_2}\Delta^2 + 5_{C_3}\Delta^3 + 5_{C_4}\Delta^4 + 5_{C_5}\Delta^5) y_0$$
$$= y_0 + 5\Delta y_0 + 10\Delta^2 y_0 + 10\Delta^3 y_0 + 5\Delta^4 y_0 + \Delta^5 y_0$$
$$= 7 + 5(24) + 10(17) + 10(1) + 5(16) + 0$$
$$= 7 + 120 + 170 + 10 + 80$$
$$6^{th} \text{ term} = y_5 = E^5 y_0 = 387$$

Example 16. Find the first term of the series whose second and subsequent terms are 36, 131, 358, 807, 1592, 2851.

Solution.

x	y	Δy	$\Delta^2 y$	$\Delta^3 y$	$\Delta^4 y$	$\Delta^5 y$
1	36					
		95				
2	131		132			
		227		90		
3	358		222		24	
		449		114		0
4	807		336		24	
		785		138		
5	1592		474			
		1259				
6	2851					

We have to find I^{st} term which is corresponding to $x = 1$. i.e.,

$$y_0 = E^{-1} y_1$$
$$= (1 + \Delta)^{-1} y_1$$
$$= (1 - \Delta + \Delta^2 - \Delta^3 + \Delta^4 - \Delta^5 + \ldots) y_1$$
$$= y_1 - \Delta y_1 + \Delta^2 y_1 - \Delta^3 y_1 + \Delta^4 y_1 - \Delta^5 y_1 + \ldots$$
$$= 36 - 95 + 132 - 90 + 24 - 0$$
$$= 7$$

Example 17. Find the sixth term of the sequence 8, 12, 19, 29, 42, ... if its third differences are zero.

Solution.

x	y	Δy	$\Delta^2 y$	$\Delta^3 y$
0	8			
		4		
1	12		3	
		7		0
2	19		3	
		10		0
3	29		3	
		13		$y_5 - 58$
4	42		$y_5 - 55$	
		$y_5 - 42$		
5	y_5			

$$\therefore \quad y_5 - 58 = 0$$
$$\Rightarrow \quad y_5 = 58$$

Example 18. Obtain the function $f(x)$ and the sixth term of the secies 0, 4, 16, 42, 88

Solution.

x	$y = f(x)$	$\Delta f(x)$	$\Delta^2 f(x)$	$\Delta^3 f(x)$	$\Delta^4 f(x)$
0	0				
		4			
1	4		8		
		12		6	
2	16		14		0
		26		6	
3	42		20		
		46			
4	88				

Now function $f(x) = y_x = E^x y_0$
$= (1 + \Delta)^x y_0$
$= (1 + {}^xC_1 \Delta + {}^xC_2 \Delta^2 + {}^xC_3 \Delta^3 + {}^xC_4 \Delta^4) y_0$
$= y_0 + {}^xC_1 \Delta y_0 + {}^xC_2 \Delta^2 y_0 + {}^xC_3 \Delta^3 y_0 + {}^xC_4 \Delta^4 y_0$
$= 0 + x(4) + \dfrac{x(x-1)}{2}(8) + \dfrac{x(x-1)(x-2)}{6}(6) + 0$
$= 4x + 4x^2 - 4x + x^3 - 3x^2 + 2x$
$= x^3 + x^2 + 2x$

$\Rightarrow \quad f(x) = x^3 + x^2 + 2x$

$\therefore \quad 6^{\text{th}}$ term $= f(5) = y_5 = (5)^3 + (5)^2 + 2(5)$
$= 160$

Example 19. Given $y_1 + y_7 = -786$, $y_2 + y_6 = 686$, $y_3 + y_5 = 1088$, and y_x is a polynomial of degree 5, find y_4.

Solution. $\because y_x$ is a polynomial of degree 5, then

$\Rightarrow \qquad \qquad \qquad \qquad \qquad \qquad \Delta^6 y_1 = 0$
$\qquad \qquad \qquad \qquad \qquad \qquad (E - 1)^6 y_1 = 0$
$\Rightarrow \qquad \qquad \qquad \qquad \qquad \qquad (1 - E)^6 y_1 = 0$
$(1 - {}^6C_1 E + {}^6C_2 E^2 - {}^6C_3 E^3 + {}^6C_4 E^4 - {}^6C_5 E^5 + {}^6C_6 E^6) y_1 = 0$
$\Rightarrow \qquad \qquad y_1 - 6y_2 + 15y_3 - 20y_4 + 15y_5 - 6y_6 + y_7 = 0$
$\Rightarrow \qquad \qquad (y_1 + y_7) - 6(y_2 + y_6) + 15(y_3 + y_5) - 20y_4 = 0$
$\qquad \qquad \qquad -786 - 6(686) + 15(1088) - 20y_4 = 0$
$\qquad \qquad \qquad \qquad \qquad 11418 - 20 y_4 = 0$
$\Rightarrow \qquad \qquad \qquad \qquad \qquad \qquad y_4 = 570.9$

EXERCISE 3.1

1. If $y = x^3 + x^2 - 2x + 1$, find the values of y for $x = 0, 1, 2, 3, 4, 5$ and form the forward difference table, find the value of y at $x = 6$ by extending the table.
2. For the function $y = \sin hx$, draw the difference table for argument 1.5, 1.6, 1.7, 1.8, ... ,2.1.
3. Form a backward difference table for $y = \log x$ given that

x :	10	20	30	40	50
y :	1	1.3010	1.4771	1.6021	1.6990

   and find values of $\nabla^3 \log 40$ and $\nabla^4 \log 50$.
4. Find the 7th term of the sequence 2, 9, 28, 65, 126, 217 and also find the general term.
5. Construct the table of differences for the data given below.

x :	0	1	2	3	4
$f(x) = y$:	1.0	1.5	2.2	3.1	4.6

   Evaluate $\Delta^3 f(1)$ and $\Delta^2 f(2)$
6. Find the first term of the series whose second and subsequent terms are 8, 3, 0, – 1, 0, ...
7. Find the fifth term of the sequence 3, 6, 11, 18.
8. Assuming that the following values of y belong to a polynomial of degree 4, compute the next three values.

x:	0	1	2	3	4	5	6	7
y:	1	–1	1	–1	1	–	–	–

9. Prove that: $\Delta^5 y_1 = y_6 - 5y_5 + 10y_4 - 10y_3 + 5y_2 - y_1$
10. Prove that : $y_3 = y_2 + \Delta y_1 + \Delta^2 y_0 + \Delta^3 y_0$
11. Find y_6 if $y_0 = 9, y_1 = 18, y_2 = 20, y_3 = 24$ given that the third differences are constants.
12. Evaluate: (a) $\Delta^n(a^{bx+c})$ (b) $\Delta^n [\log (ax + b)]$ (c) $\Delta \tan^{-1}\left(\dfrac{x-1}{x}\right)$.
13. Evaluate:
    (a) $\Delta[x^2 + 2x]$
    (b) Δab^{cx}
    (c) $\Delta[e^{ax} \log bx]$
    (d) $\Delta\left[\dfrac{1}{1+x^2}\right], h = 1$
    (e) $\Delta[\sin x \cos 3x]$
    (f) $\Delta[\tan^{-1} bx]$
    (g) $\Delta(xe^x)$
    (h) $\Delta\left[\dfrac{2^x}{x+1}\right]$

14. Evaluate:
    (a) $\Delta^3 [(1 - x)(1 - 2x)(1 - 3x)]$, $h = 1$
    (b) $\Delta^2 \left[\dfrac{1}{x^2 + 5x + 6} \right]$, $h = 1$
    (c) $\Delta^2 \left(\dfrac{1}{x} \right)$

15. Evaluate, internal of differencing being h.
    (a) $(2\Delta^2 + \Delta - 1)(x^2 + 2x + 1)$
    (b) $(\Delta + 1)(2\Delta - 1)(x^2 + 2x + 1)$

16. Prove that $\Delta \left(\dfrac{1}{f(x)} \right) = - \dfrac{\Delta f(x)}{f(x) f(x+1)}$

17. Evaluate:
    (a) $\Delta^{10}(1 - ax)(1 - bx^2)(1 - cx^3)(1 - dx^4)$
    (b) $\Delta^n (ax^n + bx^{n-1})$

18. Prove that : $\Delta^2 \left[\dfrac{a^{2x} + a^{4x}}{(a^2 - 1)^2} \right] = a^{2x} + (a^2 + 1)^2 a^{4x}$

19. Prove that:
    (a) $\delta = \Delta(1 + \Delta)^{-1/2} = \nabla(1 - \nabla)^{-1/2}$
    (b) $\nabla = 1 - e^{-hD}$ (c) $hD = \log(1 + \Delta) = -\log(1 - \nabla) = \sin^{-1}(\mu\delta)$
    (d) $E^{-1/2} = \mu - \dfrac{\delta}{2}$ (e) $\nabla \Delta = \Delta - \nabla = \delta^2$
    (f) $1 + \mu^2 \delta^2 = \left(1 + \dfrac{1}{2} \delta^2 \right)^2$ (g) $\Delta = \dfrac{1}{2} \delta^2 + \delta \sqrt{1 + \dfrac{\delta^2}{4}}$
    (h) $\mu = \dfrac{2 + \Delta}{2\sqrt{1 + \Delta}} = \sqrt{1 + \dfrac{1}{4} \delta^2}$ (i) $\Delta^r y_k = \nabla^r y_{k+r}$
    (j) $\nabla^2 = h^2 \Delta^2 - h^3 \Delta^3 + \dfrac{7}{12} h^4 \Delta^4 \ldots$

20. Prove that : $\sum\limits_{k=0}^{n-1} \Delta^2 f_k = \Delta f_n - \Delta f_0$

21. If y_x is a polynomial of degree 7 in x and $y_0 + y_8 = 1.9243$, $y_1 + y_7 = 1.9590$, $y_2 + y_6 = 1.9823$, $y_3 + y_5 = 1.9956$, find y_4.

22. Prove that :
$$\Delta \sin^{-1} x = \sin^{-1}\left[(x+1)\sqrt{1-x^2} - x\sqrt{1-(x+1)^2}\right]$$

23. Evaluate :

(a) $\Delta\left(\dfrac{e^x}{e^x + e^{-x}}\right)$ (b) $\Delta \cos a^x$

24. Find the cubic polynomial, such that
$y(0) = -5, y(1) = 1, y(2) = 9, y(3) = 25, y(4) = 55, y(5) = 105$

25. Find the cubic polynomial, such that
$y(0) = 1, y(1) = 0, y(2) = 1, y(3) = 10$. Hence or other wise find $y(4)$.

26. If $\Delta^3 u_x = 0$, prove that :
$$u_{x+1/2} = \frac{1}{2}(u_x + u_{x-1}) - \frac{1}{16}(\Delta^2 u_x + \Delta^2 u_{x+1})$$

27. Show that :
$$\mu\left[\frac{f(x)}{g(x)}\right] = \frac{\mu f(x)\,\mu g(x) - \dfrac{1}{4}\delta f(x)\,\delta g(x)}{g\left(x - \dfrac{1}{2}\right) g\left(x + \dfrac{1}{2}\right)}$$

28. Show that :
$$e^x \left(u_0 + x\,\Delta u_0 + \frac{x^2}{2!}\Delta^2 u_0 + ...\right) = u_0 + u_1 x + u_2 \frac{x^2}{2!} + ...,$$

29. Prove that :
(a) $\Delta^n e^x = e^{x+n} - n_{C_1} e^{x+n-1} + n_{C_2} e^{x+n-2} + ... + (-1)^n e^x$.
(b) $y_{x+n} = y_x + n\,\Delta y_x + n_{C_2}\,\Delta^2 y_x + ... + \Delta^n y_x$.

30. Prove that :
$$\Delta x^n - \frac{1}{2}\Delta^2 x^n + \frac{1.3}{2.4}\Delta^3 x^n - \frac{1.3.5}{2.4.6}\Delta^4 x^n + ...\ n \text{ terms} =$$
$$\left(x + \frac{1}{2}\right)^n - \left(x - \frac{1}{2}\right)^n$$

31. Using method of separation of symbols, show that :
$$\Delta^n \mu_{x-n} = \mu_x - n\mu_{x-1} + \frac{n(n-1)}{2}\mu_{x-2} + ... + (-1)^n \mu_{x-n}$$

32. Use the method of separation of symbols, prove that :

(a) $u_x + {}^xC_1\Delta^2 u_{x-1} + {}^xC_2\Delta^4 u_{x-2} + ... = u_0 + {}^xC_1\Delta u_1 + {}^xC_2\Delta^2 u_2 + ...$

(b) $u_0 + u_1 + u_2 + ... u_0 + n + {}^1C_1 u_0 + n + {}^1C_2 \Delta u_0 + n$
$$+ {}^1C_3\Delta^2 u_0 + ... + \Delta^n u_0$$

(c) $\sum_{x=0}^{\infty} u_{2x} = \frac{1}{2}\sum_{x=0}^{\infty} u_x + \frac{1}{4}\left(1 - \frac{\Delta}{2} + \frac{\Delta^2}{4} - ...\right) u_0$

33. Show that

$u_x - u_{x+1} + u_{x+2} - u_{x+3} + ... =$

$\frac{1}{2}\left[u_{x-\frac{1}{2}} - \frac{1}{8}\Delta^2 u_{x-\frac{3}{2}} + \frac{1.3}{2!}\left(\frac{1}{8}\right)^2 \Delta^4 u_{x-\frac{5}{2}} - \frac{1.3.5}{3!}\left(\frac{1}{8}\right)^3 \Delta^6 u_{x-\frac{7}{2}} + ... \right]$

34. Prove that :

$y_x + 2\Delta y_x + 3\Delta^2 y_x + ... = \frac{1}{4}\left[y_x + y_{x+1} + \frac{3}{4}y_{x+2} + \frac{1}{2}y_{x+3} + ... \right]$

35. Prove that :

$\frac{1}{3}\left[y_0 + \frac{1}{3}y_1 + \left(\frac{1}{3}\right)^2 y_2 + \left(\frac{1}{3}\right)^3 y_3 + ... \right]$

$= \frac{1}{2}\left[y_0 + \frac{1}{2}\Delta y_0 + \left(\frac{1}{2}\right)^2 \Delta^2 y_0 + ... \right]$

ANSWERS
---

1. 241.   **3.** 0.0738 and – 0.0508

4. $y_n = (n+1)^3 + 1$ and $y_6 = 344$.   **5.** 0.4 and 0.6   **6.** 15.

7. 27   **8.** $y_5 = 31, y_6 = 129, y_7 = 351$

11. 138   **12.** (a) $a^{bx+c}(a^{bh} - 1)^n$

(b) $\log\left[1 + \frac{\Delta(ax+b)}{ax+b}\right]$ (c) $\tan^{-1}\left(\frac{1}{2x^2}\right)$

13. (a) $2hx + h^2 + 2h$ (b) $ab^{cx}(b^{ch} - 1)$

(c) $e^{ax}\left[e^{ah}\log\left(1 + \frac{h}{x}\right) + (e^{ah} - 1)\log bx\right]$

(d) $-\dfrac{1+2x}{(1+x^2)[1+(1+x)^2]}$ (e) $\sin h [\cos(4x + 2h)\sin h - \cos(2x + h)]$

(f) $\tan^{-1}\dfrac{bh}{1+b^2x^2+b^2xh}$ (g) $(x+h)e^x(e^h - 1) + e^x h$

(h) $\dfrac{2^x[(x+1)(2^h+1)-h]}{(x+1)(x+h+1)}$

14. (a) -36 (b) $\dfrac{(-1)^n n!}{x(x+1)(x+2)\ldots(x+n)}$ (c) $\dfrac{(-1)^n n!}{x(x+1)(x+2)\ldots(x+n)}$

15. (a) $5x^2 + 2hx + 2h - x^2 - 2x - 1$ (b) $5h^2 + 2hx + 2h - x^2 - 2x - 1$

17. (a) $10! \, abcd \, h^{10}$ (b) $a(n)!$

21. 0.99996

23. (a) $\dfrac{e^h - e^{-h}}{(e^x + e^{-x})(e^{x+h} - e^{-x-h})}$

(b) $2 \sin \dfrac{a^x(1+a^h)}{2} \sin \dfrac{a^x(1-a^h)}{2}$

24. $x^3 - 2x^2 + 7x - 5$

25. $y = x^3 - 2x^2 + 1$, $y(4) = 33$

3.7 FACTORIAL NOTATION

A product of the form $x(x-h)(x-2)\ldots(x-\overline{n-1}h)$ is called a factorial and is denoted by $x^{(n)}$ or $[x]^n$, where n is a positive integer and h is the interval of differencing.

i.e., $\qquad x^{(n)} = x(x-h)(x-2h)\ldots(x-\overline{n-1}h)$

$x^{(n)}$ is read as x raised to the power n factorial.

In particular for $h = 1$, we have

$$x^{(1)} = x$$
$$x^{(2)} = x(x-1)$$
$$x^{(3)} = x(x-1)(x-2)$$
$$\vdots$$
$$x^{(n)} = x(x-1)(x-2)(x-3)\ldots(x-\overline{n-1})$$

3.8 RECIPROCAL OR NEGATIVE FACTORIAL NOTATION

The reciprocal factorial function is denoted by $x^{(-n)}$ and is defined as

$x^{(-n)} = \dfrac{1}{(x+h)(x+2h)\ldots(x+nh)}$, where n is a positive integer and h is interval of differencing.

Thus $\qquad x^{(-1)} = \dfrac{1}{(x+h)}$

$$x^{(-2)} = \frac{1}{(x+h)(x+2h)}$$

$$x^{(-3)} = \frac{1}{(x+h)(x+2h)(x+3h)}$$

In particular, for $h = 1$, we have

$$x^{(-n)} = \frac{1}{(x+1)(x+2)\ldots(x+n)} = \frac{1}{(x+n)^{(n)}}$$

3.9 DIFFERENCES OF A FACTORIAL FUNCTION

Theorem 1. Show that $\Delta^n x^{(n)} = n!\, h^n$

Proof.
$$\Delta x^{(n)} = (x+h)^{(n)} - x^{(n)}$$
$$= (x+h)\, x\, (x-h) \ldots (x-\overline{n-2}\, h)$$
$$\quad - x\, (x-h)\, (x-2h) \ldots (x-\overline{n-1}\, h)$$
$$= x\, (x-h)\, (x-2h) \ldots (x-\overline{n-2}\, h)\, [(x+h)$$
$$\quad - (x-\overline{n-1}\, h)\,]$$
$$\Delta x^{(n)} = x^{(n-1)}\, nh = nh\, x^{(n-1)}$$

Again
$$\Delta^2 x^{(n)} = \Delta[nh\, x^{(n-1)}]$$
$$= nh\, \Delta\, x^{(n-1)}$$
$$= nh\, [(x+h)^{(n-1)} - x^{(n-1)}]$$
$$= nh\, [(x+h)\, x\, (x-h) \ldots (x-\overline{n-3}\, h)$$
$$\quad - x\, (x-h)\, (x-2h) \ldots (x-\overline{n-2}\, h)]$$
$$= nh\, x\, (x-h)\, (x-2h) \ldots (x-\overline{n-3}\, h)\, [(x+h)$$
$$\quad - (x-\overline{n-2}\, h)\,]$$
$$= nh\, x^{(n-2)}\, (n-1)h$$
$$= n\, (n-1)\, h^2\, x^{(n-2)}$$

Proceeding like this, we get
$$\Delta^n x^{(n)} = n\, (n-1)\, (n-2) \ldots 3.2.1 h^n$$
$$= n!\, h^n,\ \text{which is a constant term.}$$

Note. In particular if $h = 1$
$$\Delta^n x^{(n)} = n!\ \text{and}\ \Delta^r x^{(n)} = 0,\ \text{if}\ r > n.$$

Theorem 2. Show that

(i) $\Delta\, x^{(-n)} = -nh\, x^{(-(n+1))}$

(ii) $\Delta^2 x^{(-n)} = (-1)^2\, n(n+1)\, h^2 x^{(-(n+2))}$

Proof. (i) $\Delta x^{(-n)} = (x+h)^{(-n)} - x^{(-n)}$

$$= \frac{1}{(x+2h)(x+3h)...(x+\overline{n+1}h)} - \frac{1}{(x+h)(x+2h)...(x+nh)}$$

$$= \frac{(x+h) - (x+\overline{n+1}h)}{(x+h)(x+2h)...(x+\overline{n+1}h)}$$

$$= \frac{-nh}{(x+h)(x+2h)...(x+\overline{n+1}h)}$$

$$= -nh \, x^{(-(n+1))}$$

(ii) $\Delta^2 x^{(-n)} = \Delta(\Delta x^{(-n)})$

$= \Delta[-nh \, x^{(-(n+1))}]$

$= -nh \, \Delta[x^{(-(n+1))}]$

$= -nh[-(n+1)h \, x^{(-(n+2))}]$

$= (-1)^2 n(n+1) h^2 x^{(-(n+2))}$

In general

$$\Delta^r x^{(-n)} = (-1)^r n(n+1)(n+2)...(n+r-1) h^r x^{(-(n+r))}$$

3.10 POLYNOMIAL IN FACTORIAL NOTATION

Any polynomial

$$f(x) = a_0 x^n + a_1 x^{n-1} + a_2 x^{n-2} + ... + a_n$$

can be expressed in the form of factorial polynomial as

$$A_0 x^{(n)} + A_1 x^{(n-1)} + A_2 x^{(n-2)} + ... + A_n$$

Now, we can find $A_0, A_1, A_2, ... A_n$ by two ways.

1. Dividing the given polynomial successively by $x, x-h, x-2h, ...$ we get the coefficients $A_n, A_{n-1}, A_{n-2}, ...$ which are nothing but the remainders in that order.

Note. If $h = 1$, divide $f(x)$ successively by $x, x-1, x-2, ...$

If $h = 2$, divide $f(x)$ successively by $x, x-2, x-4, ...$

2. Consider $a_0 x^n + a_1 x^{n-1} + a_2 x^{n-2} + ... + a_n = A_0 x^{(n)} + A_1 x^{(n-1)} + A_2 x^{(n-2)} + ... + A_n$. Now expand the R.H.S. and equate the coefficients of various powers of x on both side, we get $A_0, A_1, A_2, ... A_n$.

3.11 COMPUTATION OF MISSING TERMS

Suppose in a table, we are given with $n+1$ equidistant values of x. If out of $n+1$ values one or more values of x are missing, then it can be found with the help of differences. Let out of $n+1$ values, one value is missing. Since we are given n entries, we consider that these values represents a polynomial of degree $n-1$.

Hence, $\Delta^{n-1} y_x = $ constant

and $\Delta^n y_x = 0$ ...(1)

Since one term is missing, put $x = 0$ is (1)
$$\Delta^n y_0 = 0$$
$$\Rightarrow \quad (E - 1)^n y_0 = 0 \qquad \ldots(2)$$
with the help of (2) we can find the missing term. If in a set of $n + 2$ values, two entries are missing, then put $x = 0$ and $x = 1$ in (1) and we get
$$\Delta^n y_0 = 0 \text{ and } \Delta^n y_1 = 0 \qquad \ldots(3)$$
simplifying (3), we get the missing terms.

ILLUSTRATIVE EXAMPLES

Example 1. Express (i) $2x^3 - 3x^2 + 4x - 8$, if $h = 1$
(ii) $3x^4 + 8x^3 + 3x^2 - 27x + 9$, if $h = 1$
(iii) $x^3 - 3x^2 + 5x + 7$, if $h = 2$
as a factorial polynomial and find their successive differences.

Solution (i) Let $\quad f(x) = 2x^3 - 3x^2 + 4x - 8$
$$= A\, x^{(3)} + B\, x^{(2)} + C\, x^{(1)} + D$$
$$= Ax(x-1)(x-2) + Bx(x-1) + Cx + D$$

Now, put $x = 0$; we get $D = -8$
put $x = 1$; we get $C = 3$
put $x = 2$; we get $B = 3$
comparing coefficients of x^3 on both sides, we get $A = 2$.

$\therefore \qquad f(x) = 2x^{(3)} + 3x^{(2)} + 3x^{(1)} - 8$

Now, $\quad \Delta f(x) = 2.3\, x^{(2)} + 3.2\, x^{(1)} + 3.1 = 6x^{(2)} + 6x^{(1)} + 3$
$\Delta^2 f(x) = 6.2\, x^{(1)} + = 6 = 12\, x^{(1)} + 6$
$\Delta^3 f(x) = 12$
$\Delta^4 f(x) = 0$.

(ii) Let $\quad f(x) = 3x^4 + 8x^3 + 3x^2 - 27 + 9$
$$= A\, x^{(4)} + B\, x^{(3)} + C\, x^{(2)} + D\, x^{(1)} + E$$
$$= Ax(x-1)(x-2)(x-3)$$
$$\quad + Bx(x-1)(x-2) + Cx(x-1) + Dx + E$$

Now, put $x = 0$; $E = 9$
put $x = 1$; $D = -13$
put $x = 2$; $C = 48$
put $x = 3$; $B = 26$
comparing the coefficients of x^4 on both sides, we get $A = 3$

$\therefore \qquad f(x) = 3\, x^{(4)} + 26\, x^{(3)} + 48\, x^{(2)} - 13\, x^{(1)} + 9$

Now, $\quad \Delta f(x) = 12\, x^{(3)} + 78\, x^{(2)} + 96\, x^{(1)} - 13$
$\Delta^2 f(x) = 36\, x^{(2)} + 156\, x^{(1)} + 96$

$$\Delta^3 f(x) = 72\, x^{(1)} + 156$$
$$\Delta^4 f(x) = 72$$
$$\Delta^5 f(x) = 0$$

(iii) Let $f(x) = x^3 - 3x^2 + 5x + 7$
$$= A\, x^{(3)} + B\, x^{(2)} + C\, x^{(1)} + D$$
$$= A\, x(x-2)(x-4) + B\, x(x-2) + C\, x + D$$
$$[\because h = 2]$$

put $x = 0$; $D = 7$
put $x = 2$; $C = 3$
put $x = 4$; $B = 3$
and $A = 1$

$\therefore \qquad f(x) = x^{(3)} + 3x^{(2)} + 3x^{(1)} + 7 \qquad$ where $h = 2$

Now, $\quad \Delta f(x) = 3h\, x^{(2)} + 6h\, x^{(1)} + 3h = 6x^{(2)} + 12\, x^{(1)} + 6$
$$\Delta^2 f(x) = 2.6.2\, x^{(1)} + 12.2.1 = 24\, x^{(1)} + 24$$
$$\Delta^3 f(x) = 24.2 = 48$$
$$\Delta^4 f(x) = 0$$

Example 2. Express $f(x) = x^4 - 12x^3 + 24x^2 - 30x + 9$ as a factorial polynomial interval of differencing being unity. Also prove that $\Delta^4 f(x) = 24$.

Solution. Let $\quad f(x) = x^4 - 12x^3 + 24x^2 - 30x + 9$
$$= A\, x^{(4)} + B\, x^{(3)} + C\, x^{(2)} + D\, x^{(1)} + E$$

Now, dividing the given polynomial successively by $x, x-1, x-2, x-3$ by synthetic division method.

0	1	−12	24	−30	9
		0	0	0	0
1	1	−12	24	−30	9 = E
		1	−11	13	
2	1	−11	13	−17 = D	
		2	−18		
3	1	−9	−5 = C		
		3			
	1	−6 = B			
	1 = A				

Hence, $\quad f(x) = x^{(4)} - 6\, x^{(3)} - 5\, x^{(2)} - 17\, x^{(1)} + 9$
$$\Delta f(x) = 4\, x^{(3)} - 18\, x^{(2)} - 10\, x^{(1)} - 17$$
$$\Delta^2 f(x) = 12\, x^{(2)} - 36\, x^{(1)} - 10$$

$$\Delta^3 f(x) = 24 x^{(1)} - 36$$
$$\Delta^4 f(x) = 24$$

Example 3 Obtain the function whose first difference is
  (i) $x^3 + 4x^2 + 9x + 12$
  (ii) $3x^2 + 13x - 7$

Solution. (i) Let $f(x)$ be the required function
then,
$$\Delta f(x) = x^3 + 4x^2 + 9x + 12$$
$$= A x^{(3)} + B x^{(2)} + C x^{(1)} + D$$
$$= A x (x - 1) (x - 2) + B x (x - 1) + C x + D$$

Now, putting $x = 0$ on both sides, we get $D = 12$
Putting $x = 1$
$$C + D = 1 + 4 + 9 + 12 = 26 \quad \Rightarrow \quad C = 14$$
Putting $x = 2$
$$2B + 2C + D = 8 + 16 + 18 + 12 \quad \Rightarrow \quad B = 7$$

Equating the coefficients of x^3 on both sides, we get $A = 1$.
Thus,
$$\Delta f(x) = x^{(3)} + 7 x^{(2)} + 14 x^{(1)} + 12$$
$$f(x) = \frac{x^{(4)}}{4} + \frac{7x^{(3)}}{3} + \frac{14x^{(2)}}{2} + 12x^{(1)} + k,$$

where k is a constant

$$= \frac{x(x-1)(x-2)(x-3)}{4} + \frac{7x(x-1)(x-2)}{3}$$
$$+ 7 x(x - 1) + 12 x + k$$

$$f(x) = \frac{1}{4}(x^4 - 6x^3 - 7x^2 - 6x) + \frac{7}{3}(x^3 - 3x^2 + 2x)$$
$$+ 7(x^2 - x) + 12x + k$$

$$\Rightarrow \quad f(x) = \frac{x^4}{4} + \frac{5}{6}x^3 - \frac{7}{4}x^2 + \frac{49}{6}x + k$$

(ii) Let $\Delta f(x) = 3x^2 + 13x - 7$
$$= A x^{(2)} + B x^{(1)} + C = A x(x - 1) + Bx + C$$

put $x = 0$; $C = -7$
put $x = 1$; $B = 16$
Equating the coefficients of x^3 ; $A = 3$
$$\therefore \quad \Delta f(x) = 3 x^{(2)} + 16 x^{(1)} - 7$$
$$f(x) = \frac{3 x^{(3)}}{3} + \frac{16 x^{(2)}}{2} - 7 x^{(1)} + k$$
$$= x^{(3)} + 8 x^{(2)} - 7 x^{(1)} + k$$

$$= x(x-1)(x-2) + 8x(x-1) - 7x + k$$
$$= x^3 - 3x^2 + 2x + 8x^2 - 8x - 7x + k$$
$$= x^3 + 5x^2 - 13x + k$$

Example 4. Obtain the missing term, from the following table:

x :	0	1	2	3	4
y :	1	3	9	—	81

Solution. Since only 4 values of y are given, so we get a polynomial of degree 3, hence fourth differences are zero.

i.e.,
$$\Delta^4 y_0 = 0$$
$$(E-1)^4 y_0 = 0$$
$$(E^4 - 4E^3 + 6E^2 - 4E + 1) y_0 = 0$$
$$E^4 y_0 - 4E^3 y_0 + 6E^2 y_0 - 4E y_0 + y_0 = 0$$
$$y_4 - 4y_3 + 6y_2 - 4y_1 + y_0 = 0$$
$$81 - 4y_3 + 6(9) - 4(3) + 1 = 0$$
$$4y_3 = 124$$
$$\Rightarrow \quad y_3 = 31$$

Example 5. Estimate the production for 1964 and 1966 from the following data:

Year :	1961	1962	1963	1964	1965	1966	1967
Production :	200	220	260	—	350	—	430

Solution. Since five values are given, so we get a polynomial of degree four, hence fifth differences are zero.

$$\Delta^5 y_k = 0$$
$$(E-1)^5 y_k = 0$$
$$(E^5 - 5E^4 + 10E^3 - 10E^2 + 5E - 1) y_k = 0$$
$$E^5 y_k - 5E^4 y_k + 10E^3 y_k - 10E^2 y_k + 5E y_k - y_k = 0$$
$$y_{k+5} - 5y_{k+4} + 10y_{k+3} - 10y_{k+2} + 5y_{k+1} - y_k = 0 \qquad \ldots(1)$$

put $k = 0$
$$y_5 - 5y_4 + 10y_3 - 10y_2 + 5y_1 - y_0 = 0$$
$$y_5 - 5(350) + 10y_3 - 10(260) + 5(220) - 200 = 0$$
$$y_5 + 10y_3 = 3450 \qquad \ldots(2)$$

put $k = 1$
$$y_6 - 5y_5 + 10y_4 - 10y_3 + 5y_2 - y_1 = 0$$
$$430 - 5y_5 + 10(350) - 10y_3 + 5(260) - 220 = 0$$
$$5y_5 + 10y_3 = 5010 \qquad \ldots(3)$$

Solving (2) and (3), we get
$$y_3 = 306 \text{ and } y_5 = 390$$

∴ Production for year 1964 = 306
and for year 1966 = 390

Example 6. Find the cubic polynomial from the given data :

x	:	0	1	2	3	4
y	:	-5	-10	-9	4	35

Solution.

x	y	Δy	$\Delta^2 y$	$\Delta^3 y$	$\Delta^4 y$
0	-5				
		-5			
1	-10		6		
		1		6	
2	-9		12		0
		13		6	
3	4		18		
		31			
4	35				

$$y_x = E^x y_0$$
$$= (1 + \Delta)^x y_0$$
$$= (1 + {}^xC_1\Delta + {}^xC_2\Delta^2 + {}^xC_3\Delta^3 + {}^xC_4\Delta^4 + {}^xC_5\Delta^5 + \dots) y_0$$
$$= y_0 + {}^xC_1\Delta y_0 + {}^xC_2\Delta^2 y_0 + {}^xC_3\Delta^3 y_0 + {}^xC_4\Delta^4 y_0 + {}^xC_5\Delta^5 y_0 + \dots$$
$$= -5 + x(-5) + \frac{x(x-1)}{2!} \times 6 + \frac{x(x-1)(x-2)}{3!} \times 6 + 0 + 0 + \dots$$
$$= -5 - 5x + 3x^2 - 3x + x(x-1)(x-2)$$
$$= -5 - 5x + 3x^2 - 3x + x^3 - 3x^2 + 2x$$
$$y_x = x^3 - 6x - 5$$

i.e., $f(x) = x^3 - 6x - 5$

Example 7. Find $\Delta^2 f(x)$ if

$$f(x) = \frac{1}{x(x+4)(x+8)}$$

Solution.
$$f(x) = \frac{1}{x(x+4)(x+8)}$$
$$= (x-4)^{(-3)} \qquad \text{[where } h = 4\text{]}$$
$$\Delta f(x) = \Delta (x-4)^{(-3)}$$
$$= (-3) \cdot 4 \, (x-4)^{(-4)}$$
$$\Delta^2 f(x) = (-3)(-4) \cdot 4^2 \, (x-4)^{(-5)}$$
$$= 192 \, (x-4)^{(-5)}$$
$$= \frac{192}{x(x+4)(x+8)(x+12)(x+16)}$$

Example 8. if $y = (3x + 1)(3x + 4) \dots (3x + 22)$, prove that
$$\Delta^4 y = 136080 \, (3x + 13)(3x + 16)(3x + 19)(3x + 22)$$

Solution. Clearly $y = (3x + 1)(3x + 4) \dots (3x + 22)$ has 8 factors

$$\therefore \quad y = 3^8\left(x+\frac{1}{3}\right)(x+4/3)\ldots\left(x+\frac{22}{3}\right)$$

$$y = 3^8\left(x+\frac{22}{3}\right)^{(8)}, \text{ where } h = 1$$

$$\therefore \quad \Delta y = 8.3^8\left(x+\frac{22}{3}\right)^{(7)}$$

$$\Delta^2 y = 8.7.3^8\left(x+\frac{22}{3}\right)^{(6)}$$

$$\Delta^3 y = 8.7.6.3^8\left(x+\frac{22}{3}\right)^{(5)}$$

$$\Delta^4 y = 8.7.6.5.3^8\left(x+\frac{22}{3}\right)^{(4)}$$

$$= 56 \times 30 \times 3^8 \left(x+\frac{22}{3}\right)\left(x+\frac{19}{3}\right)\left(x+\frac{16}{3}\right)\left(x+\frac{13}{3}\right)$$

$$= 56 \times 30 \times 3^4 (3x+22)(3x+19)(3x+16)(3x+13)$$

$$\Delta^4 y = 136080 \, (3x+13)(3x+16)(3x+19)(3x+22)$$

Example 9. if $y = \dfrac{1}{(2x+1)(2x-1)(2x-3)(2x-5)}$ find $\Delta^3 y$.

Solution.
$$y = \frac{1}{(2x+1)(2x-1)(2x-3)(2x-5)}$$

$$= \frac{1}{16\left(x+\dfrac{1}{2}\right)\left(x-\dfrac{1}{2}\right)(x-3/2)(x-5/2)}$$

$$= \frac{1}{16}\left(x+\frac{3}{2}\right)^{(-4)} \qquad \text{[where } h = -1\text{]}$$

$$\Delta y = \frac{1}{16}(-4)(-1)(x+3/2)^{(-5)}$$

$$\Delta^2 y = \frac{1}{16}(-4)(-5)(-1)^2\left(x+\frac{3}{2}\right)^{(-6)}$$

$$\Delta^3 y = \frac{1}{16}(-4)(-5)(-6)(-1)^3 (x+3/2)^{(-7)}$$

$$= \frac{120}{16} \frac{1}{\left(x+\dfrac{1}{2}\right)\left(x-\dfrac{1}{2}\right)\left(x-\dfrac{3}{2}\right)\left(x-\dfrac{5}{2}\right)\left(x-\dfrac{7}{2}\right)\left(x-\dfrac{9}{2}\right)\left(x-\dfrac{11}{2}\right)}$$

$$= \frac{120}{16} \cdot 2^7 \cdot \frac{1}{(2x+1)(2x-1)(2x-3)(2x-5)(2x-7)(2x-9)(2x-11)}$$

Example 10. Prove that $y_4 = y_3 + \Delta y_2 + \Delta^2 y_1 + \Delta^3 y_1$

Solution.
$$\Delta y_2 = y_3 - y_2$$
$$\Delta^2 y_1 = \Delta(\Delta y_1)$$
$$= \Delta(y_2 - y_1)$$
$$= y_3 - y_2 - y_2 + y_1$$
$$\Delta^2 y_1 = y_3 - 2y_2 + y_1$$
$$\Delta^3 y_1 = \Delta(\Delta^2 y_1)$$
$$= \Delta(y_3 - 2y_2 + y_1)$$
$$= y_4 - y_3 - 2y_3 + 3y_2 - y_1$$
$$= y_4 - 3y_3 + 3y_2 - y_1$$
$$\therefore \quad y_3 + \Delta y_2 + \Delta^2 y_1 + \Delta^3 y_1 = y_3 + y_3 - y_2 + y_3 - 2y_2 + y_1 + y_4$$
$$\qquad - 3y_3 + 3y_2 - y_1$$
$$= y_4$$

EXERCISE 3.2

1. Find the missing term in the following tables :

   (a)
x :	1	2	3	4	5	6	7
y :	2	4	8	—	32	64	128

   (b)
x :	10	15	20	25	30	35	40
y :	270	—	222	200	—	164	148

2. Express the following functions in terms of factorial polynomials and hence find their successive differences :

   (a) $7x^4 + 12x^3 - 6x^2 + 5x - 3$, if $h = 2$
   (b) $2x^3 - 3x^2 + 3x - 10$, if $h = 1$
   (c) $11x^4 + 5x^3 + 2x^2 + x - 15$, if $h = 1$
   (d) $x^3 + x^2 + x + 1$, if $h = 2$
   (e) $x^3 + 3x^2 + 5x + 12$, if $h = 1$

3. Obtain the function whose first difference is

   (a) $x^3 + 3x^2 + 5x + 12$ 　　(b) $2x^3 + 3x^2 - 5x + 4$

4. If $\Delta f(x) = 2x^3 - 6x^2 + 7x + 10$, find $f(x)$.

5. If $y = \dfrac{1}{(3x+1)(3x+4)(3x+7)}$, prove that
$$\Delta^2 y = \dfrac{108}{(3x+1)(3x+4)(3x+7)(3x+10)(3x+13)}$$

6. If $y = \dfrac{1}{x(x+3)(x+6)}$, prove that
$$\Delta^2 y = \dfrac{108}{x(x+3)(x+6)(x+9)(x+12)}$$

118 *Numerical and Statistical Techniques*

ANSWERS

1. (a) 16.1 (b) 246 and 180.8
2. (a) $7 x^{(4)} + 96 x^{(3)} + 262 x^{(2)} + 97 x^{(1)} - 3$
   (b) $2 x^{(3)} + 3 x^{(2)} + 2 x^{(1)} - 10 x^{(0)}$
   (c) $11 x^{(4)} + 71 x^{(3)} + 94 x^{(2)} + 19 x^{(1)} - 15 x^{(0)}$
   (d) $x^{(3)} + 7x^{(2)} + 7x^{(1)} + 1$
   (e) $x^{(3)} + 6x^{(2)} + 5x^{(1)} + 12$
3. (a) $\frac{1}{4}(x^4 + 2x^3 + 11x^2 + 40x) + c$
   (b) $\frac{x^{(4)}}{2} + 3 x^{(2)} + 4 x^{(1)} + c$
4. $f(x) = \frac{x^{(4)}}{2} + \frac{3}{2} x^{(2)} + 10 x^{(1)} + c$

3.12 FINITE INTEGRATION (OR INVERSE OPERATOR Δ^{-1})

If $\Delta y_x = u_x$ then $y_x = \Delta^{-1} u_x$, where Δ^{-1} is called finite integration operator or inverse of operator Δ.

If $C(x)$ is a periodic function of period h which is equal to the interval of differencing, then

$$\Delta C(x) = C(x + h) - C(x) \text{ (By definition of } \Delta)$$
$$= C(x) - C(x) \qquad (\therefore C(x) \text{ is periodic function})$$
$$= 0$$

This shows that, if $C(x)$ is periodic function whose period and interval of differencing is same h, then $\Delta C(x) = 0$.

Hence if $\quad \Delta y(x) = u(x)$
then $\Delta(y(x) + C(x)) = \Delta y(x) + \Delta C(x)$
$$= \Delta y(x) + 0$$
$$= \Delta y(x) = u(x)$$
$\therefore \qquad \Delta^{-1} u(x) = y(x) + C(x)$

where $C(x)$ is the periodic function of period h. The following inverse operator results can be remembered from the corresponding forward operator results.

(i) $\Delta^{-1}(u_x + v_x) = \Delta^{-1} u_x + \Delta^{-1} v_x$
(ii) $\Delta^{-1}(C u_x) = C \Delta^{-1} u_x$
(iii) $\Delta^{-1}(e^{ax+b}) = \dfrac{e^{ax+b}}{e^{ah} - 1}$

Inparticular $\Delta^{-1}(e^x) = \dfrac{e^x}{e^h - 1}$

(iv) $\Delta^{-1}(a^x) = \dfrac{a^x}{a^h - 1}$, $a \neq 1$

(v) $\Delta^{-1} x^{(n)} = \dfrac{x^{(n+1)}}{n+1}$, $n \neq -1$ and $h = 1$

(vi) $\Delta^{-1}(a + bx)^{(n)} = \dfrac{(a+bx)^{(n+1)}}{(n+1)hb}$, $n \neq -1$

ILLUSTRATIVE EXAMPLES

Example 1. Find $\Delta^{-1} x(x+1)(x+2)$

Solution. We have $(x+2)(x+1)x = (x+2)^{(3)}$, if $h = 1$

$\therefore \quad \Delta^{-1}(x+2)(x+1)x = \Delta^{-1}(x+2)^{(3)}$

$$= \dfrac{(x+2)^{(4)}}{4} + C(x)$$

$$= \dfrac{(x+2)(x+1)x(x-1)}{4} + C(x)$$

where $C(x)$ is a periodic function of period 1.

Example 2. Find $\Delta^{-1} \dfrac{1}{x(x+1)(x+2)}$

Solution. $\dfrac{1}{x(x+1)(x+2)} = (x-1)^{(-3)}$

$\therefore \quad \Delta^{-1}\left[\dfrac{1}{x(x+1)(x+2)}\right] = \Delta^{-1}\left[(x-1)^{(-3)}\right]$

$$= \dfrac{(x-1)^{(-2)}}{(-2)} + C(x)$$

$$= -\dfrac{1}{2x(x+1)} + C(x)$$

3.13 SUMMATION OF SERIES

Finite differences are used in finding the sum of series.

Let us find the sum of the series

$$u_1 + u_2 + u_3 + \cdots + u_n$$

Let the xth term be u_x, such that

120 Numerical and Statistical Techniques

$$u_x = \Delta y_x$$

∴ $u_x = y_{x+1} - y_x \ (h = 1)$

Hence
$u_1 = y_2 - y_1$
$u_2 = y_3 - y_2$
$u_3 = y_4 - y_3$
$\vdots$
$u_n = y_{n+1} - y_n$

Adding vertically, we get

$$S_n = u_1 + u_2 + u_3 + \cdots + u_n = y_{n+1} - y_1 = (y_x)_1^{n+1}$$

$$\Rightarrow \boxed{S_n = \sum_{x=1}^{n} u_x = \left[\Delta^{-1} u_x\right]_1^{n+1}}$$

3.14 MONTMORT'S THEOREM

$$u_0 + xu_1 + x^2 u_2 + x^3 u_3 + \ldots \infty = \frac{u_0}{1-x} + \frac{x\Delta u_0}{(1-x)^2} + \frac{x^2 \Delta^2 u_0}{(1-x)^3} + \cdots \text{ to } \infty$$

Proof: L.H.S $= u_0 + xu_1 + x^2 u_2 + x^3 u_3 + \ldots \infty$

$= u_0 + x E u_0 + x^2 E^2 u_0 + x^3 E^3 u_0 + \ldots \infty$

$= (1 + xE + x^2 E^2 + x^3 E^3 + \ldots \infty) u_0$

$= \dfrac{1}{1 - xE} u_0$ \hspace{1em} [∵ The series is in G.P.]

$= \dfrac{1}{1 - x(1 + \Delta)} u_0$

$= \dfrac{1}{1 - x - x\Delta} u_0$

$= \dfrac{1}{(1-x)\left[1 - \dfrac{x\Delta}{1-x}\right]} u_0$

$= \dfrac{1}{1-x} \left[1 - \dfrac{x\Delta}{1-x}\right]^{-1} u_0$

$= \dfrac{1}{1-x} \left[1 + \dfrac{x\Delta}{1-x} + \dfrac{x^2 \Delta^2}{(1-x)^2} + \cdots \infty\right] u_0$

$= \dfrac{u_0}{1-x} + \dfrac{x\Delta u_0}{(1-x)^2} + \dfrac{x^2 \Delta^2 u_0}{(1-x)^3} + \cdots \infty$

$=$ R.H.S.

ILLUSTRATIVE EXAMPLES

Example 1. Sum the series:
$$2\cdot 3 + 3\cdot 4 + 4\cdot 5 + \cdots + (n+1)(n+2)$$

Solution. x^{th} term; $u_x = (x+1)(x+2) = (x+2)^{(2)}$

$\therefore \quad S_n = $ Sum to n terms

$$= \left[\Delta^{-1} u_x\right]_1^{n+1}$$

$$= \left[\Delta^{-1}(x+2)^{(2)}\right]_1^{n+1}$$

$$= \left[\frac{(x+2)^{(3)}}{3}\right]_1^{n+1}$$

$$= \frac{(n+3)^{(3)}}{3} - \frac{3^{(3)}}{3}$$

$$= \frac{(n+3)(n+2)(n+1) - 3\cdot 2\cdot 1}{3}$$

$$= \frac{n^3 + 6n^2 + 11n}{3} = \frac{n(n^2 + 6n + 11)}{3}$$

Example 2. Sum the series to n terms of
$$1\cdot 2\cdot 3 + 2\cdot 3\cdot 4 + 3\cdot 4\cdot 5 + \ldots$$

Solution. x^{th} term $= u_x = x(x+1)(x+2) = (x+2)^{(3)}$

$S_n = $ Sum to n terms $= \left[\Delta^{-1} u_x\right]_1^{n+1}$

$$S_n = \left[\Delta^{-1}(x+2)^{(3)}\right]_1^{n+1}$$

$$= \left[\frac{(x+2)^{(4)}}{4}\right]_1^{n+1}$$

$$= \frac{1}{4}\left[(n+3)^{(4)} - 3^{(4)}\right]$$

$$= \frac{1}{4}\left[(n+3)(n+2)(n+1)n - 3\cdot 2\cdot 1\cdot 0\right]$$

$$S_n = \frac{n(n+1)(n+2)(n+3)}{4}$$

Example 3. Sum to n terms of $\dfrac{1}{2\cdot 3}+\dfrac{1}{3\cdot 4}+\dfrac{1}{4\cdot 5}+\cdots$

Solution. x^{th} term $= u_x = \dfrac{1}{(x+1)(x+2)} = (x)^{(-2)}$

$\therefore \quad S_n = \left[\Delta^{-1} u_x\right]_1^{n+1}$

$ = \left[\Delta^{-1} x^{(-2)}\right]_1^{n+1}$

$ = \left[\dfrac{x^{(-1)}}{-1}\right]_1^{n+1}$

$ = \dfrac{(n+1)^{(-1)}}{-1} - \dfrac{1^{(-1)}}{-1}$

$ = -\dfrac{1}{n+2} + \dfrac{1}{2}$

$ = \dfrac{1}{2} - \dfrac{1}{n+2}$

Example 4. Sum to n terms of the series

$$\dfrac{1}{1\cdot 2\cdot 3}+\dfrac{1}{2\cdot 3\cdot 4}+\dfrac{1}{3\cdot 4\cdot 5}+\cdots$$

Solution. x^{th} term $= u_x = \dfrac{1}{x(x+1)(x+2)} = (x-1)^{(-3)}$

$\therefore \quad S_n = $ Sum to n term $= \left[\Delta^{-1} u_x\right]_1^{n+1}$

$\therefore \quad S_n = \left[\Delta^{-1}(x-1)^{(-3)}\right]_1^{n+1}$

$ = \left[\dfrac{(x-1)^{(-2)}}{-2}\right]_1^{n+1}$

$ = -\dfrac{1}{2}\left[n^{(-2)} - 0^{(-2)}\right]$

$ = -\dfrac{1}{2}\left[\dfrac{1}{(n+1)(n+2)} - \dfrac{1}{1\cdot 2}\right]$

$S_n = \dfrac{1}{2}\left[\dfrac{1}{2} - \dfrac{1}{(n+1)(n+2)}\right]$

Example 5. Sum the series $1^2 + 2^2 x + 3^2 x^2 + \ldots \infty$

Solution. Here $u_0 = 1^2 = 1, u_1 = 2^2 = 4, u_2 = 3^2 = 9, u_3 = 4^2 = 16$

The difference table is

u	Δu	$\Delta^2 u$	$\Delta^3 u$
1			
	3		
4		2	
	5		0
9		2	
	7		
16			

∴ By Montmort's theorem

$$1^2 + 2^2 x + 3^2 x^2 + 4^2 x^3 + \ldots \infty$$
$$= u_0 + u_1 x + u_2 x^2 + u_3 x^3 + \ldots \infty$$
$$= \frac{u_0}{1-x} + \frac{x \Delta u_0}{(1-x)^2} + \frac{x^2 \Delta^2 u_0}{(1-x)^3} + \ldots$$
$$= \frac{1}{1-x} + \frac{x \cdot 3}{(1-x)^2} + \frac{2 \cdot x^2}{(1-x)^3}$$
$$= \frac{(1-x)^2 + 3x(1-x) + 2x^2}{(1-x)^3}$$
$$= \frac{1 + x^2 - 2x + 3x - 3x^2 + 2x^2}{(1-x)^3} = \frac{x+1}{(1-x)^3}$$

Example 6. Sum the series $1 \cdot 2 + 2 \cdot 3x + 3 \cdot 4 x^2 + \ldots$ to ∞

Solution. Here $u_0 = 1 \cdot 2 = 2, u_1 = 2 \cdot 3 = 6, u_2 = 3 \cdot 4 = 12, u_3 = 4 \cdot 5 = 20$

The difference table is

u	Δu	$\Delta^2 u$	$\Delta^3 u$
2			
	4		
6		2	
	6		0
12		2	
	8		
20			

∴
$$1 \cdot 2 + 2 \cdot 3 x + 3 \cdot 4 x^2 + 4 \cdot 5 x^3 + \ldots \infty$$
$$= u_0 + u_1 x + u_2 x^2 + u_3 x^3 + \ldots \infty$$
$$= \frac{u_0}{1-x} + \frac{x \Delta u_0}{(1-x)^2} + \frac{x^2 \Delta^2 u_0}{(1-x)^3} + \frac{x^3 \Delta^3 u_0}{(1-x)^4} + \ldots \infty$$
$$= \frac{2}{1-x} + \frac{x \cdot 4}{(1-x)^2} + \frac{x^2 \cdot 2}{(1-x)^3}$$

$$= \frac{2(1-x)^2 + 4x(1-x) + 2x^2}{(1-x)^3}$$

$$= \frac{2(1+x^2-2x) + 4x - 4x^2 + 2x^2}{(1-x)^3}$$

$$= \frac{2 + 2x^2 - 4x + 4x - 4x^2 + 2x^2}{(1-x)^3}$$

$$= \frac{2}{(1-x)^3}$$

Example 7. Using Montmort's theorem, sum the series
$$1\cdot 3 + 3\cdot 5x + 5\cdot 7x^2 + 7\cdot 9x^3 + \cdots \infty$$

Solution. Here $u_0 = 1\cdot 3 = 3$, $u_1 = 3\cdot 5 = 15$, $u_2 = 5\cdot 7 = 35$, $u_3 = 7\cdot 9 = 63$

The difference table is

u	Δu	$\Delta^2 u$	$\Delta^3 u$
3			
	12		
15		8	
	20		0
35		8	
	28		
63			

$\therefore \quad 1\cdot 3 + 3\cdot 5x + 5\cdot 7x^2 + 7\cdot 9x^3 + \cdots \infty$

$$= u_0 + u_1 x + u_2 x^2 + u_3 x^3 + \cdots \infty$$

$$= \frac{u_0}{1-x} + \frac{x \Delta u_0}{(1-x)^2} + \frac{x^2 \Delta^2 u_0}{(1-x)^3} + \cdots$$

$$= \frac{3}{1-x} + \frac{x\cdot 12}{(1-x)^2} + \frac{x^2 \cdot 8}{(1-x)^3} + 0 \cdots$$

$$= \frac{3(1-x)^2 + 12x(1-x) + 8x^2}{(1-x)^3}$$

$$= \frac{3 + 3x^2 - 6x + 12x - 12x^2 + 8x^2}{(1-x)^3}$$

$$= \frac{3 + 6x - x^2}{(1-x)^3}, \quad x \neq 1$$

Example 8. Using the method of separation of symbols show that
$$(u_1 - u_0) - x(u_2 - u_1) + x^2(u_3 - u_2) - \cdots \text{ to } \infty$$

$$= \frac{\Delta u_0}{1+x} - x\frac{\Delta^2 u_0}{(1+x)^2} + x^2 \frac{\Delta^3 u_0}{(1+x)^3} + \cdots \infty$$

Solution. L.H.S
$$= (u_1 - u_0) - x(u_2 - u_1) + x^2(u_3 - u_2) \ldots \infty$$
$$= \Delta u_0 - x\Delta u_1 + x^2 \Delta u_2 - \ldots$$
$$= \Delta u_0 - x\Delta E u_0 + x^2 \Delta E^2 u_0 - \ldots$$
$$= \Delta(1 - xE + x^2E^2 - \ldots)u_0$$
$$= \left[\Delta(1 + xE)^{-1}\right] u_0$$
$$= \left(\frac{\Delta}{1 + xE}\right) u_0$$
$$= \left[\frac{\Delta}{1 + x(1 + \Delta)}\right] u_0$$
$$= \frac{\Delta}{(1 + x)\left[1 + \frac{x\Delta}{1 + x}\right]} u_0$$
$$= \frac{1}{1 + x}\left[1 - \frac{x\Delta}{1 + x} + \frac{x^2 \Delta^2}{(1 + x)^2} - \cdots\right] \Delta u_0$$
$$= \frac{\Delta u_0}{1 + x} - \frac{x\Delta^2 u_0}{(1 + x)^2} + \frac{x^2 \Delta^3 u_0}{(1 + x)^3} - \cdots$$
$$= \text{R.H.S}$$

Example 9. Using the method of separation of symbols, prove that
$$y_x = y_{x-1} + \Delta y_{x-2} + \ldots\ldots + \Delta^{n-1} y_{x-n} + \Delta^n y_{x-n}$$

Solution. R.H.S
$$= y_{x-1} + \Delta y_{x-2} + \ldots\ldots + \Delta^{n-1} y_{x-n} + \Delta^n y_{x-n}$$
$$= E^{-1} y_x + \Delta E^{-2} y_x + \ldots + \Delta^{n-1} E^{-n} y_x + \Delta^n E^{-n} y_x$$
$$= E^{-1}\left[1 + \frac{\Delta}{E} + \frac{\Delta^2}{E^2} + \cdots + \frac{\Delta^{n-1}}{E^{n-1}}\right] y_x + \frac{\Delta^n}{E^n} y_x$$
$$= E^{-1}\left[\frac{\left(\frac{\Delta}{E}\right)^n - 1}{\frac{\Delta}{E} - 1}\right] y_x + \frac{\Delta^n}{E^n} y_x$$
$$= E^{-1}\left[\frac{\Delta^n - E^n}{\Delta - E} \cdot \frac{1}{E^{n-1}}\right] y_x + \frac{\Delta^n}{E^n} y_x$$
$$= E^{-1}\left[\frac{\Delta^n - E^n}{E - 1 - E} \cdot \frac{1}{E^{n-1}}\right] y_x + \frac{\Delta^n}{E^n} y_x$$

$$= \left[-\frac{(\Delta^n - E^n)}{E^n}\right] y_x + \frac{\Delta^n}{E^n} y_x$$

$$= \left[1 - \frac{\Delta^n}{E^n} + \frac{\Delta^n}{E^n}\right] y_x = y_x = \text{L.H.S.}$$

EXERCISE 3.3

1. Sum to n terms the series
   (a) $2 \cdot 5 + 5 \cdot 8 + 8 \cdot 11 + 11 \cdot 14 + \ldots$
   (b) $1^2 + 2^2 + 3^2 + \ldots + n^2$
   (c) $1 \cdot 3 \cdot 5 + 2 \cdot 4 \cdot 6 + 3 \cdot 5 \cdot 7 + \ldots$

2. Sum to n terms of the series
$$\frac{1}{1 \cdot 2 \cdot 3} + \frac{1}{2 \cdot 3 \cdot 4} + \frac{1}{3 \cdot 4 \cdot 5} + \cdots$$

3. Find the sum to n terms of the series whose n^{th} terms is
   (i) $2n - 1$ (ii) $n(n - 1)(n - 2)$
   (iii) $(n + 1)(n + 2)(n + 3)(n + 4)$

4. Using Montmort's theorem, sum the series
$$1 \cdot 3 + 3 \cdot 5\, x + 5 \cdot 7\, x^2 + 7 \cdot 9\, x^3 + \ldots \infty$$

5. Sum the series
   (a) $1^2 + 2^2 x + 3^2 x^2 + \ldots$ to ∞
   (b) $1 \cdot 2 + 2 \cdot 3\, x + 3 \cdot 4\, x^2 + \ldots$ to ∞
   (c) $5 + \frac{4x}{1!} + \frac{5x^2}{2!} + \frac{14x^3}{3!} + \frac{37x^4}{4!} + \frac{80x^5}{5!} + \cdots$ to ∞

6. Show that $\sum_{x=1}^{n} u_x = u_1 + u_2 + \cdots + u_n$

$$= n_{c_1} u_1 + n_{c_2} \Delta u_1 + n_{c_3} \Delta^2 u_1 + \cdots \Delta^{n-1} u_1$$

and hence sum the series $1^2 + 2^2 + 3^2 + \ldots + n^2$.

7. Prove that
$$u_0 - u_1 + u_2 - u_3 + \cdots = \frac{1}{2} u_0 - \frac{1}{4} \Delta u_0 + \frac{1}{8} \Delta^2 u_0 - \frac{1}{16} \Delta^3 u_0 + \cdots$$

ANSWERS

1. (a) $n(3n^2 + 6n + 1)$ (b) $\dfrac{1}{6} n(n+1)(2n+1)$ (c) $\dfrac{n(n+1)(n+4)(n+5)}{4}$

2. $\dfrac{1}{2}\left[\dfrac{1}{2} - \dfrac{1}{(n+1)(n+2)}\right]$

3. (i) n^2 (ii) $\dfrac{n^{(4)}}{4}$ (iii) $\dfrac{n(n+1)(n+2)(n+3)(n+4)}{5}$

4. $\dfrac{3 + 6x - x^2}{(1-x)^3}$

5. (a) $\dfrac{1+x}{(1-x)^3}$ (b) $\dfrac{2}{(1-x)^3}$ (c) $e^x(x^3 + x^2 - x + 5)$

Chapter 4

Interpolation

4.1 INTERPOLATION

Interpolation means the process of computing intermediate values of a function from a given set of tabular values of the function.

Suppose we are given the following values of $y = f(x)$ for a set of values of x;

x	x_0	x_1	x_2	...	x_n
y	y_0	y_1	y_2	...	y_n

Thus the process of finding the value of y corresponding to any value of $x = x_i$ between x_0 and x_n is called interpolation. The process of finding the value of a function outside the interval (x_0, x_n) is called extrapolation. But in general the word interpolation is used in both processes.

If the form of the function $f(x)$ is known, we can find the value of $f(x)$ for any value of x by simple substitution. But in most of the practical problems, the exact form of $f(x)$ is not known. In such cases we replace $f(x)$ by a simpler function $\phi(x)$ which has same values as $f(x)$ for different values of x. The function $\phi(x)$ is called interpolation function or smoothing function or interpolating polynomial and the process is known as polynomial interpolation.

Usually, polynomial interpolation is preferred due to the following reasons:

(*i*) They are free from singularities.

(*ii*) They are easy to manipulate

(*iii*) They are easy to differentiate and integrate.

Even though there are other methods like graphical method and method of curve fitting, in this chapter we will study polynomial interpolation using the calculus of finite differences.

4.2 NEWTON'S-GREGORY FORWARD INTERPOLATION FORMULA

Let $y = f(x)$ be the function which takes the values $y_0, y_1, y_2, \ldots y_n$ for $(n+1)$ vales $x_0, x_1, x_2, \ldots x_n$ of argument x. Let these arguments are equally spaced with interval h, then

$$x_i = x_0 + ih, \text{ for } i = 1, 2, 3, \ldots, n.$$

$\therefore \quad x_1 = x_0 + h, \; x_2 = x_0 + 2h, \text{ etc.}$

Let $y_n(x)$ be a polynomial of nth degree in x, such that,

$$y_i = f(x_i) = y_n(x_i), \quad i = 0, 1, 2, \ldots, n.$$

Let
$$y_n(x) = a_0 + a_1(x - x_0) + a_2(x - x_0)(x - x_1)$$
$$+ a_3(x - x_0)(x - x_1)(x - x_2) + \ldots$$
$$+ a_n(x - x_0)(x - x_1) \ldots (x - x_{n-1}) \qquad \ldots(1)$$

where a_i, $i = 0, 1, 2, 3, \ldots, n$ are $(n+1)$ unknowns which are to be found.

Put $\qquad x = x_0$ in (1), then
$$y_n(x_0) = a_0 = y_0$$

Put $\qquad x = x_1$, then
$$y_n(x_1) = a_0 + a_1(x_1 - x_0)$$
$\Rightarrow \qquad y_1 = a_0 + a_1 h.$

$$a_1 = \frac{y_1 - a_0}{h} = \frac{y_1 - y_0}{h}$$

$$a_1 = \frac{\Delta y_0}{h}$$

Put $\qquad x = x_2$, then
$$y_n(x_2) = a_0 + a_1(x_2 - x_0) + a_2(x_2 - x_0)(x_2 - x_1)$$

$\Rightarrow \qquad y_2 = y_0 + \dfrac{\Delta y_0}{h} \cdot 2h + a_2 \cdot 2h \cdot h.$

$$= y_0 + 2 \Delta y_0 + 2 a_2 h^2$$

$$a_2 = \frac{y_2 - y_0 - 2\Delta y_0}{2h^2} = \frac{y_2 - 2y_1 + y_0}{2h^2}$$

$$a_2 = \frac{\Delta^2 y_0}{2! h^2}$$

Continuing in this way, we get

$$a_3 = \frac{\Delta^3 y_0}{3! h^3}, \; a_4 = \frac{\Delta^4 y_0}{4! h^4} \text{ and so on.}$$

Thus $\qquad a_n = \dfrac{\Delta^n y_0}{n! h^n}$

putting these values in (1)

$$y_n(x) = y_0 + (x - x_0)\frac{\Delta y_0}{1!h} + (x - x_0)(x - x_1)\frac{\Delta^2 y_0}{2!h^2} + (x - x_0)(x - x_1)(x - x_2)\frac{\Delta^3 y_0}{3!h^3}$$
$$+ \cdots + (x - x_0)(x - x_1) \cdots (x - x_{n-1})\frac{\Delta^n y_0}{n!h^n}$$

Put $x = x_0 + uh \Rightarrow u = \frac{x - x_0}{h}$, where u is a real number, we have

$$y_n(x_0 + uh) = y_0 + (hu)\Delta y_0 + (hu)(hu - h)\frac{\Delta^2 y_0}{2!h^2}$$
$$+ (hu)(hu - h)(hu - 2h)\frac{\Delta^3 y_0}{3!h^3} + \cdots +$$
$$(hu)(hu - h)(hu - 2h) \cdots (hu - \overline{n - 1}h)\frac{\Delta^n y_0}{n!h^n}$$

$$\Rightarrow y_n(x_0 + uh) = y_0 + u\Delta y_0 + \frac{u(u-1)}{2!}\Delta^2 y_0 + \frac{u(u-1)(u-2)}{3!}\Delta^3 y_0$$
$$+ \cdots + \frac{u(u-1)(u-2) \cdots (u-n+1)}{n!}\Delta^n y_0 \qquad \ldots(2)$$

where $u = \frac{x - x_0}{h}$.

Equation (2) is called Newton's Gregory Forward Interpolation Formula.

Aliter. We can also derive the above formula by using separation of symbols method.

$$y_n(x) = y_n(x_0 + uh) = E^u y_n(x_0).$$
$$= E^u y_0$$
$$= (1 + \Delta)^u y_0.$$
$$= [1 + u_{c_1}\Delta + u_{c_2}\Delta^2 + u_{c_3}\Delta^3 + \cdots u_{c_n}\Delta^n + \cdots]y_0$$
$$= y_0 + u\Delta y_0 + \frac{u(u-1)}{2!}\Delta^2 y_0 + \frac{u(u-1)(u-2)}{3!}\Delta^3 y_0 + \cdots$$
$$\cdots + \frac{u(u-1)(u-2) \cdots (u-n+1)}{n!}\Delta^n y_0 + \cdots$$

where $u = \frac{x - x_0}{h}$

If $y = f(x)$ is a polynomial of nth degree, then $\Delta^{n+1} y_0, \ldots$ are zero. Hence,

$$y_n(x) = y_n(x_0 + uh) = y_0 + u\Delta y_0 + \frac{u(u-1)}{2!}\Delta^2 y_0 + \frac{u(u-1)(u-2)}{3!}\Delta^3 y_0 + \ldots$$
$$+ \frac{u(u-1)(u-2) \ldots (u-n+1)}{n!}\Delta^n y_0$$

we can also remember this formula as,
$$y_n(x) = y_n(x_0 + uh)$$
$$= y_0 + u_{c_1}\Delta y_0 + u_{c_2}\Delta^2 y_0 + u_{c_3}\Delta^3 y_0 + \cdots + u_{c_n}\Delta^n y_0.$$

Note:
1. If we put $n = 1$, then we get,
$$y_1(x) = y_0 + u\,\Delta y_0$$
which is formula for linear interpolation.
2. If we put $n = 2$, we get,
$$y_2(x) = y_0 + u\Delta y_0 + \frac{u(u-1)}{2}\Delta^2 y_0$$
which is formula for parabolic interpolation.
3. This formula is called forward interpolation formula due to the fact that this formula contains y_0 and its forward difference. This is used mainly to interpolate the value of y near to the beginning value of the table.
4. This formula is applicable only if the interval of differencing h is constant.
5. This formula can also be written as.
$$y_n(x) = y_0 + \frac{u^{(1)}}{1!}\Delta y_0 + \frac{u^{(2)}}{2!}\Delta^2 y_0 + \frac{u^{(3)}}{3!}\Delta^3 y_0 + \cdots + \frac{u^{(n)}}{n!}\Delta^n y_0$$
where $\quad u^{(n)} = u(u-1)(u-2)(u-3)\ldots(u-n+1)$

4.3 ERROR IN POLYNOMIAL INTERPOLATION

If $y = f(x)$ is the exact curve and $y = y_n(x)$ is the interpolating polynomial curve, then the error in polynomial interpolation is given by
$$\text{Error} = f(x) - y_n(x) = \frac{(x - x_0)(x - x_1)\cdots(x - x_n)}{(n+1)!}f^{n+1}(c)$$
for any value of x such that $x_0 < x < x_n$ and $x_0 < c < x_n$.

4.4 ERROR IN NEWTON'S GREGORY FORWARD INTERPOLATION FORMULA

$$\text{Error} = f(x) - y_n(x) = \frac{u(u-1)(u-2)\cdots(u-n)}{(n+1)!}h^{n+1}f^{n+1}(c)$$
where $\quad u = \dfrac{x - x_0}{h}$

ILLUSTRATIVE EXAMPLES

Example 1. The population of a town was as given below. Estimate the population for the year 1896.

132 *Numerical and Statistical Techniques*

Year(x)	1893	1903	1913	1923	1933
Population(y) (in thousands)	48	69	83	91	98

Solution: Since $x = 1896$ is near to the beginning of the table, we use Newton's forward formula.

Here, $\qquad x = 1896, x_0 = 1893$ and $h = 10$

$\therefore \qquad u = \dfrac{x - x_0}{h} = \dfrac{1896 - 1893}{10} = 0.3$

The difference table is

x	y	Δy	$\Delta^2 y$	$\Delta^3 y$	$\Delta^4 y$
1893	48				
		21			
1903	69		−7		
		14		1	
1913	83		−6		4
		8		5	
1923	91		−1		
		7			
1933	98				

Now, applying Newton's forward Difference formula

$y(1896) = y_0 + u\Delta y_0 + \dfrac{u(u-1)}{2!}\Delta^2 y_0 + \dfrac{u(u-1)(u-2)}{3!}\Delta^3 y_0$

$\qquad\qquad + \dfrac{u(u-1)(u-2)(x-3)}{4!}\Delta^4 y_0$

$\qquad = 48 + (0.3)(21) + \dfrac{(0.3)(0.3-1)}{2!}(-7)$

$\qquad\quad + \dfrac{(0.3)(0.3-1)(0.3-2)}{3!}(1)$

$\qquad\quad + \dfrac{(0.3)(0.3-1)(0.3-2)(0.3-3)}{4!}(4)$

$\qquad = 48 + 6.3 + 0.735 + 0.0595 - 0.16065$

$\qquad = 54.93$ thousands.

Example 2. A cubic polynomial passes through the points (0, −1), (1, 1), (2, 1) and (3, −2). Find the polynomial.

Solution. The forward difference table is,

x	y	Δy	$\Delta^2 y$	$\Delta^3 y$
0	– 1			
		2		
1	1		– 2	
		0		– 1
2	1		– 3	
		– 3		
3	– 2			

By Newton's Forward Difference formula, taking

$$u = \frac{x - x_0}{h} = \frac{x - 0}{1} = x$$

$$y(x) = y_0 + x\Delta y_0 + \frac{x(x-1)}{2!}\Delta^2 y_0 + \frac{x(x-1)(x-2)}{3!}\Delta^3 y_0$$

$$= -1 + x(2) + \frac{x(x-1)}{2}(-2) + \frac{x(x-1)(x-2)}{6}(-1)$$

$$= -1 + 2x - x^2 + x + \left(-\frac{1}{6}\right)(x^3 - 3x^2 + 2x)$$

$$= -\frac{1}{6}x^3 - \frac{1}{2}x^2 + \frac{8}{3}x - 1$$

Example 3. The hourly declination of the moon on a day is given below. Find the declination at $35^m\ 15^s$

Hour	0	1	2	3	4
Dec.	8°29'53.7"	8°18'19.4"	8°6'43.5"	7°55'6.1"	7°43'27.2"

Solution. The difference table is

x	y	Δy	$\Delta^2 y$	$\Delta^3 y$	$\Delta^4 y$
0	8°29'53.7"				
		–11'34.3"			
1	8°18'19.4"		–1.6"		
		–11'35.9"		0.1"	
2	8°6'43.5"		–1.5"		–0.1"
		–11'37.4"		0	
3	7°55'6.1"		–1.5"		
		–11'38.9"			
4	7°43'27.2"				

134 *Numerical and Statistical Techniques*

Since $35^m\ 15^s$ is near to the beginning of the table, we use Newton's forward formula.

$$\therefore \quad u = \frac{x - x_0}{h} = \frac{35^m 15^s - 0^h}{1^h} = \frac{2115^s}{3600^s} = 0.5875$$

Now, using Newton's Forward Interpolation formula;

$$y(35^m\ 15^s) = 8°29'53.7'' + (0.5875)(-11'34.3'')$$
$$+ \frac{(0.5875)(0.5875 - 1)}{2!}(-1.6'') + \frac{(0.5875)(0.5875 - 1)(0.5875 - 2)}{3!}(0.1'')$$
$$+ \frac{(0.5875)(0.5875 - 1)(0.5875 - 2)(0.5875 - 3)}{4!}(-0.1'')$$
$$= 8°29'53.7'' - 6'47.9'' + 0.19'' + 0.0057'' + 0.0034''$$
$$= 8°29'53.89'' - 6'47.9''$$
$$= 8°23'5.99''$$

Example 4. The table below gives the values of tan x for $0.10 \le x \le 0.30$.

x	0.10	0.15	0.20	0.25	0.30
$y = \tan x$	0.1003	0.1511	0.2027	0.2553	0.3093

find tan 0.13, using Newton's Forward difference formula.

Solution. The difference table is

x	y	Δy	$\Delta^2 y$	$\Delta^3 y$	$\Delta^4 y$
0.10	0.1003				
		0.0508			
0.15	0.1511		0.0008		
		0.0516		0.0002	
0.20	0.2027		0.0010		0.0002
		0.0526		0.0004	
0.25	0.2553		0.0014		
		0.0540			
0.30	0.3093				

For $\tan 0.13,\ x = 0.13,\ h = 0.05$

$$\therefore \quad u = \frac{x - x_0}{h} = \frac{0.13 - 0.10}{0.05} = 0.6$$

$\therefore$ By Newton's Forward formula,

$$y(0.13) = y_0 + u\Delta y_0 + \frac{u(u-1)}{2!}\Delta^2 y_0 + \frac{u(u-1)(u-2)}{3!}\Delta^3 y_0$$

$$+ \frac{u(u-1)(u-2)(u-3)}{4!}\Delta^4 y_0$$

$$= 0.1003 + (0.6)(0.0508) + \frac{0.6(0.6-1)}{2}(0.0008)$$

$$+ \frac{0.6(0.6-1)(0.6-2)}{6}(0.0002)$$

$$+ \frac{0.6(0.6-1)(0.6-2)(0.6-3)}{24}(0.0002)$$

$$= 0.1003 + 0.03048 - 0.000096 + 0.000011 - 0.000007$$

$$= 0.1307.$$

Example 5. From the data given below find the number of students who secured marks between 45 and 50.

Marks range	30-40	40-50	50-60	60-70	70-80
No. of students	35	49	72	41	23

Solution. The difference table is

x (Marks)	y (No. of students)	Δy	$\Delta^2 y$	$\Delta^3 y$	$\Delta^4 y$
Below 40	35				
		49			
Below 50	84		23		
		72		−54	
Below 60	156		−31		67
		41		13	
Below 70	197		−18		
		23			
Below 80	220				

Now for $\quad x = 45, u = \dfrac{45-40}{10} = 0.2$

∴ By Newton's Forward difference formula;

$$y(45) = 35 + 0.2(49) + \frac{0.2(0.2-1)}{2!}(23) + \frac{(0.2)(0.2-1)(0.2-2)}{3!}(-54)$$

$$+ \frac{(0.2)(0.2-1)(0.2-2)(0.2-3)}{4!}(67)$$

$$= 35 + 9.8 - 1.84 - 2.592 - 2.251$$

$$= 38.117 \approx 38 \text{ students}$$

Hence, no. of students who secured marks less than 45 = 38
No. of students who got marks between 40 & 45 = 38 – 35 = 3.

Example 6. Find the annual premium at the age of 23 given,

Age	21	25	29	33	37
Premium (in rupees)	14.27	15.81	17.72	19.96	21.23

Solution. The difference table is

x	y	Δy	$\Delta^2 y$	$\Delta^3 y$	$\Delta^4 y$
21	14.27				
		1.54			
25	15.81		0.37		
		1.91		– 0.04	
29	17.72		0.33		– 1.26
		2.24		– 1.3	
33	19.96		– 0.97		
		1.27			
37	21.23				

Hence $\quad u = \dfrac{23 - 21}{4} = 0.5$

Now, by Newton's Forward formula,

$$y(23) = 14.27 + (0.5)(1.54) + \frac{(0.5)(-0.5)}{2!}(0.37)$$

$$+ \frac{(0.5)(-0.5)(-1.5)}{3!}(-0.04)$$

$$+ \frac{(0.5)(-0.5)(-1.5)(-2.5)}{4!}(-1.26)$$

$$= 14.27 + 0.77 - 0.04625 - 0.0025 + 0.0492$$
$$= 15.04$$

Hence, premium at the age of 23 is Rs 15.04

Example 7. Estimate $e^{-1.19}$ from the given data :

x	1.00	1.25	1.50	1.75	2.00
e^{-x}	0.3679	0.2865	0.2231	0.1738	0.1353

Solution. Here, $u = \dfrac{x - x_0}{h} = \dfrac{1.19 - 1.00}{0.25} = 0.76$

The forward difference table is

x	$y=e^{-x}$	$10^4 y$	$20^4 \Delta y$	$10^4 \Delta^2 y$	$10^4 \Delta^3 y$	$10^4 \Delta^4 y$
1.00	0.3679	3679				
			−814			
1.25	0.2865	2865		180		
			−634		−39	
1.50	0.2231	2231		141		6
			−493		−33	
1.75	0.1738	1738		108		
			−385			
2.00	0.1353	1353				

Now, applying Newton's forward Interpolation formula, we have

$$10^4 y(1.19) = 3679 + (0.76)(-814) + \frac{(0.76)(0.76-1)}{2!}(180)$$

$$+ \frac{(0.76)(0.76-1)(0.76-2)}{3!}(-39)$$

$$+ \frac{(0.76)(0.76-1)(0.76-2)(0.76-3)}{4!}(6)$$

$$= 3679 - 618.64 - 16.416 - 1.470 - 0.127$$

$$10^4 y(1.9) = 3042.34$$

$$\therefore \quad y(1.19) = 0.3042$$

Example 8. Calculate $\sqrt{5.5}$, given $\sqrt{5} = 2.236$, $\sqrt{6} = 2.449$, $\sqrt{7} = 2.646$ and $\sqrt{8} = 2.828$.

Solution. The difference table is

x	$y = \sqrt{x}$	Δy	$\Delta^2 y$	$\Delta^3 y$
5	2.236			
		0.213		
6	2.449		−0.016	
		0.197		0.001
7	2.646		−0.015	
		0.182		
8	2.828			

Here, $x = 5.5$, $x_0 = 5$, $h = 1$.

$$\therefore \quad u = \frac{x - x_0}{h} = 0.5$$

By Newton's Forward formula,

$$y(x) = y_0 + u\Delta y_0 + \frac{u(u-1)}{2!}\Delta^2 y_0 + \frac{u(u-1)(u-2)}{3!}\Delta^3 y_0$$

$$y(5.5) = 2.236 + (0.5)(0.213) + \frac{(0.5)(-0.5)}{2!}(-0.016)$$

$$+ \frac{(0.5)(-0.5)(-1.5)}{3!}(0.001)$$

$$= 2.236 + 0.1065 + 0.002 + 0.0000625$$

$$\therefore \quad \sqrt{5.5} = 2.345$$

Example 9. Find the number of men getting wages between Rs.20 and Rs.25 from the following table:

wages(Rs.)	0–19	20–39	40–59	60–79	80–99
Frequency	41	62	65	50	17

Solution. The difference table is

x	y	Δy	$\Delta^2 y$	$\Delta^3 y$	$\Delta^4 y$
Below 20	41				
		62			
Below 40	103		3		
		65		−18	
Below 60	168		−15		0
		50		−18	
Below 80	218		−33		
		17			
Below 100	235				

Now, to find number of men getting wages between Rs.20 and Rs. 25 we find no. of men getting wages below 25

So, $\quad u = \dfrac{x - x_0}{h} = \dfrac{25 - 20}{20} = 0.25$

By Newton's Forward formula,

$$y(25) = 41 + (0.25)(62) + \frac{(0.25)(0.25-1)}{2!}(3)$$

$$\frac{(0.25)(0.25-1)(0.25-2)}{3!}(-18) + 0$$

$$= 41 + 15.5 - 0.28125 - 0.98438$$

$$= 55.23437 \approx 55 \text{ men.}$$

$\therefore$ No. of men getting wages below Rs.25 = 55

No. of men getting wages below Rs.20 = 41

∴ No. of men getting wages between Rs.20
and   Rs. 25 = 55 – 41
              = 14 men

Example 10. Find a polynomial which takes the values

x	1	3	5	7	9	11
y	3	14	19	21	23	28

Solution. Here, $u = \dfrac{x - x_0}{h} = \dfrac{x-1}{2}$

The difference table is

x	y	Δy	Δ²y	Δ³y	Δ⁴y
1	3				
		11			
3	14		–6		
		5		3	
5	19		–3		0
		2		3	
7	21		0		0
		2		3	
9	23		3		
		5			
11	28				

Using, Newton's forward Interpolation formula,

$$y(x) = y_0 + u\Delta y_0 + \frac{u(u-1)}{2!}\Delta^2 y_0 + \frac{u(u-1)(u-2)}{3!}\Delta^3 y_0$$

$$= 3 + \left(\frac{x-1}{2}\right)(11) + \left(\frac{x-1}{2}\right)\left(\frac{x-1}{2}-1\right)\frac{(-6)}{2!}$$

$$+ \left(\frac{x-1}{2}\right)\left(\frac{x-1}{2}-1\right)\left(\frac{x-1}{2}-2\right)\frac{(3)}{3!}$$

$$= 3 + \frac{11x-11}{2} + \left(\frac{x-1}{2}\right)\left(\frac{x-3}{2}\right)(-3) + \left(\frac{x-1}{2}\right)\left(\frac{x-3}{2}\right)\left(\frac{x-5}{2}\right)\frac{1}{2}$$

$$= \frac{1}{16}[x^3 - 21x^2 + 159x - 91].$$

Example 11. The following temperatures were taken on a day:

Time	2 a.m.	6 a.m.	10 a.m.	14 a.m.
Temp.	40.2°	42.4°	51.0°	72.4°

Find the temperature at 3 a.m.

Solution. The difference table is

Time(x)	Temp. (y)	Δy	$\Delta^2 y$	$\Delta^3 y$
2	40.2			
		2.2		
6	42.4		6.4	
		8.6		6.4
10	51.0		12.8	
		21.4		
14	72.4			

Here, $u = \dfrac{x - x_0}{h} = \dfrac{3-2}{4} = \dfrac{1}{4}$

Now, using Newton's Forward Difference Interpolation formula

$$y(3) = 40.2 + \frac{1}{4}(2.2) + \frac{1}{4}\left(\frac{1}{4}-1\right)\left(\frac{6.4}{2!}\right) + \frac{1}{4}\left(\frac{1}{4}-1\right)\left(\frac{1}{4}-2\right)\left(\frac{6.4}{3!}\right)$$

$$= 40.2 + 0.55 - 0.6 + 0.35$$

$$= 40.5°$$

∴ The temperature at 3 a.m. is 40.5°

Example 12. Find $y(1.15)$ from the following table :

x	1.0	1.1	1.2	1.3	1.4	1.5	1.6	1.7	1.8
y	0.84147	0.89121	0.93204	0.96356	0.98545	0.99749	0.99957	0.99385	0.97385

Solution. Here, $x = 1.15$, $x_0 = 1.0$ and $h = 0.1$

∴ $u = \dfrac{x - x_0}{h} = \dfrac{1.15 - 1.0}{0.1} = 1.5$

The difference table is

x	y	Δy	$\Delta^2 y$	$\Delta^3 y$	$\Delta^4 y$
1.0	0.84147				
		0.04974			
			−0.00891		
1.1	0.89121			−0.00040	
		0.04083			0.00008
1.2	0.93204		−0.00931		
		0.03152		−0.00032	
1.3	0.96356		−0.00963		0.00010
		0.02189		−0.00022	
1.4	0.98545		−0.00985		0.00011
		0.01204		−0.00011	
1.5	0.99749		−0.00996		0.01371
		0.00208		0.0136	
1.6	0.99957		−0.00364		−0.03568
		0.00572		−0.02208	
1.7	0.99385		−0.02572		
		−0.02000			
1.8	0.97385				

Neglecting higher difference because it does not effect the answer much.

Here, $$u = \frac{x - x_0}{h} = \frac{1.15 - 1.0}{0.1} = 1.5$$

Now, using Newton's Forward difference Interpolation formula

$$y(1.15) = (0.84147) + (1.5)(0.04974) + \frac{(1.5)(1.5-1)}{2!}(-0.00891)$$

$$+ \frac{(1.5)(1.5-1)(1.5-2)}{3!}(-0.00040)$$

$$+ \frac{(1.5)(1.5-1)(1.5-2)(1.5-3)}{4!}(0.00008)$$

$$= 0.84147 + 0.07461 - 0.003341 + 0.000025 + 0.000002$$

$$= 0.9128$$

Example 13. Use Newton's method to find a polynomial $y(x)$ of lowest degree such that $y(x) = 2^x$ for $x = 0, 1, 2, 3, 4$.

Hence, find the value of $y(1.5)$

Solution. The difference table for $y = 2^x$ for $x = 0, 1, 2, 3, 4$ is

142 Numerical and Statistical Techniques

x	$y = 2^x$	Δy	$\Delta^2 y$	$\Delta^3 y$	$\Delta^4 y$
0	1				
		1			
1	2		1		
		2		1	
2	4		2		1
		4		2	
3	8		4		
		8			
4	16				

Here, $\quad u = \dfrac{x - x_0}{h} = \dfrac{x - 0}{1} = x$

$\therefore \quad y(x) = 1 + x(1) + \dfrac{x(x-1)}{2!}(1) + \dfrac{x(x-1)(x-2)}{3!}(1)$

$\qquad\qquad + \dfrac{x(x-1)(x-2)(x-3)}{4!}(1)$

$\qquad = 1 + x + \dfrac{x^2 - x}{2} + \dfrac{x^3 - 3x^2 + 2x}{6} + \dfrac{x^4 - 6x^3 + 11x^2 - 6x}{24}$

$\qquad = \dfrac{1}{24}[x^4 - 2x^3 + 11x^2 + 14x + 24]$

$\qquad y(x) = \dfrac{x^4}{24} - \dfrac{x^3}{12} + \dfrac{11}{24}x^2 + \dfrac{7}{12}x + 1$.

$\therefore \quad y(1.5) = \dfrac{(1.5)^4}{24} - \dfrac{(1.5)^3}{12} + \dfrac{11}{24}(1.5)^2 + \dfrac{7}{12}(1.5) + 1 = 2.8359$

4.5 NEWTON'S GREGORY BACKWARD INTERPOLATION FORMULA

Newton's forward interpolation formula is inconvenient for interpolating near to the end of the table. For this purpose we derive backward interpolation formula.

Let $y = f(x)$ be a function which takes the values $y_0, y_1, y_2, ..., y_n$ for $(n + 1)$ values of arguments, say $x_0, x_1, x_2, ..., x_n$. Let values of x are equally spaced, with interval h, then

$$x_i = x_0 + ih \text{ for } i = 1, 2, 3, ..., n.$$

Let $y_n(x)$ be the polynomial of nth degree in x, such that,

$$y_i = f(x_i) = y_n(x_i), i = 0, 1, 2, ..., n.$$

Now, assume,
$$y_n(x) = a_0 + a_1(x - x_n) + a_2(x - x_n)(x - x_{n-1}) + a_3(x - x_n)(x - x_{n-1})(x - x_{n-2})$$
$$+ \ldots + a_n(x - x_n)(x - x_{n-1}) \ldots (x - x_1) \qquad \ldots(1)$$
where $a_0, a_1, a_2, \ldots, a_n$ are $(n + 1)$ unknowns whose values are to be found.

Put $\qquad x = x_n$ is (1), we get
$$y_n(x_n) = a_0 = y_n$$

Put $\qquad x = x_{n-1}$, we get,
$$y_n(x_{n-1}) = a_0 + a_1(x_{n-1} - x_n)$$
$$y_{n-1} = y_n + a_1(-h) \qquad [\because x_n - x_{n-1} = h]$$
$$\Rightarrow \qquad a_1 = \frac{y_n - y_{n-1}}{h}$$
$$a_1 = \frac{\nabla y_n}{h}$$

Put $\qquad x = x_{n-2}$, we get,
$$y_n(x_{n-2}) = a_0 + a_1(x_{n-2} - x_n) + a_2(x_{n-2} - x_n)(x_{n-2} - x_{n-1})$$
$$y_{n-2} = a_0 + a_1(-2h) + a_2(-2h)(-h)$$
$$\Rightarrow \qquad y_{n-2} = y_n - 2a_1h + 2a_2h^2$$
$$\Rightarrow \qquad a_2 = \frac{-y_n + y_{n-2} + 2\nabla y_n}{2!h^2}$$
$$a_2 = \frac{-y_n + y_{n-2} + 2y_n - 2y_{n-1}}{2!h^2}$$
$$= \frac{y_n - 2y_{n-1} + y_{n-2}}{2!h^2}$$
$$\Rightarrow \qquad a_2 = \frac{(y_n - y_{n-1}) - (y_{n-1} - y_{n-2})}{2!h^2}$$
$$= \frac{\nabla y_n - \nabla y_{n-1}}{2!h^2}$$
$$\Rightarrow \qquad a_2 = \frac{\nabla^2 y_n}{2!h^2}$$

Continuing, we get
$$a_3 = \frac{\nabla^3 y_n}{3!h^3}, \quad a_4 = \frac{\nabla^4 y_n}{4!h^4}, \quad \ldots, \quad a_n = \frac{\nabla^n y_n}{n!h^n}$$

Putting all these values in (1)

$$y_n(x) = y_n + \frac{\nabla y_n}{h}(x - x_n) + (x - x_n)(x - x_{n-1})\frac{\nabla^2 y_n}{2!h^2}$$

$$+ (x - x_n)(x - x_{n-1})(x - x_{n-2})\frac{\nabla^3 y_n}{3!h^3} + \cdots$$

$$+ (x - x_n)(x - x_{n-1}) \cdots (x - x_1)\frac{\nabla^n y_n}{n!h^n} \qquad \ldots(2)$$

Now, Let $\quad x = x_n + uh$, where u is a real number

$\therefore \quad x - x_{n-1} = x - x_n + h = uh + h = (u + 1)h$.

$\quad x - x_{n-2} = x - x_n + 2h = uh + 2h = (u + 2)h$, etc.

Substituting these values in (2), we get

$$y_n(x_n + uh) = y_n + u\nabla y_n + \frac{u(u+1)}{2!}\nabla^2 y_n + \frac{u(u+1)(u+2)}{3}\nabla^3 y_n +$$

$$\cdots + \frac{u(u+1)(u+2)\cdots(u+n-1)}{n!}\nabla^n y_n \qquad \ldots(3)$$

Equation (2) is called Newton's Gregory Backward Interpolation formula.

Aliter. We can also derive the above formula by symbolic operator methods.

$$y_n(x) = y_n(x + uh)$$

$$= E^u y_n(x_n) = E^u y_n$$

$$= (1 - \nabla)^{-u} y_n \qquad [\therefore E = (1 - \nabla)^{-1}]$$

$$= \left[1 + u\nabla + \frac{u(u+1)}{2!}\nabla^2 + \frac{u(u+1)(u+2)}{3!}\nabla^3 \right.$$

$$\left. + \cdots + \frac{u(u+1)(u+2)\cdots(u+n-1)}{n!} + \cdots\right] y_n$$

$\Rightarrow \quad y_n(x) = y_n + u\nabla y_n + \dfrac{u(u+1)}{2!}\nabla^2 y_n + \dfrac{u(u+1)(u+2)}{3!}\nabla^3 y_n +$

$$\cdots + \frac{u(u+1)(u+2)\cdots(u+n-1)}{n!}\nabla^n y_n + \cdots$$

where $\quad u = \dfrac{x - x_n}{h}$

If $y = f(x)$ is a polynomial of degree n, then $\nabla^{n+1} y_n$, $\nabla^{n+2} y_n$,....vanishes.

Hence,

$$y_n(x) = y_n + u\nabla y_n + \frac{u(u+1)}{2!}\nabla^2 y_n + \frac{u(u+1)(u+2)}{3!}\nabla^3 y_n + \cdots$$
$$\cdots + \frac{u(u+1)(u+2)\cdots(u+n-1)}{n!}\nabla^n y_n$$

Note: 1. Since this formula involves the backward difference Operator, it is called Backward Interpolation formula.

2. This formula is used to interpolate the values of y nearer to the end of the table.

4.6 ERROR IN NEWTON'S GREGORY BACKWARD INTERPOLATION FORMULA

$$\text{Error} = f(x) - y_n(x)$$
$$= \frac{u(u+1)(u+2)\cdots(u+n)h^{n+1}}{(n+1)!} f^{n+1}(c)$$

where $u = \dfrac{x - x_n}{h}$

ILLUSTRATIVE EXAMPLES

Example 1. From the following table of half-yearly premium for policies maturing at different ages, find the premium for policies maturing at 64.

Age x (years)	46	51	56	61	66
Premium y (Rupees)	116	97	83	74	68

Solution. The difference table is

x	y	∇y	$\nabla^2 y$	$\nabla^3 y$	$\nabla^4 y$
46	116				
		–19			
51	97		5		
		–14		0	
56	83		5		–2
		–9		–2	
61	74		3		
		–6			
66	68				

For $x = 64$, we use Newton's Backward interpolation formula

Here, $x = 64, x_n = 66, u = \dfrac{x - x_n}{h} = \dfrac{64 - 66}{5} = -0.4$

$$\therefore \quad y(x) = y_n + u\nabla y_n + \frac{u(u+1)}{2!}\nabla^2 y_n + \frac{u(u+1)(u+2)}{3!}\nabla^3 y_n$$
$$+ \frac{u(u+1)(u+2)(u+3)}{4!}\nabla^4 y_n$$

$$\Rightarrow \quad y(64) = 68 + (-0.4)(-6) + \frac{(-0.4)(-0.4+1)}{2!}(3) + \frac{(-0.4)(-0.4+1)(-0.4+2)}{3!}(-2)$$
$$+ \frac{(-0.4)(-0.4+1)(-0.4+2)(-0.4+3)}{4!}(-2)$$
$$= 68 + 2.4 - 0.36 + 0.128 + 0.0832$$
$$y(64) = 70.25$$

∴ The premium at the age of 64 is Rs 70.25.

Example 2. Find y at $x = 42$ from the following data:

x	20	25	30	35	40	45
y	354	332	291	260	231	204

Solution. For $x = 42$, which is near to the end of the table we use Newton's Backward formula.

Here, $$u = \frac{x - x_n}{h} = \frac{42 - 45}{5} = -0.6$$

The difference table is

x	y	∇y	$\nabla^2 y$	$\nabla^3 y$	$\nabla^4 y$	$\nabla^5 y$
20	354					
		−22				
25	332		−19			
		−41		29		
30	291		10		−37	
		−31		−8		45
35	260		2		8	
		−29		0		
40	231		2			
		−27				
45	204					

∴ By Newton's Backward Interpolation formula,

$$y(x) = y_n + u\nabla y_n + \frac{u(u+1)}{2!}\nabla^2 y_n + \frac{u(u+1)(u+2)}{3!}\nabla^3 y_n$$
$$+ \frac{u(u+1)(u+2)(u+3)}{4!}\nabla^4 y_n + \frac{u(u+1)(u+2)(u+3)(u+4)}{5!}\nabla^5 y_n$$

$$y(42) = 204 + (-0.6)(-27) + \frac{(-0.6)(0.4)}{2!}(2) + \frac{(-0.6)(0.4)(1.4)}{3!}(0)$$
$$+ \frac{(-0.6)(0.4)(1.4)(2.4)}{4!}(8) + \frac{(-0.6)(0.4)(1.4)(2.4)(3.4)}{5!}(45)$$
$$= 204 + 16.2 - 0.24 + 0 - 0.2688 - 1.0282$$
$$y(42) = 218.663$$

Example 3. The population of a town in the census is given below. Estimate the increase in the population during the period 1896 to 1916.

year	1891	1901	1911	1921	1931
Population (in Thousands)	46	66	81	93	101

Solution. The difference table is

x	y	Δy	$\Delta^2 y$	$\Delta^3 y$	$\Delta^4 y$
1891	46				
		20			
1901	66		−5		
		15		2	
1911	81		−3		−3
		12		−1	
1921	93		−4		
		8			
1931	101				

For $x = 1896$, $u = \dfrac{1896 - 1891}{10} = 0.2$; and for this we use Newton's forward interpolation formula.

$$y(1896) = y_0 + u\Delta y_0 + \frac{u(u-1)}{2!}\Delta^2 y_0 + \frac{u(u-1)(u-2)}{3!}\Delta^3 y_0$$
$$+ \frac{u(u-1)(u-2)(u-3)}{4!}\Delta^4 y_0$$

$$= 46 + (0.2)(20) + \frac{(0.2)(-0.8)}{2!}(-5) + \frac{(0.2)(-0.8)(-1.8)}{3!}(2)$$
$$+ \frac{(0.2)(-0.8)(-1.8)(-2.8)}{4!}(-3)$$

$$= 46 + 4 + 0.4 + 0.096 + 0.1008$$
$$= 50.597 \text{ thousands.}$$

For $x = 1916$, $u = \dfrac{x - x_n}{h} = \dfrac{1916 - 1931}{10} = -1.5$; and for this we use Newton's Backward interpolation formula.

$$y(1916) = 101 + (-1.5)(8) + \frac{(-1.5)(-0.5)}{2!}(-4) + \frac{(-1.5)(-0.5)(0.5)}{3!}(-1)$$
$$+ \frac{(-1.5)(-0.5)(0.5)(1.5)}{4!}(-3)$$
$$= 101 - 12 - 1.5 - 0.0625 - 0.0703$$
$$= 87.367 \text{ thousands}$$

Therefore, increase in population during the period 1896 to 1916
$$= 87.367 - 50.597 = 36.77 \text{ thousands.}$$

Example 4. Construct a polynomial for the given data :

x	0	1	2	3	4
y	3	6	11	18	27

Solution. To find a polynomial we can used forward or backward formula. we apply Newton's backward interpolation formula,

For this, $u = \dfrac{x - x_n}{h} = \dfrac{x-4}{1} = x - 4$

The difference table is

x	y	∇y	$\nabla^2 y$	$\nabla^3 y$	$\nabla^4 y$
0	3				
		3			
1	6		2		
		5		0	
2	11		2		0
		7		0	
3	18		2		
		9			
4	27				

∴ By Newton's Backward Interpolation formula.

$$y(x) = 27 + (x-4)(9) + \frac{(x-4)(x-4+1)}{2!}(2) + 0 + 0.$$
$$= 27 + 9x - 36 + x^2 - 7x + 12$$
$$y(x) = x^2 + 2x + 3.$$

Note: We see that, since 2nd difference are constant we get a polynomial of degree 2.

Example 5. Given sin 45° = 0.7071 sin 50° = 0.7660, sin 55° = 0.8192 sin 60° = 0.8660, and sin 65° = 0.9063, find sin 57°.

Solution. The difference table is

x	y	∇y	$\nabla^2 y$	$\nabla^3 y$	$\nabla^4 y$
45°	0.7071				
		0.0589			
50°	0.7660		−0.0057		
		0.0532		−0.0007	
55°	0.8192		−0.0064		0.0006
		0.0468		−0.0001	
60°	0.8660		−0.0065		
		0.0403			
65°	0.9063				

Now, $u = \dfrac{x - x_n}{h} = \dfrac{57° - 65°}{5°} = -1.6$

$\therefore\ y(57°) = (0.9063) + (-1.6)(0.0403) + \dfrac{(-1.6)(-1.6+1)}{2!}(-0.0065)$

$\qquad + \dfrac{(-1.6)(-1.6+1)(-1.6+2)}{3!}(-0.0001)$

$\qquad + \dfrac{(-1.6)(-1.6+1)(-1.6+2)(-1.6+3)}{4!}(0.0006)$

$\qquad = 0.9063 - 0.06448 - 0.00312 - 0.000006 + 0.000013$

$\therefore \qquad \sin 57° = 0.8387.$

Example 6. Evaluate log 58.75 from the following table:

x	40	45	50	55	60	65
$y = \log x$	1.60206	1.65321	1.69897	1.74036	1.77815	1.81291

Solution. The difference table is:

x	y	∇y	$\nabla^2 y$	$\nabla^3 y$	$\nabla^4 y$	$\nabla^5 y$
40	1.60206					
		0.05115				
45	1.65321		−0.00539			
		0.04576		0.00102		
50	1.69897		−0.00437		−0.00025	
		0.04139		0.00077		0.00005
55	1.74036		−0.00360		−0.00020	
		0.03779		0.00057		
60	1.77815		−0.00303			
		0.03476				
65	1.81291					

For $x = 58.75$ we use Newton's backward difference interpolation formula

Here, $\quad u = \dfrac{x - x_n}{h} = \dfrac{58.75 - 65}{5} = -1.25$

$\therefore \quad y(58.75) = \log 58.75$

$$= y_n + u\nabla y_n + \frac{u(u+1)}{2!}\nabla^2 y_n + \frac{u(u+1)(u+2)}{3!}\nabla^3 y_n$$

$$+ \frac{u(u+1)(u+2)(u+3)}{4!}\nabla^4 y_n + \frac{u(u+1)(u+2)(u+3)(u+4)}{5!}\nabla^5 y_n$$

$\log 58.75 = (1.81291) + (-1.25)(0.03476) + \dfrac{(-1.25)(-0.25)}{2!}(-0.00303)$

$$+ \frac{(-1.25)(-0.25)(0.75)}{3!}(0.00057)$$

$$+ \frac{(-1.25)(-0.25)(0.75)(1.75)}{4!}(-0.00020)$$

$$+ \frac{(-1.25)(-0.25)(0.75)(1.75)(2.75)}{5!}(0.00005)$$

$\quad = 1.81291 - 0.04345 - 0.00047 + 0.00002$

$\log 58.75 = 1.76901$

Example 7. Given

x	1	2	3	4	5	6	7	8
$y = f(x)$	1	8	27	64	125	216	343	512

Find $f(7.5)$.

Solution. Here, $\quad u = \dfrac{x - x_n}{h} = \dfrac{7.5 - 8}{1} = -0.5$

The difference table is

x	$y = f(x)$	∇y	$\nabla^2 y$	$\nabla^3 y$	$\nabla^4 y$
1	1				
		7			
2	8		12		
		19		6	
3	27		18		0
		37		6	
4	64		24		0
		61		6	
5	125		30		0
		91		6	
6	216		36		0
		127		6	
7	343		42		
		169			
8	512				

Now, by Newton's Backward formula,

$$y(7.5) = f(7.5) = y_n + u\nabla y_n + \frac{u(u+1)}{2!}\nabla^2 y_n + \frac{u(u+1)(u+2)}{3!}\nabla^3 y_n$$

$$= 512 + (-0.5)(169) + \frac{(-0.5)(0.5)}{2!}(42) + \frac{(-0.5)(0.5)(1.5)}{3!}(6)$$

$$= 512 - 84.5 - 5.25 - 0.375$$

$$\therefore f(7.5) = 421.875$$

Example 8. The following data are taken from the steam table:

Temp.°C	140	150	160	170	180
Pressure kg f/cm^2	3.685	4.854	6.302	8.076	10.225

Find the pressure at temperature $t = 175°$.

Solution. The difference table is

t	p	∇p	$\nabla^2 p$	$\nabla^3 p$	$\nabla^4 p$
140	3.685				
		1.169			
150	4.854		0.279		
		1.448		0.047	
160	6.302		0.326		0.002
		1.774		0.049	
170	8.076		0.375		
		2.149			
180	10.225				

$$u = \frac{t - t_n}{h} = \frac{175 - 180}{10} = -0.5$$

$$p(t = 175) = p_n + u\nabla p_n + \frac{u(u+1)}{2!}\nabla^2 p_n + \frac{u(u+1)(u+2)}{3!}\nabla^3 p_n$$

$$+ \frac{u(u+1)(u+2)(u+3)}{4!}\nabla^4 p_n$$

$$= 10.225 + (-0.5)(2.149) + \frac{(-0.5)(0.5)}{2!}(0.375) + \frac{(-0.5)(0.5)(1.5)}{3!}(0.049)$$

$$+ \frac{(-0.5)(0.5)(1.5)(2.5)}{4!}(0.002)$$

$$= 10.225 - 1.0745 - 0.046875 - 0.0030625 - 0.0000781$$

$$= 9.1004844 \approx 9.100$$

EXERCISE 4.1

1. From the following table of values of x and y, find the value of y at $x = 0.23$ and $x = 0.29$

x:	0.20	0.22	0.24	0.26	0.28	0.30
y:	1.6596	1.6698	1.6804	1.6912	1.7024	1.7139

2. Find the area of the circle of diameter 82 cm. from the following table:

d(cm):	80	85	90	95	100
A(cm^2):	5026	5674	6362	7088	7854

3. If l_x represents the number of persons living at age x in a life table, find as accurately as the data will permit the value of l_{47}. Given that $l_{20} = 512$, $l_{30} = 439$, $l_{40} = 346$ and $l_{50} = 243$.

4. Find a polynomial of degree four which takes the values:

x:	2	4	6	8	10
y:	0	0	1	0	0

5. Find the values of y at $x = 21$ and $x = 28$ from the following table:

x:	20	23	26	29
y:	0.3420	0.3907	0.4384	0.4848

6. From the following table, find $\sin 53°$:

$x°$:	15°	20°	25°	30°	35°	40°	45°	50°	55°
$\sin x°$:	0.2588	0.3420	0.4226	0.5000	0.5736	0.6428	0.7071	0.7660	0.8192

7. From the following table find the value of $\tan 45° \ 15'$:

$x°$:	45	46	47	48	49	50
$\tan x°$:	1.00000	1.03553	1.07237	1.11061	1.15037	1.19175

8. The population of a town is as follows:

Year x:	1941	1951	1961	1971	1981	1991
Population in lakhs y:	20	24	29	36	46	51

   Estimate the population in crease during the period 1946 to 1976.

9. From the following table, find θ at $x = 43$ and $x = 84$.

x:	40	50	60	70	80	90
θ:	184	204	226	250	276	304

   Also express θ in terms of x.

10. From the following table, find the value of $e^{0.24}$

x:	0.1	0.2	0.3	0.4	0.5
e^x:	1.10517	1.22140	1.34986	1.49182	1.64872

11. The following table gives the marks secured by 100 students in the paper of Mathematics:

Marks Range:	30-40	40-50	50-60	60-70	70-80
No. of students:	25	35	22	11	7

   Find the number of students
   (i) who got more than 55 marks.
   (ii) who got marks in the range of 36-45.

12. The number of deaths in four successive ten year age groups are given:

Age group:	25-35	35-45	45-55	55-65
Deaths:	13229	18139	24225	31496

   Find the number of deaths between 45 and 50.

13. Find the number of men getting wages between Rs.10 and Rs.15 from the following table:

Wages(in Rs):	0-10	10-20	20-30	30-40
Frequency:	9	30	35	42

14. Using Newton's forward interpolation formula, obtain the interpolating polynomial satisfying the following data:

x:	1	2	3	4
f(x):	26	18	4	1

15. The marks obtained by 492 candidates in physics examination are given in the following table

Marks	No. of candidates
0-40	210
40-45	43
45-50	54
50-55	74
55-60	32
60-65	79

   Find the number of candidates who got marks more than 48 but not more than 50.

16. Find $y(32)$ if $y(10) = 35.3$, $y(15) = 32.4$, $y(20) = 29.2$, $y(25) = 26.1$, $y(30) = 23.2$ and $y(35) = 20.5$.

17. The following table gives the values of the probability integral
$$f(x) = \frac{1}{\sqrt{2\pi}} \int_0^x e^{-x^2} dx$$ for certain equidistant values of x.

Find the value of $\dfrac{1}{\sqrt{2\pi}}\int_0^{0.543} e^{-x^2}\, dx$ using the data given below

x:	0.51	0.52	0.53	0.54	0.55
$f(x)$:	0.5292	0.5379	0.5465	0.5549	0.5633

18. Estimate the value of tan 17°, from the following table:

$\theta°$:	0	4	8	12	16	20	24
$\tan\theta°$:	0	0.0699	0.1405	0.2126	0.2167	0.3640	0.4402

19. Find the value of $e^{2.25}$ by Newton's Interpolation formula from the following table:

x:	1.7	1.8	1.9	2.0	2.1	2.2	2.3
e^x:	5.474	6.050	6.686	7.389	8.166	9.025	9.974

20. The following table gives the values of density of saturated water vapours for various temperatures of saturated steam:

Temperature °C:	100	150	200	250	300
Density:	958	917	865	794	712

find the density when the temperature is 130° C and 275°C.

21. Using Newton's backward difference formula, find the polynomial of degree four passing through the points (1, 1), (2, –1), (3, 1), (4, –1) and (5, 1).

22. Given $u_0 = 10$, $u_2 = 8$, $u_4 = 10$, $u_6 = 50$, find u_5.

23. Given that $\sqrt{12500} = 111.8034$, $\sqrt{12510} = 111.8481$, $\sqrt{12520} = 111.8928$, $\sqrt{12530} = 111.9375$ and $\sqrt{12540} = 111.982$. find $\sqrt{12536}$.

24. Find the value of $\log_{10} 2.91$, using the given table:

x:	2.0	2.2	2.4	2.6	2.8	3.0
$\log_{10} x$:	0.30103	0.34242	0.38021	0.41497	0.44716	0.47721

25. The following data are part of a table for $g(x) = \dfrac{\sin x}{x^2}$:

x:	0.1	0.2	0.3	0.4	0.5
$g(x)$:	9.9833	4.9667	3.2836	2.4339	1.9177

find the value of $g(0.45)$.

26. The table gives the distance in nautical miles of the visible horizon for the given heights above the earths surface, find the distance when height is 410 ft.

x = height:	100	150	200	250	300	350	400
y = distance:	10.63	13.03	15.04	16.81	18.42	19.90	21.27

27. Find $f(1.1)$ from the following table:

x:	1	2	3	4	5
$f(x)$:	7	12	29	64	123

28. Find y at $x = 2.5$, given:

x:	0	1	2	3
y:	560	556	520	385

29. Estimate log 62.37 from the following table:

x:	45	50	55	60	65
$\log x$:	1.65321	1.69897	1.74036	1.77815	1.81291

30. Find $f(1.0428)$ and $f(1.9675)$, from the following table:

x:	1.0	1.1	1.2	1.3	1.4	1.5	1.6
$f(x)$:	0.5652	0.6375	0.7147	0.7973	0.8861	0.9817	1.0848
	1.7	1.8	1.9	2.0			
	1.1964	1.3172	1.0428	1.5906			

ANSWERS
---

1. 1.6751 ; 1.7081 **2.** 5280.10 cm^2 **3.** 276.353

4. $\dfrac{1}{64}[x^4 - 24x^3 + 196x^2 - 624x + 640]$ **5.** 0.3583 ; 0.4695

6. 0.7986 **7.** 1.00876 **8.** 19.119 lakhs.

9. 189.79 ; 286.96 ; $\theta = 0.01 x^2 + 1.1 x + 124$ **10.** 1.27125

11. (i) 28 (ii) 36

12. 11278   **13.** 15

14. $\dfrac{17}{6}x^3 - 20x^2 + \dfrac{193}{6}x + 11$ **15.** 27

16. 22.0948   **17.** 0.55743   **18.** 0.3057

19. 9.4877   **20.** 935 ; 759

21. $\dfrac{1}{3}(2x^4 - 24x^3 + 100x^2 - 168x + 93)$ **22.** 23.125

23. 111.964   **24.** 0.46389   **25.** 2.1479

26. 21.53 nautical miles   **27.** 7.13   **28.** 469.06

29. 1.79498.   **30.** 0.5956 ; 1.5433

4.7 CENTRAL DIFFERENCE INTERPOLATION FORMULA

We have already discussed Newton's forward and backward interpolation formula which are applicable for interpolation near the beginning and end of the tabulated values respectively. They are not applicable to interpolate near the centre of the table. For this, we use central difference interpolation formula which are most suited for interpolation near the middle of a table. We have already discussed the central difference operator δ.

156 Numerical and Statistical Techniques

The various central difference interpolation formula are
(i) Gauss's Forward interpolation formula.
(ii) Gauss's Backward interpolation formula.
(iii) Stirling's Formula.
(iv) Bessel's Formula.
(v) Laplace Everett's Formula.

4.8 GAUSS'S FORWARD INTERPOLATION FORMULA

Newton's-Gregory forward interpolation formula is given by,

$$y_n(x) = y_n(x_0 + uh) = y_0 + u\Delta y_0 + \frac{u(u-1)}{2!}\Delta^2 y_0$$

$$+ \frac{u(u-1)(u-2)}{3!}\Delta^3 y_0 + \cdots \quad \ldots(1)$$

where $\quad u = \dfrac{x - x_0}{h}$

Now,
$$\Delta^2 y_0 = \Delta^2 E y_{-1} = \Delta^2(1+\Delta)y_{-1}$$
$$= \Delta^2(y_{-1} + \Delta y_{-1}) = \Delta^2 y_{-1} + \Delta^3 y_{-1} \quad \ldots(a)$$
$$\Delta^3 y_0 = \Delta^3 E y_{-1} = \Delta^3(1+\Delta)y_{-1} = \Delta^3 y_{-1} + \Delta^4 y_{-1} \quad \ldots(b)$$
$$\Delta^4 y_0 = \Delta^4 E y_{-1} = \Delta^4(1+\Delta)y_{-1} = \Delta^4 y_{-1} + \Delta^5 y_{-1} \quad \ldots(c)$$
$$\Delta^3 y_{-1} = \Delta^3 E y_{-2} = \Delta^3(1+\Delta)y_{-2} = \Delta^3 y_{-2} + \Delta^4 y_{-2} \quad \ldots(d)$$
$$\Delta^4 y_{-1} = \Delta^4 y_{-2} + \Delta^5 y_{-2} \text{ and so on.}$$

Substituting these values in equation (1), we get

$$y_n(x) = y_0 + u\Delta y_0 + \frac{u(u-1)}{2!}(\Delta^2 y_{-1} + \Delta^3 y_{-1})$$

$$+ \frac{u(u-1)(u-2)}{3!}(\Delta^3 y_{-1} + \Delta^4 y_{-1}) + \frac{u(u-1)(u-2)(u-3)}{4!}(\Delta^4 y_{-1} + \Delta^5 y_{-1}) + \cdots$$

$$= y_0 + u\Delta y_0 + \frac{u(u-1)}{2!}\Delta^2 y_{-1} + \left[\frac{u(u-1)}{2!} + \frac{u(u-1)(u-2)}{3!}\right]\Delta^3 y_{-1}$$

$$+ \left[\frac{u(u-1)(u-2)}{3!} + \frac{u(u-1)(u-2)(u-3)}{4!}\right]\Delta^4 y_{-1} + \cdots$$

$$= y_0 + u\Delta y_0 + \frac{u(u-1)}{2!}\Delta^2 y_{-1} + \frac{(u+1)u(u-1)}{3!}\Delta^3 y_{-1}$$

$$+ \frac{(u+1)u(u-1)(u-2)}{4!}\Delta^4 y_{-1} + \frac{(u+1)u(u-1)(u-2)(u-3)}{5!}\Delta^5 y_{-1} + \cdots$$

$$= y_0 + u\Delta y_0 + \frac{u(u-1)}{2!}\Delta^2 y_{-1} + \frac{(u+1)u(u-1)}{3!}\Delta^3 y_{-1}$$

$$+ \frac{(u+1)u(u-1)(u-2)}{4!}(\Delta^4 y_{-2} + \Delta^5 y_{-2})$$

$$+ \frac{(u+1)u(u-1)(u-2)(u-3)}{5!}[\Delta^5 y_{-2} + \Delta^6 y_{-2}] + \cdots \text{ [Using equation (d)]}$$

$$y_n(x) = y_0 + u\Delta y_0 + \frac{u(u-1)}{2!}\Delta^2 y_{-1} + \frac{(u+1)u(u-1)}{3!}\Delta^3 y_{-1}$$

$$+ \frac{(u+1)u(u-1)(u-2)}{4!}\Delta^4 y_{-2}$$

$$+ \frac{(u+2)(u+1)u(u-1)(u-2)}{5!}\Delta^5 y_{-2} + \cdots \quad \ldots(2)$$

Equation (2) is called Gauss's forward interpolation formula. This formula can also be written as,

$$y_n(x) = y_n(x_0 + uh)$$

$$= y_0 + \binom{u}{1}\Delta y_0 + \binom{u}{2}\Delta^2 y_{-1} + \binom{u+1}{3}\Delta^3 y_{-1} + \binom{u+1}{4}\Delta^4 y_{-2}$$

$$+ \binom{u+2}{5}\Delta^5 y_{-2} + \cdots$$

where
$$\binom{u}{r} = \frac{u(u-1)(u-2)(u-3)\cdots(u-r+1)}{r!}$$

or
$$y_n(x) = y_n(x_0 + uh)$$

$$= y_0 + u_{C_1}\Delta y_0 + u_{C_2}\Delta^2 y_{-1} + u+1_{C_3}\Delta^3 y_{-1}$$

$$+ u+1_{C_4}\Delta^4 y_{-2} + u+2_{C_5}\Delta^5 y_{-2} + \cdots$$

Note: 1. This formula involves odd differences below The central line and even differences on the central line

Central line: y_0 , $\Delta^2 y_{-1}$, $\Delta^4 y_{-2}$, $\Delta^6 y_{-3}$

Next line: Δy_0 , $\Delta^3 y_{-1}$, $\Delta^5 y_{-2}$

2. This formula is used for $0 < u < 1$.

3. This formula can be written easily with the help of the following table:

Coefficients:	1	$^u C_1$	$^u C_2$	$^{u+1} C_3$	$^{u+1} C_4$	$^{u+2} C_5$
Differences:	y_0	Δy_0	$\Delta^2 y_{-1}$	$\Delta^3 y_{-1}$	$\Delta^4 y_{-2}$	$\Delta^5 y_{-2}$

4.9 GAUSS'S BACKWARD INTERPOLATION FORMULA

Newton's Gregory forward difference interpolation formula is

$$y_n(x) = y_n(x_0 + uh)$$

$$= y_0 + u\Delta y_0 + \frac{u(u-1)}{2!}\Delta^2 y_0 + \frac{u(u-1)(u-2)}{3!}\Delta^3 y_0 + \cdots \quad \ldots(1)$$

where $\quad u = \dfrac{x - x_0}{h}$

Now, $\quad \Delta y_0 = \Delta E y_{-1} = \Delta(1 + \Delta)y_{-1} = \Delta y_{-1} + \Delta^2 y_{-1}$

$\quad\quad\quad \Delta^2 y_0 = \Delta^2 y_{-1} + \Delta^3 y_{-1}$

$\quad\quad\quad \Delta^3 y_0 = \Delta^3 y_{-1} + \Delta^4 y_{-1}$

$\quad\quad\quad \Delta^3 y_{-1} = \Delta^3 y_{-2} + \Delta^4 y_{-2}$

$\quad\quad\quad \Delta^4 y_{-1} = \Delta^4 y_{-2} + \Delta^5 y_{-2}$ and so on

from (1)

$$y_n(x) = y_0 + u(\Delta y_{-1} + \Delta^2 y_{-1}) + \frac{u(u-1)}{2!}(\Delta^2 y_{-1} + \Delta^3 y_{-1})$$

$$+ \frac{u(u-1)(u-2)}{3!}(\Delta^3 y_{-1} + \Delta^4 y_{-1})$$

$$+ \frac{u(u-1)(u-2)(u-3)}{4!}(\Delta^4 y_{-1} + \Delta^5 y_{-1}) + \cdots$$

$$= y_0 + u\Delta y_{-1} + \left[u + u\frac{(u-1)}{2!}\right]\Delta^2 y_{-1}$$

$$+ \left[\frac{u(u-1)}{2!} + \frac{u(u-1)(u-2)}{3!}\right]\Delta^3 y_{-1}$$

$$+ \left[\frac{u(u-1)(u-2)}{3!} + \frac{u(u-1)(u-2)(u-3)}{4!}\right]\Delta^4 y_{-1} + \cdots$$

$$= y_0 + u\Delta y_{-1} + \frac{(u+1)u}{2!}\Delta^2 y_{-1} + \frac{(u+1)u(u-1)}{3!}\Delta^3 y_{-1}$$

$$+ \frac{(u+1)u(u-1)(u-2)}{4!}\Delta^4 y_{-1} + \cdots$$

$$= y_0 + u\Delta y_{-1} + \frac{(u+1)u}{2!}\Delta^2 y_{-1} + \frac{(u+1)u(u-1)}{3!}(\Delta^3 y_{-2} + \Delta^4 y_{-2})$$

$$+ \frac{(u+1)u(u-1)(u-2)}{4!}(\Delta^4 y_{-2} + \Delta^5 y_{-2}) + \cdots$$

$$= y_0 + u\Delta y_{-1} + \frac{(u+1)u}{2!}\Delta^2 y_{-1} + \frac{(u+1)u(u-1)}{3!}\Delta^3 y_{-2}$$

$$+ \frac{(u+2)(u+1)u(u-1)}{4!}\Delta^4 y_{-2} + \cdots \quad \ldots(2)$$

Equation (2) is known as Gauss's Backward difference interpolation formula.

This formula can also be written as.

$$y_n(x) = y_n(x_0 + uh) = y_0 + \binom{u}{1}\Delta y_{-1} + \binom{u+1}{2}\Delta^2 y_{-1} + \binom{u+1}{3}\Delta^3 y_{-2}$$

$$+ \binom{u+2}{4}\Delta^4 y_{-2} + \binom{u+2}{5}\Delta^5 y_{-3} + \cdots$$

or $y_n(x) = y_n(x_0 + uh) = y_0 + u_{C_1}\Delta y_{-1} + u+1_{C_2}\Delta^2 y_{-1} + u+1_{C_3}\Delta^3 y_{-2}$

$$+ u+2_{C_4}\Delta^4 y_{-2} + u+2_{C_5}\Delta^5 y_{-3} + \cdots$$

Note: 1. This formula involves odd differences above the central line and even differences on the central line

Previous line: ——— Δy_{-1} ——— $\Delta^3 y_{-2}$ ——— $\Delta^5 y_{-3}$ ———

Central line: ——— y_0 ——— $\Delta^2 y_{-1}$ ——— $\Delta^4 y_{-2}$ ———

2. This formula is used for $-1 < u < 0$.

3. This backward formula can be written with the help of the following table:

Coefficients:	1	u_{C_1}	$u+1_{C_2}$	$u+1_{C_3}$	$u+2_{C_4}$	$u+2_{C_5}$
Differences:	y_0	Δy_{-1}	$\Delta^2 y_{-1}$	$\Delta^3 y_{-2}$	$\Delta^4 y_{-2}$	$\Delta^5 y_{-3}$

ILLUSTRATIVE EXAMPLES

Example 1. From the following table, find the value of $e^{1.17}$ using Gauss's forward formula:

| x: | 1.00 | 1.05 | 1.10 | 1.15 | 1.20 | 1.25 | 1.30 |
| e^x: | 2.7183 | 2.8577 | 3.0042 | 3.1582 | 3.3201 | 3.4903 | 3.6693 |

Solution. Here $h = 0.05$ and $x = 1.17$

Taking origin at 1.15, we have

$$u = \frac{x - x_0}{h} = \frac{1.17 - 1.15}{0.05} = 0.4 \in [0, 1]$$

Now, we form the difference table:

160 Numerical and Statistical Techniques

u	x	$y = e^x$	Δy	$\Delta^2 y$	$\Delta^3 y$	$\Delta^4 y$	$\Delta^5 y$	$\Delta^6 y$
-3	1.00	2.7183						
			0.1394					
-2	1.05	2.8577		0.0071				
			0.1465		0.0004			
-1	1.10	3.0042		0.0075		0		
			0.1540		0.0004		0	
0	1.15	3.1582		0.0079		0		0.0001
			0.1619		0.0004		0.0001	
1	1.20	3.3201		0.0083		0.0001		
			0.1702		0.0005			
2	1.25	3.4903		0.0088				
			0.1790					
3	1.30	3.6693						

Using Gauss's forward formula

$$e^{1.17} = y_o + u\Delta y_0 + \frac{u(u-1)}{2!}\Delta^2 y_{-1} + \frac{(u+1)u(u-1)}{3!}\Delta^3 y_{-1}$$
$$+ \frac{(u+1)u(u-1)(u-2)}{4!}\Delta^4 y_{-2} + \frac{(u+2)(u+1)u(u-1)(u-2)}{5!}\Delta 5 y_{-2} + \ldots$$

$$= 3.1582 + (0.4)(0.1619) + \frac{(0.4)(0.4-1)}{2!}(0.0079)$$
$$+ \frac{(0.4+1)(0.4)(0.4-1)}{3!}(0.0004) + 0 \ldots$$

$$= 3.1582 + 0.06476 - 0.0009 - 0.0000224$$
$$e^{1.17} = 3.2220$$

Example 2. Given $\sqrt{100} = 10$, $\sqrt{110} = 10.489$, $\sqrt{120} = 10.954$, $\sqrt{130} = 11.402$ and $\sqrt{140} = 11.832$, find $\sqrt{122}$ using Gauss's forward formula.

Solution. The difference table is

u	x	$y = \sqrt{x}$	Δy	$\Delta^2 y$	$\Delta^3 y$	$\Delta^4 y$
-2	100	10				
			0.489			
-1	110	10.489		-0.024		
			0.465		0.007	
0	120	10.954		-0.017		-0.008
			0.448		-0.001	
1	130	11.402		-0.018		
			0.430			
2	140	11.832				

Taking 120 as origin,
$$u = \frac{x - x_0}{h} = \frac{122 - 120}{10} = 0.2$$
By using Gauss's forward formula.
$$\sqrt{122} = 10.954 + (0.2)(0.448) + \frac{(0.2)(0.2 - 1)}{2!}(-0.017)$$
$$+ \frac{(0.2 + 1)(0.2)(0.2 - 1)}{3!}(-0.001) + \frac{(0.2 + 1)(0.2)(0.2 - 1)(0.2 - 2)}{4!}(-0.008)$$
$$= 10.954 + 0.0896 + 0.00136 + 0.00032 - 0.000115$$
$$\sqrt{122} = 11.045$$

Example 3. The population of a town is given as:

Year:	1931	1941	1951	1961	1971
Population (in thousands):	15	20	27	39	52

Estimate the population of the town in the year 1946. Using Gauss's Backward formula.

Solution. Here $h = 10$, $x = 1946$.

Taking $x_0 = 1951$ as origin, we have
$$x = \frac{x - x_0}{h} = \frac{1946 - 1951}{10} = -\frac{1}{2} = 0.5$$

The difference table is

u	x	y	Δy	$\Delta^2 y$	$\Delta^3 y$	$\Delta^4 y$
-2	1931	15				
			5			
-1	1941	20		2		
			7		3	
0	1951	27		5		-7
			12		-4	
1	1961	39		1		
			13			
2	1971	52				

Using, Gauss's backward formula.
$$y(x) = y_0 + u\Delta y_{-1} + \frac{(u+1)u}{2!}\Delta^2 y_{-1} + \frac{(u+1)u(u-1)}{3!}\Delta^3 y_{-2}$$
$$+ \frac{(u+1)u(u-1)(u+2)}{4!}\Delta^4 y_{-2}$$
$$y(1946) = 27 + (-0.5)(7) + \frac{(-0.5+1)(-0.5)}{2!}(5) + \frac{(-0.5+1)(-0.5)(-0.5-1)}{3!}(3)$$
$$+ \frac{(-0.5+2)(-0.5+1)(-0.5)(-0.5-1)}{4!}(-7)$$

= 27 − 3.5 − 0.625 + 0.1875 − 0.16406
= 22.898 Thousands = 22898

Example 4. Use Gauss's backward formula to find tan 51° 42′ given

x:	50	51	52	53	54
tan x:	1.1918	1.2349	1.2799	1.3270	1.3764

Solution. Here $x = 51°42'$, $h = 1° = 60'$

Taking 52° as origin,

$$u = \frac{x - x_0}{h} = \frac{51°42' - 52°}{1°} = -\frac{18'}{60'} = -0.3$$

The difference table is

u	x	y = tan x	Δy	Δ²y	Δ³y	Δ⁴y
−2	50	1.1918				
			0.0431			
−1	51	1.2349		0.0019		
			0.0450		0.0002	
0	52	1.2799		0.0021		0
			0.0471		0.0002	
1	53	1.3270		0.0023		
			0.0494			
2	54	1.3764				

Using Gauss's backward formula,

$$\tan 51° 42' = y_0 + u\Delta y_{-1} + \frac{(u+1)u}{2!}\Delta^2 y_{-1} + \frac{(u+1)u(u-1)}{3!}\Delta^3 y_{-2}$$

$$= 1.2799 + (-0.3)(0.0450) + \frac{(-0.3+1)(-0.3)}{2!}(0.0021)$$

$$+ \frac{(-0.3+1)(-0.3)(-0.3-1)}{3!}(0.0002)$$

= 1.2799 − 0.0135 − 0.0002205 + 0.0000091.

tan 51° 42′ = 1.2661704 = 1.2662

Example 5. Find the value of cos 51° 42′ by Gauss's backward formula. Given that

x:	50°	51°	52°	53°	54°
cos x:	0.6428	0.6293	0.6157	0.6018	0.5878

Solution. Here $h = 1$. Taking origin at 52° i.e., $x_0 = 52°$

$$\therefore \quad u = \frac{x - x_0}{h} = \frac{51°42' - 52°}{1°} = \frac{-18'}{60'} = -0.3$$

The difference table is

u	x	y	Δy	$\Delta^2 y$	$\Delta^3 y$	$\Delta^4 y$
-2	50°	0.6428				
			-0.0135			
-1	51°	0.6293		-0.0001		
			-0.0136		-0.0002	
0	52°	0.6157		-0.0003		0.0004
			-0.0139		0.0002	
1	53°	0.6018		-0.0001		
			-0.0140			
2	54°	0.5878				

By using, Gauss's backward formula,

$$y(x) = y_0 + \frac{u}{1!}\Delta y_{-1} + \frac{(u+1)u}{2!}\Delta^2 y_{-1} + \frac{(u+1)u(u-1)}{3!}\Delta^3 y_2$$

$$+ \frac{(u+2)(u+1)u(u-1)}{4!}\Delta^4 y_{-2} + \cdots$$

$$= 0.6157 + (-0.3)(-0.0136) + \frac{(0.7)(-0.3)}{2!}(-0.0003)$$

$$+ \frac{(0.7)(-0.3)(-1.3)}{3!}(-0.0002) + \frac{(1.7)(0.7)(-0.3)(-1.3)}{4!}(0.0004)$$

$$= 0.6157 + 0.00408 + 0.0000315 - 0.0000091 + 0.000007735$$

$$= 0.619810135$$

Hence cos 51° 42′ = 0.619810135

Example 6 Using Gauss's backward interpolation formula, find the population for the year 1936 given that

Year (x):	1901	1911	1921	1931	1941	1951
Population(y):	12	15	20	27	39	52
(in thousand)						

Solution. Here $h = 10$, Taking origin at 1941 i.e., $x_0 = 1941$.

$$\therefore \quad u = \frac{x - x_0}{h} = \frac{1936 - 1941}{10} = -0.5$$

The difference table is

u	x	y	Δy	Δ²y	Δ³y	Δ⁴y	Δ⁵y
−4	1901	12					
			3				
−3	1911	15		2			
			5		0		
−2	1921	20		2		3	
			7		3		−10
−1	1931	27		5		−7	
			12		−4		
0	1941	39		1			
			13				
1	1951	52					

By Gauss's backward formula

$$y(x) = y_0 + \frac{u}{1!}Dy_{-1} + \frac{(u+1)u}{2!}\Delta^2 y_{-1} + \frac{(u+1)u(u-1)}{3!}\Delta^3 y_{-2}$$

$$+ \frac{(u+2)(u+1)u(u-1)}{4!}\Delta^4 y_{-2} + \frac{(u+2)(u+1)u(u-1)(u-2)}{5!}\Delta^5 y_{-3} + \cdots$$

$$= 39 + (-0.5)(12) + \frac{(0.5)(-0.5)}{2} \times 1 + \frac{(0.5)(-0.5)(-1.5)}{6} \times -4$$

$$y(x) = 39 - 6 - \frac{1}{8} - \frac{1}{4} = 32.625 \text{(thousands)} = 32625$$

EXERCISE 4.2

1. Apply Gauss's forward formula to obtain $f(x)$ at $x = 3.5$ form the following table:

x :	2	3	4	5
f(x) :	2.626	3.454	4.784	6.986

2. Use Gauss's interpolation formula to get y_{16} for the following data:

x :	5	10	15	20	25
y :	26.782	19.951	14.001	8.762	4.163

3. Find $f\left(\frac{1}{2}\right)$ for the following data by using Gauss's forward formula:
   $f(2) = 10, f(1) = 8, f(0) = 5, f(-1) = 10$

4. Find $f(25)$ for the following data by using Gauss's formula:
   $f(20) = 14, f(24) = 32, f(28) = 35$, and $f(32) = 40$

5. The population of a town is given below. Apply Gauss's backward formula to get the population in 1926.

   year(x):   1911   1921   1931   1941   1951
   Population (in thousands)y: 15   20   27   39   52

6. If $\sqrt{12500} = 111.803399$, $\sqrt{12510} = 111.848111$, $\sqrt{12520} = 111.892805$, $\sqrt{12530} = 111.937483$, find $\sqrt{12516}$ by Gauss's backward formula.

7. Use Gauss's backward interpolation formula to obtain the sales of a concern for the year 1976 for the following data:

   year   : 1940   1950   1960   1970   1980   1990
   Sales(in Lakhs of Rs.) : 17   20   27   32   36   38

ANSWERS

1. 4.033125   2. 12.901   3. 0.06
4. 33.41   5. 22.898   6. 111.8749294
7. 30.114368

4.10 STIRLING'S FORMULA

By Gauss's forward formula, we have

$$y = y_0 + u\Delta y_0 + \frac{u(u-1)}{2!}\Delta^2 y_{-1} + \frac{(u+1)u(u-1)}{3!}\Delta^3 y_{-1}$$

$$+ \frac{(u+1)u(u-1)(u-2)}{4!}\Delta^4 y_{-2} + \cdots \quad \ldots(1)$$

By Gauss's Backward formula, we have

$$y = y_0 + u\Delta y_{-1} + \frac{(u+1)u}{2!}\Delta^2 y_{-1} + \frac{(u+1)u(u-1)}{3!}\Delta^3 y_{-2}$$

$$+ \frac{(u+2)(u+1)u(u-1)}{4!}\Delta^4 y_{-2} + \cdots \quad \ldots(2)$$

Taking Mean of (1) and (2), we get

$$y = y_0 + u\left(\frac{\Delta y_0 + \Delta y_{-1}}{2}\right) + \frac{u^2}{2!}\Delta^2 y_{-1} + \frac{u(u+1)(u-1)}{3!}\left(\frac{\Delta^3 y_{-1} + \Delta^3 y_{-2}}{2}\right)$$

$$+ \frac{u^2(u^2-1^2)}{4!}\Delta^4 y_{-2} + \cdots$$

or $$y = y_0 + \frac{u}{1!}\left(\frac{\Delta y_0 + \Delta y_{-1}}{2}\right) + \frac{u^2}{2!}\Delta^2 y_{-1} + \frac{u(u^2-1^2)}{3!}\left(\frac{\Delta^3 y_{-1} + \Delta^3 y_{-2}}{2}\right)$$

$$+ \frac{u^2(u^2-1^2)}{4!}\Delta^4 y_{-2} + \cdots \quad \ldots(3)$$

This is called stirling's formula and it is the average of two Gauss's formula. This is useful when $|u| < \dfrac{1}{2}$ or $-\dfrac{1}{2} < u < \dfrac{1}{2}$. It gives best estimate when $-\dfrac{1}{4} < u < \dfrac{1}{4}$

Note: 1. The above formula involves the means of the odd differences just above and just below the central line and even difference on the central line.

    2. The formula can be remembered with the help of the following table.

Coefficients	1	u	$\dfrac{u^2}{2}$	$\dfrac{u(u^2-1^2)}{3!}$	$\dfrac{u^2(u^2-1^2)}{4!}$
Differences	y_0	$\dfrac{1}{2}(\Delta y_0 + \Delta y_{-1})$	$\Delta^2 y_{-1}$	$\dfrac{1}{2}(\Delta^3 y_{-1} + \Delta^3 y_{-2})$	$\Delta^4 y_{-2}$

4.11 BESSEL'S FORMULA

By Gauss's forward formula, we have

$$y = y(x) = y(x_0 + uh) = y_0 + \frac{u}{1!}\Delta y_0 + \frac{u(u-1)}{2!}\Delta^2 y_{-1}$$

$$+ \frac{(u+1)u(u-1)}{3!}\Delta^3 y_{-1} + \frac{(u+1)u(u-1)(u-2)}{4!}\Delta^4 y_{-2}$$

$$+ \frac{(u+2)(u+1)u(u-1)(u-2)}{5!}\Delta^5 y_{-2} + \cdots \qquad \ldots(1)$$

we know that $\quad \Delta y_0 = y_1 - y_0$

$\therefore \qquad\qquad y_0 = y_1 - \Delta y_0 \qquad\qquad\qquad\qquad\qquad \ldots(2)$

$\qquad\qquad\quad y_{-1} = y_0 - \Delta y_{-1}$

$\therefore \qquad\qquad \Delta^2 y_{-1} = \Delta^2 y_0 - \Delta^3 y_{-1} \qquad\qquad\qquad \ldots(3)$

Also $\qquad\quad \Delta^4 y_{-2} = \Delta^4 y_{-1} - \Delta^5 y_{-2}$ etc

Hence equation (1) is rewritten as,

$$y = y(x) = y(x_0 + uh) = \left(\frac{y_0}{2} + \frac{y_0}{2}\right) + u\Delta y_o + \frac{1}{2}\frac{u(u-1)}{2!}\Delta^2 y_{-1}$$

$$+ \frac{1}{2}\frac{u(u-1)}{2!}\Delta^2 y_{-1} + \frac{(u+1)u(u-1)}{3!}\Delta^3 y_{-1} + \cdots \qquad \ldots(4)$$

Using equation (2) and (3) in (4), we get

$$y = y(x) = y(x_0 + uh)$$

$$= \frac{y_0}{2} + \frac{1}{2}(y_1 - \Delta y_0) + u\Delta y_0 + \frac{1}{2}\frac{u(u-1)}{2!}\Delta^2 y_{-1}$$

$$+ \frac{1}{2}\cdot\frac{u(u-1)}{2!}\left(\Delta^2 y_0 - \Delta^3 y_{-1}\right) + \frac{(u+1)u(u-1)}{3!}\Delta^3 y_{-1} + \cdots$$

$$= \frac{y_0 + y_1}{2} + \left(u - \frac{1}{2}\right)\Delta y_0 + \frac{1}{2}\frac{u(u-1)}{2!}(\Delta^2 y_{-1} + \Delta^2 y_0)$$

$$+ \frac{u(u-1)}{2!}\left(-\frac{1}{2} + \frac{u+1}{3}\right)\Delta^3 y_{-1} + \cdots$$

$$y = y(x) = y(x_0 + uh)$$

$$= \frac{y_0 + y_1}{2} + \left(u - \frac{1}{2}\right)\Delta y_0 + \frac{u(u-1)}{2!}\left(\frac{\Delta^2 y_{-1} + \Delta^2 y_0}{2}\right)$$

$$+ \frac{\left(u - \frac{1}{2}\right)u(u-1)}{3!}\Delta^3 y_{-1} + \frac{(u+1)u(u-1)(u-2)}{4!}\left(\frac{\Delta^4 y_{-2} + \Delta^4 y_{-1}}{2}\right) + \cdots$$

...(5)

This is called Bessel's formula. It is useful when $u = \frac{1}{2}$. It gives better results when $\frac{1}{4} < u < \frac{3}{4}$.

Note: If $u = \frac{1}{2}$, then the coefficients of all odd order differences are zero and above formula becomes

$$y\left(u = \frac{1}{2}\right) = \frac{y_0 + y_1}{2} - \frac{1}{8}\left(\frac{\Delta^2 y_{-1} + \Delta^2 y_0}{2}\right) + \frac{3}{128}\left(\frac{\Delta^4 y_{-2} + \Delta^4 y_{-1}}{2}\right) + \cdots$$

This formula is suited to compute the values of the function midway between two given values. This is also known as formula for interpolating to halves.

4.12 LAPLACE-EVERETT FORMULA

Gauss's forward formula

$$y(x) = y(x_0 + uh) = y_0 + u\Delta y_0 + \frac{u(u-1)}{2!}\Delta^2 y_{-1} + \frac{(u+1)u(u-1)}{3!}\Delta^3 y_{-1} +$$

$$\frac{(u+1)u(u-1)(u-2)}{4!}\Delta^4 y_{-2} + \frac{(u+2)(u+1)u(u-1)(u-2)}{5!}\Delta^5 y_{-2} + \cdots$$

...(1)

We know that $\Delta y_0 = y_1 - y_0$

$\therefore \quad \Delta^3 y_{-1} = \Delta^2 y_0 - \Delta^2 y_{-1}$

$\Delta^5 y_{-2} = \Delta^4 y_{-1} - \Delta^4 y_{-2}$

Substituting these values in equation (1), we get

$$y(x) = y_0 + u(y_1 - y_0) + \frac{u(u-1)}{2!}\Delta^2 y_{-1} + \frac{(u+1)u(u-1)}{3!}(\Delta^2 y_0 - \Delta^2 y_{-1})$$

$$+\frac{(u+1)u(u-1)(u-2)}{4!}\Delta^4 y_{-2}+\frac{(u+2)(u+1)u(u-1)(u-2)}{5!}$$

$$(\Delta^4 y_{-1}-\Delta^4 y_{-2})+\cdots$$

$$y(x)=(1-u)y_0+uy_1+\frac{(u+1)u(u-1)}{3!}\Delta^2 y_0-\frac{u(u-1)(u-2)}{3!}\Delta^2 y_{-1}$$

$$+\frac{(u+2)(u+1)u(u-1)(u-2)}{5!}\Delta^4 y_{-1}-\frac{(u+1)u(u-1)(u-2)(u-3)}{5!}\Delta^4 y_{-2}+\cdots$$

$$=\left\{uy_1+\frac{(u+1)u(u-1)}{3!}\Delta^2 y_0+\frac{(u+2)(u+1)u(u-1)(u-2)}{5!}\Delta^4 y_{-1}+\cdots\right\}$$

$$+\left\{(1-u)y_0+\frac{(1-u+1)(1-u)(1-u-1)}{3!}\Delta^2 y_{-1}+\right.$$

$$\left.\frac{(1-u+2)(1-u+1)(1-u)(1-u-1)(1-u-2)}{5!}\Delta^4 y_{-2}+\cdots\right\}$$

$$=\left\{uy_1+\frac{(u+1)u(u-1)}{3!}\Delta^2 y_0+\frac{(u+2)(u+1)u(u-1)(u-2)}{5!}\Delta^4 y_{-1}+\cdots\right\}$$

$$+\left\{vy_0+\frac{(v+1)v(v-1)}{3!}\Delta^2 y_{-1}+\frac{(v+2)(v+1)v(v-1)(v-2)}{5!}\Delta^4 y_{-2}+\cdots\right\}$$

where $v = 1-u$

or $$y(x)=\left\{uy_1+\frac{u(u^2-1^2)}{3!}\Delta^2 y_0+\frac{u(u^2-1^2)(u^2-2^2)}{5!}\Delta^4 y_{-1}+\cdots\right\}$$

$$+\left\{vy_0+\frac{v(v^2-1^2)}{3!}\Delta^2 y_{-1}+\frac{v(v^2-1^2)(v^2-2^2)}{5!}\Delta^4 y_{-2}+\cdots\right\}$$

This is called Laplace-Everett's formula.

Note: This formula involves even differences on and below the central line.

This gives best estimation when $u > \frac{1}{2}$. This can be used if $0 < u < 1$.

4.13 RELATION BETWEEN BESSEL'S AND EVERETT'S FORMULA

These two formulae are very closely related and it is possible to deduce one formula from the other by a suitable rearrangement. To see this we start with Bessel's formula

$$y=y(x_0+uh)=\frac{y_0+y_1}{2}+\left(u-\frac{1}{2}\right)\Delta y_0+\frac{u(u-1)}{2!}\left(\frac{\Delta^2 y_{-1}+\Delta^2 y_0}{2}\right)$$

$$+\frac{\left(u-\frac{1}{2}\right)u(u-1)}{3!}\Delta^3 y_{-1}+\frac{(u+1)u(u-1)(u-2)}{4!}\left(\frac{\Delta^4 y_{-2}+\Delta^4 y_{-1}}{2}\right)+\cdots$$

...(1)

$$= \frac{y_0 + y_1}{2} + \left(u - \frac{1}{2}\right)\Delta y_0 + \frac{u(u-1)}{2}\left(\frac{\Delta^2 y_{-1} + \Delta^2 y_0}{2}\right)$$

$$+ \frac{\left(u - \frac{1}{2}\right)u(u-1)}{3!}\Delta^3 y_{-1} + \cdots \qquad \ldots(2)$$

Now we express the odd order differences in terms of the corresponding lower even order differences. That is,

$$\Delta^3 y_{-1} = \Delta^2 y_0 - \Delta^2 y_{-1} \text{ etc}$$

Then the equation (2) reduces to,

$$y = y(x_0 + uh) = \frac{y_0 + y_1}{2} + \left(u - \frac{1}{2}\right)(y_1 - y_0) + \frac{u(u-1)}{2}\left(\frac{\Delta^2 y_{-1} + \Delta^2 y_0}{2}\right)$$

$$+ \frac{\left(u - \frac{1}{2}\right)u(u-1)}{3!}(\Delta^2 y_0 - \Delta^2 y_{-1}) + \cdots$$

$$= (1-u)y_0 + uy_1 + \binom{u+1}{3}\Delta^2 y_0 - \binom{u}{3}\Delta^2 y_{-1} + \cdots$$

$$= (1-u)y_0 + uy_1 + \frac{(u+1)u(u-1)}{3!}\Delta^2 y_0 - \frac{u(u-1)(u-2)}{3!}\Delta^2 y_{-1} + \cdots$$

$$= vy_0 + uy_1 + \frac{u(u^2 - 1^2)}{3!}\Delta^2 y_0 + \frac{(v-1)v(1+v)}{3!}\Delta^2 y_{-1} + \cdots$$

$$= \left[vy_0 + \frac{v(v^2 - 1^2)}{3!}\Delta^2 y_{-1} + \cdots\right] + \left[uy_1 + \frac{u(u^2 - 1^2)}{3!}\Delta^2 y_0 + \cdots\right] \qquad \ldots(3)$$

where $v = 1 - u$

Equation (3) is nothing but the Everett's formula upto second differences.

This shows that Bessel's formula truncated after the third differences is the same as Everett's formula truncated after the second differences. In a similar way, we can also start from Everett's formula and obtain Bessel's formula.

4.14 ADVANTAGES OF CENTRAL DIFFERENCE INTERPOLATION FORMULA

In the preceding sections, we have derived some interpolation formulae of great practical importance. A natural question is: Which one of these formulae gives the most accurate result?

In general, the coefficients in the central difference interpolation formulae are smaller and rapidly convergent than those in Newton-Gregory formulae. The coefficients in stirling's formula decrease more rapidly than those of Bessel's formula. The coefficients in Bessel's formula decrease more rapidly than those in Newton-Gregory formulae. Therefore, it is advisable to use stirling's or Bessel's central difference formula instead of Newton-Gregory formulae. One can have the following rules in selecting the interpolation formulae.

(i) If interpolation is required near the beginning of the tabular values, then use Newton-Gregory forward interpolation formula.

(ii) If interpolation is required near the end of the tabular values, then use Newton-Gregory backward interpolation formula.

(iii) If interpolation is required near the middle values of the table, then use either stirling's or Bessel's formula.

(iv) If $-\frac{1}{4} \leq u \leq \frac{1}{4}$, then use stirling's formula.

(v) If $\frac{1}{4} \leq u \leq \frac{3}{4}$, then use Bessel's formula.

ILLUSTRATIVE EXAMPLES

Example 1. Use stirling's formula to find y_{28}, given
$y_{20} = 49225$, $y_{25} = 48316$, $y_{30} = 47236$, $y_{35} = 45926$, $y_{40} = 44306$.

Solution. Here $h = 5$. Take the origin x_0 as 30.

$$\therefore \quad u = \frac{x - x_0}{h} = \frac{28 - 30}{5} = \frac{-2}{5} = -0.4$$

The difference table is

u	x	y	Δy	$\Delta^2 y$	$\Delta^3 y$	$\Delta^4 y$
-2	20	49225				
			-909			
-1	25	48316		-171		
			-1080		-59	
0	30	47236		-230		-21
			-1310		-80	
1	35	45926		-310		
			-1620			
2	40	44306				

By stirling's formula,

$$y(x) = y(x_0 + uh) = y_0 + \frac{u}{1}\left(\frac{\Delta y_0 + \Delta y_{-1}}{2}\right) + \frac{u^2}{2}\Delta^2 y_{-1}$$
$$+ \frac{u(u^2 - 1)}{3!}\left(\frac{\Delta^3 y_{-1} + \Delta^3 y_{-2}}{2}\right) + \frac{u(u^2 - 1^2)}{4!}\Delta^4 y_{-2} + \cdots$$

$$y(28) = 47236 + (-0.4)\left(\frac{-1080-1310}{2}\right) + \frac{(-0.4)^2}{2}(-230)$$
$$+ \frac{(-0.4 \times -0.84)}{6}\left(\frac{-59-80}{2}\right) + \frac{(-0.4)\times(-0.84)}{24} \times -21$$
$$y(28) = 47907.414$$

Example 2. Use Stirling's formula to find y_{35}, given $y_{20} = 512$, $y_{30} = 439$, $y_{40} = 346$ and $y_{50} = 243$.

Solution. Here $h = 10$, Take the origin x_0 as 30.

$$\therefore \quad u = \frac{x-x_0}{h} = \frac{35-30}{10} = 0.5$$

The difference table is

u	x	y	Δy	$\Delta^2 y$	$\Delta^3 y$
-1	20	512			
			-73		
0	30	439		-20	
			-93		10
1	40	346		-10	
			-103		
2	50	243			

By Stirling's formula

$$y(x) = y_0 + u\left(\frac{\Delta y_0 + \Delta y_{-1}}{2}\right) + \frac{u^2}{2}\Delta^2 y_{-1} + \frac{u(u^2-1)}{3!}\left(\frac{\Delta^3 y_{-1} + \Delta^3 y_{-2}}{2}\right) + \cdots$$

$$y(25) = 439 + 0.5\left(\frac{-93-73}{2}\right) + \frac{(0.5)^2}{2} \times -20 + \frac{0.5 \times -0.75}{6} \times \left(\frac{10}{2}\right)$$

$$y(25) = 394.6875$$

Example 3. Apply Bessel's formula to get the value of y_{25} given $y_{20} = 2854$, $y_{24} = 3162$, $y_{28} = 3544$, $y_{32} = 3992$.

Solution. Here $h = 4$, Take 24 as origin

$$\therefore \quad u = \frac{x-x_0}{h} = \frac{25-24}{4} = \frac{1}{4} = 0.25$$

The difference table is

u	x	y	Δy	$\Delta^2 y$	$\Delta^3 y$
-1	20	2854			
			308		
0	24	3162		74	
			382		-8
1	28	3544		66	
			448		
2	32	3992			

By Bessel's formula, we have

$$y(x) = \frac{y_0 + y_1}{2} + \left(u - \frac{1}{2}\right)\Delta y_0 + \frac{u(u-1)}{2!}\left(\frac{\Delta^2 y_{-1} + \Delta^2 y_0}{2}\right) + \frac{\left(u - \frac{1}{2}\right)u(u-1)}{3!}\Delta^3 y_{-1} + \ldots$$

$$y(25) = \frac{3162 + 3544}{2} + (-0.25) \times 382$$
$$+ \frac{0.25 \times -0.75}{2}\left(\frac{74 + 66}{2}\right) + \frac{(-0.25 \times .25 \times -0.75)}{6} \times -8$$
$$= 3353 - 95.5 - 6.5625 - 0.0625$$
$$y(25) = 3250.875$$

Example 4. From the table given below, find $y(5)$ by using Bessel's formula.

x:	0	4	8	12
y:	143	158	177	199

Solution. Here $h = 4$. Take 4 as origin.

$$\therefore \quad u = \frac{x - x_0}{h} = \frac{5 - 4}{4} = 0.25$$

The difference table is

u	x	y	Δy	$\Delta^2 y$	$\Delta^3 y$
-1	0	143			
			15		
0	4	158		4	
			19		-1
1	8	177		3	
			22		
2	12	199			

By Bessel's formula, we have

$$y(x) = \frac{y_0 + y_1}{2} + \left(u - \frac{1}{2}\right)\Delta y_0 + \frac{u(u-1)}{2!}\left(\frac{\Delta^2 y_{-1} + \Delta^2 y_0}{2}\right) + \frac{\left(u - \frac{1}{2}\right)u(u-1)}{3!}\Delta^3 y_{-1} + \cdots$$

$$= \frac{158 + 177}{2} + (0.25 - 0.5)(19) + \frac{(0.25)(0.25 - 1)}{2}$$

$$\left(\frac{4+3}{2}\right) + \frac{(0.25-1)(0.25)(0.25-0.5)}{6} \times -1$$

$$y(5) = 167.5 - 4.75 - 0.328125 - 0.0078125$$

$$y(5) = 162.4140625$$

Example 5. Given $y_{20} = 24$, $y_{24} = 32$, $y_{28} = 35$ and $y_{32} = 40$, find y_{25} by using Bessel's interpolation formula.

Solution. Here $h = 4$, Take origin as 24.

$$\therefore \quad u = \frac{x - x_0}{h} = \frac{25 - 24}{4} = \frac{1}{4} = 0.25$$

The difference table is

u	x	y	Δy	$\Delta^2 y$	$\Delta^3 y$
-1	20	24			
			8		
0	24	32		-5	
			3		7
1	28	35		2	
			5		
2	32	40			

By Bessel's formula, we have

$$y(x) = \frac{y_0 + y_1}{2} + \left(u - \frac{1}{2}\right)\Delta y_0 + \frac{u(u-1)}{2}$$

$$\left(\frac{\Delta^2 y_{-1} + \Delta^2 y_0}{2}\right) + \frac{(u-1)u\left(u - \frac{1}{2}\right)}{3!} \Delta^3 y_{-1} + \cdots$$

$$y(25) = \frac{32 + 35}{2} + (0.25 - 0.5)(3) + \frac{0.25(0.25-1)}{2}$$

$$\left(\frac{-5+2}{2}\right) + \frac{(0.25-1)(0.25)(0.25-0.5)}{6} \times 7$$

$$y(25) = 32.9453125$$

Example 6. Apply Bessel's formula to find the value of $y(27.4)$ from the following table:

x:	25	26	27	28	29	30
y:	4.000	3.846	3.704	3.571	3.448	3.333

Solution. Take 27 as origin, $h = 1$

$$\therefore \quad u = \frac{x - x_0}{h} = \frac{27.4 - 27}{1} = 0.4$$

The difference table is

u	x	y	Δy	$\Delta^2 y$	$\Delta^3 y$	$\Delta^4 y$	$\Delta^5 y$
-2	25	4.000					
			-0.154				
-1	26	3.846		0.012			
			-0.142		-0.003		
0	27	3.704		0.009		0.004	
			-0.133		0.001		-0.007
1	28	3.571		0.010		-0.003	
			-0.123		-0.002		
2	29	3.448		0.008			
			-0.115				
3	30	3.333					

Bessel's formula is

$$y(x) = \frac{y_0 + y_1}{2} + \left(u - \frac{1}{2}\right)\Delta y_0 + \frac{u(u-1)}{2}\left(\frac{\Delta^2 y_{-1} + \Delta^2 y_0}{2}\right)$$

$$+ \frac{(u-1)u\left(u - \frac{1}{2}\right)}{3!}\Delta^3 y_{-1} + \frac{(u+1)u(u-1)(u-2)}{4!}\left(\frac{\Delta^4 y_{-1} + \Delta^4 y_{-2}}{2}\right)$$

$$+ \frac{(u-2)(u-1)\left(u - \frac{1}{2}\right)u(u+1)}{5!}\Delta^5 y_{-2} + \cdots$$

$$y(27.4) = \frac{3.704 + 3.571}{2} + (0.4 - 0.5) \times -0.133 + \frac{0.4(0.4-1)}{2}\left(\frac{0.009 + 0.010}{2}\right)$$

$$+ \frac{(0.4-1)(0.4)(0.4-0.5)}{6} \times 0.001$$

$$+ \frac{(0.4+1)(0.4)(0.4-1)(0.4-2)}{24}\left(\frac{0.004 - 0.003}{2}\right)$$

$$+ \frac{(0.4-2)(0.4-1)(0.4-0.5)0.4(0.4+1)}{120} \times -0.007$$

$$y(27.4) = 3.649678336$$

Example 7. Using Everett's formula, evaluate $y(30)$ if $y(20) = 2854$, $y(28) = 3162$, $y(36) = 7088$, $y(44) = 7984$.

Solution. Taking 28 as origin. Here $h = 8$

$$\therefore \quad u = \frac{x - x_0}{h} = \frac{30 - 28}{8} = \frac{2}{8} = 0.25$$

Also $v = 1 - u = 1 - 0.25 = 0.75$

The difference table is

u	x	y	Δy	$\Delta^2 y$	$\Delta^3 y$
-1	20	2854			
			308		
0	28	3162		3618	
			3926		-6648
1	36	7088		-3030	
			896		
2	44	7984			

By Laplace-Everett's formula, we have

$$y(x) = \left[vy_0 + \frac{v(v^2 - 1^2)}{3!} \Delta^2 y_{-1} + \frac{v(v^2 - 1^2)(v^2 - 2^2)}{5!} \Delta^4 y_{-2} + \cdots \right]$$

$$+ \left[uy_1 + \frac{u(u^2 - 1^2)}{3!} \Delta^2 y_0 + \frac{u(u^2 - 1^2)(u^2 - 2^2)}{5!} \Delta^4 y_{-1} + \cdots \right]$$

$$y(30) = \left[(0.75)(3162) + \frac{(0.75)(-0.4375)}{6} \times (3618) + \cdots \right]$$

$$+ \left[(0.25)(7088) + \frac{(0.25)(-0.9375)}{6} \times (-3030) + \cdots \right]$$

$y(30) = 4064$

Example 8. Apply Laplace-Everett's formula to find the values of y_{25}, given that

$$y_{20} = 2854, \ y_{24} = 3162, \ y_{28} = 3544 \text{ and } y_{32} = 3992$$

Solution. Taking 24 as origin, $h = 4$

$$\therefore \quad u = \frac{x - x_0}{h} = \frac{25 - 24}{4} = \frac{1}{4} = 0.25$$

$$v = 1 - u = 1 - 0.25 = 0.75$$

The difference table is

u	x	y	Δy	$\Delta^2 y$	$\Delta^3 y$
−1	20	2854			
			308		
0	24	3162		74	
			382		−8
1	28	3544		66	
			448		
2	32	3992			

By Laplace-Everett's formula, we have

$$y(x) = \left[vy_0 + \frac{v(v^2-1^2)}{3!}\Delta^2 y_{-1} + \frac{v(v^2-1^2)(v^2-2^2)}{5!}\Delta^4 y_{-2} + \cdots \right]$$

$$+ \left[uy_1 + \frac{u(u^2-1^2)}{3!}\Delta^2 y_0 + \frac{u(u^2-1^2)(u^2-2^2)}{5!}\Delta^4 y_{-1} + \cdots \right]$$

$$y(25) = \left[(0.75)(3162) + \frac{(0.75)(-0.4375)}{6} \times (74) + \cdots \right]$$

$$+ \left[(0.25)(3544) + \frac{(0.25)(-0.9375)}{6} \times (66) + \cdots \right]$$

$$= [2371.5 - 4.046875] + [886 - 2.578125]$$

$$= 3257.5 - 6.625 = 3250.875$$

EXERCISE 4.3

1. Apply central difference formula to find $f(12)$ from the following data:

x:	5	10	15	20
$f(x)$:	54.14	60.54	67.72	75.88

2. Using Bessel's formula to find $y(62.5)$ from the following data:

x:	60	61	62	63	64	65
y:	7782	7853	7924	7993	8062	8129

3. Apply Bessel's formula to get the value of $y(45)$ for the following data:

x:	40	44	48	52
y:	51.08	63.24	70.88	79.84

4. Estimate $\sqrt{1.12}$ using Stirling's formula from the following table:

x:	1.0	1.05	1.10	1.15	1.20	1.25	1.30
$f(x)$:	1.00000	1.02470	1.04881	1.07238	1.09544	1.11803	1.14017

5. Find $y(12)$, if $y(0) = 0$, $y(10) = 43214$, $y(20) = 86002$, $y(30) = 128372$ by using Everett's formula.

6. Given the table:

x:	21	22	23	24	25	26
$\log x$:	1.3222	1.3424	1.3617	1.3802	1.3979	1.4150

   Apply Laplace-Everett's formula to find the value of log 2375.

7. Apply Everett's formula to find the value of $f(26)$ and $f(27)$ from the following table:

x:	15	20	25	30	35	40
$f(x)$:	12.849	16.351	19.524	22.396	24.999	27.356

8. Using Bessel's formula obtain the value of $y(5)$ from the following table:

x:	0	4	8	12
y:	14.27	15.81	17.72	19.96

9. Using Everett's formula to find $f(34)$ from the following table:

x:	20	25	30	35	40
$y = f(x)$:	11.4699	12.7834	13.7648	14.4982	15.0463

10. Apply Bessel's formula to find $y(2.73)$ from the following table:

x:	2.5	2.6	2.7	2.8	2.9	3.0
y:	0.4938	0.4953	0.4965	0.4974	0.4981	0.4987

ANSWERS
---

1. 63.30
2. 7957
3. 65.0175
4. 1.05830
5. 56266
6. 3.3756
7. 20.121, 20.707
8. 16.25
9. 14.368214
10. 0.4968

4.15 INTERPOLATION WITH UNEQUAL INTERVALS

In the previous sections on interpolation we had the intervals of differencing to be a constant h, that is $x_i - x_{i-1} = h$ constant, for $i = 1, 2, \ldots n$. If the values of x's are given at unequal intervals, our Newton's forward, backward formula and central difference interpolation formula will not hold good. That is why we introduce a new idea of divided differences. Let the function $y = f(x)$ assume the values $f(x_0), f(x_1) \ldots, f(x_n)$ corresponding to the arguments $x_0, x_1, x_2, \ldots, x_n$ respectively, where the intervals $x_1 - x_0, x_2 - x_1, \ldots, x_n - x_{n-1}$ need not be equal. In this section we shall study two such formulae.

1. Newton's divided difference interpolation formula.
2. Lagrange's interpolation formula.

4.16 DIVIDED DIFFERENCES

The first divided difference of $f(x)$ **for the arguments** x_0, x_1 **is defined as** $\dfrac{f(x_1) - f(x_0)}{x_1 - x_0}$ **and is denoted by** $f(x_0, x_1)$ **or** $[x_0, x_1]$ **or** $\underset{x_1}{\Delta} f(x_0)$. That is,

$$f(x_0, x_1) = [x_0, x_1] = \underset{x_1}{\Delta} f(x_0) = \frac{f(x_1) - f(x_0)}{x_1 - x_0}$$

similarly, for the other arguments, we have

$$f(x_1, x_2) = [x_1, x_2] = \underset{x_2}{\Delta} f(x_1) = \frac{f(x_2) - f(x_1)}{x_2 - x_1}$$

and $\quad f(x_{n-1}, x_n) = [x_{n-1}, x_n] = \underset{x_n}{\Delta} f(x_{n-1})$

$$= \frac{f(x_n) - f(x_{n-1})}{x_n - x_{n-1}}, n = 1, 2, 3, ..., n.$$

The second divided difference of $f(x)$ **for three arguments** x_0, x_1, x_2 **is defined as**

$$f(x_0, x_1, x_2) = [x_0, x_1, x_2]$$
$$= \underset{x_1, x_2}{\Delta^2} f(x_0) = \frac{f(x_1, x_2) - f(x_0, x_1)}{x_2 - x_0}$$

In the same way, we define the **third divided difference** of $f(x)$ for the four arguments x_0, x_1, x_2, x_3 as

$$f(x_0, x_1, x_2, x_3) = [x_0, x_1, x_2, x_3]$$
$$= \underset{x_1, x_2, x_3}{\Delta^3} f(x_0) = \frac{f(x_1, x_2, x_3) - f(x_0, x_1, x_2)}{x_3 - x_0}$$

Now we will see below the divided difference table

Argument x	Entry $f(x)$	First divided difference $\Delta f(x)$	Second divided difference $\Delta^2 f(x)$	Third divided difference $\Delta^3 f(x)$	Fourth divided difference $\Delta^4 f(x)$
x_0	$f(x_0)$				
		$f(x_0, x_1)$			
x_1	$f(x_1)$		$f(x_0, x_1, x_2)$		
		$f(x_1, x_2)$		$f(x_0, x_1, x_2, x_3)$	
x_2	$f(x_2)$		$f(x_1, x_2, x_3)$		$f(x_0, x_1, x_2, x_3, x_4)$
		$f(x_2, x_3)$		$f(x_1, x_2, x_3, x_4)$	
x_3	$f(x_3)$		$f(x_2, x_3, x_4)$		
		$f(x_3, x_4)$			
x_4	$f(x_4)$				

Interpolation

ILLUSTRATIVE EXAMPLES

Example 1. Form the divided difference table for the following data:

x:	2	4	9	10
$f(x)$:	4	56	711	980

Solution. The divided difference table is

x	$f(x)$	$\Delta f(x)$	$\Delta^2 f(x)$	$\Delta^3 f(x)$
2	4			
		$\dfrac{56-4}{4-2} = 26$		
4	56		$\dfrac{131-26}{9-2} = 15$	
		$\dfrac{711-56}{9-4} = 131$		$\dfrac{23-15}{10-2} = 1$
9	711		$\dfrac{269-131}{10-4} = 23$	
		$\dfrac{980-711}{10-9} = 269$		
10	980			

Example 2. Construct the divided difference table for the following data:

x:	0	2	3	4	7	9
$f(x)$:	4	26	58	112	466	922

Solution. The divided difference table is

x	$f(x)$	$\Delta f(x)$	$\Delta^2 f(x)$	$\Delta^3 f(x)$	$\Delta^4 f(x)$
0	4				
		11			
2	26		7		
		32		1	
3	58		11		0
		54		1	
4	112		16		0
		118		1	
7	466		22		
		228			
9	922				

4.17 PROPERTIES OF DIVIDED DIFFERENCES

1. The value of any divided difference is independent of the order of the argument *i.e.*, the divided differences are symmetrical in all their arguments.

$$f(x_0, x_1) = \frac{f(x_1) - f(x_0)}{x_1 - x_0} = \frac{f(x_0) - f(x_1)}{x_0 - x_1} = f(x_1, x_0) \qquad \ldots(1)$$

Also, $f(x_0, x_1) = \dfrac{f(x_0)}{x_0 - x_1} - \dfrac{f(x_1)}{x_0 - x_1} = \dfrac{f(x_0)}{x_0 - x_1} + \dfrac{f(x_1)}{x_1 - x_0}$ ...(2)

In the same way,

$$f(x_1, x_0) = \dfrac{f(x_1)}{x_1 - x_0} + \dfrac{f(x_0)}{x_0 - x_1} \qquad ...(3)$$

From (2) and (3), we have
$$f(x_0, x_1) = f(x_1, x_0)$$

Similarly, $f(x_0, x_1, x_2) = \dfrac{f(x_0)}{(x_0 - x_1)(x_0 - x_2)} + \dfrac{f(x_1)}{(x_1 - x_0)(x_1 - x_2)}$

$$+ \dfrac{f(x_2)}{(x_2 - x_0)(x_2 - x_1)}$$

$f(x_0, x_1, x_2) = f(x_2, x_0, x_1)$ or $f(x_1, x_2, x_0)$ or $f(x_1, x_0, x_2)$

This shows that $f(x_0, x_1, x_2)$ is independent of the order of the arguments.
By Mathematical induction, we can prove that

$$f(x_0, x_1, x_2, ... x_n) = \dfrac{f(x_0)}{(x_0 - x_1)(x_0 - x_2) \cdots (x_0 - x_n)}$$

$$+ \dfrac{f(x_1)}{(x_1 - x_0)(x_1 - x_2) ... (x_1 - x_n)} + \dfrac{f(x_2)}{(x_2 - x_0)(x_2 - x_1) ... (x_2 - x_n)}$$

$$+ \cdots \dfrac{f(x_n)}{(x_n - x_0)(x_n - x_1) \cdots (x_n - x_{n-1})}$$

Thus the divided differences are symmetrical w.r.t. any two arguments.

2. The nth divided differences of a polynomial of nth degree are constant.
If the arguments $x_0, x_1, x_2, ... x_n$ are equally spaced, then we have

$$x_1 - x_0 = x_2 - x_1 = ... = x_n - x_{n-1} = h$$

$$f(x_0, x_1) = [x_0, x_1] = \dfrac{f(x_1) - f(x_0)}{x_1 - x_0} = \dfrac{\Delta f(x_0)}{h}$$

$f(x_0, x_1, x_2) = [x_0, x_1, x_2]$

$$= \dfrac{f(x_1, x_2) - f(x_0, x_1)}{x_2 - x_0} = \dfrac{1}{2h}\left(\dfrac{\Delta f(x_1)}{h} - \dfrac{\Delta f(x_0)}{h}\right)$$

$$= \dfrac{1}{2h^2}\Delta^2 f(x_0) = \dfrac{1}{2!h^2}\Delta^2 f(x_0)$$

Similarly

$$f(x_0, x_1, x_2, ... x_n) = [x_0, x_1, x_2, ..., x_n] = \dfrac{1}{n!h^n}\Delta^n f(x_0)$$

If tabulated function is a nth degree polynomial
$\therefore \qquad \Delta^n f(x_0) = $ constant
$\therefore$ nth divided differences will also be constant.

4.18 RELATION BETWEEN DIVIDED DIFFERENCES AND FORWARD DIFFERENCES

If the arguments $x_0, x_1, x_2,...,x_n$ are equally spaced, then we have $x_1 - x_0 = x_2 - x_1 = x_3 - x_2 = ... = x_n - x_{n-1} = h$.

Now $f(x_0, x_1) = [x_0, x_1]$

$$= \underset{x_1}{\Delta} f(x_0) = \frac{f(x_1) - f(x_0)}{x_1 - x_0} = \frac{\Delta f(x_0)}{h}$$

$f(x_0, x_1, x_2) = [x_0, x_1, x_2]$

$$= \underset{x_1, x_2}{\Delta^2} f(x_0) = \frac{f(x_1, x_2) - f(x_0, x_1)}{x_2 - x_0}$$

$$= \frac{1}{2h}\left(\frac{\Delta f(x_1)}{h} - \frac{\Delta f(x_0)}{h}\right)$$

$$= \frac{1}{2h^2}\Delta^2 f(x_0) = \frac{1}{2!h^2}\Delta^2 f(x_0)$$

Similarly $\Delta^3 f(x_0) = \frac{1}{3!h^3}\Delta^3 f(x_0)$

$$\Delta^n f(x_0) = \frac{1}{n!h^n}\Delta^n f(x_0) = \frac{\Delta^n f(x_0)}{n!h^n}$$

4.19 NEWTON'S GENERAL INTERPOLATION FORMULA (FOR UNEQUAL INTERVALS) OR NEWTON'S DIVIDED DIFFERENCE INTERPOLATION FORMULA

Let the function $y = f(x)$ takes the values $f(x_0), f(x_1),...f(x_n)$ corresponding to the arguments $x_0, x_1, x_2,...,x_n$.

From the definition of divided difference, we have

$$f(x, x_0) = \frac{f(x) - f(x_0)}{x - x_0}$$

$\therefore \quad f(x) = f(x_0) + (x - x_0) f(x, x_0)$ ...(1)

Again, $f(x, x_0, x_1) = \frac{f(x, x_0) - f(x_0, x_1)}{x - x_1}$

$\therefore \quad f(x, x_0) = f(x_0, x_1) + (x - x_1) f(x, x_0, x_1)$ ...(2)

From (1) and (2), we have

$f(x) = f(x_0) + (x - x_0) f(x_0, x_1) + (x - x_0)(x - x_1) f(x, x_0, x_1)$ ...(3)

Also, $f(x, x_0, x_1, x_2) = \frac{f(x, x_0, x_1) - f(x_0, x_1, x_2)}{x - x_2}$

$\therefore \quad f(x, x_0, x_1) = f(x_0, x_1, x_2) + (x - x_2) f(x, x_0, x_1, x_2)$ ...(4)

From (3) and (4), we get

182 Numerical and Statistical Techniques

$$f(x) = f(x_0) + (x - x_0) f(x_0, x_1) + (x - x_0)(x - x_1) f(x_0, x_1, x_2)$$
$$+ (x - x_0)(x - x_1)(x - x_2) f(x, x_0, x_1, x_2) \quad \ldots(5)$$

Proceeding like this, we get

$$f(x) = f(x_0) + (x - x_0)f(x_0, x_1)$$
$$+ (x - x_0)(x - x_1)f(x_0, x_1, x_2) + (x - x_0)(x - x_1)(x - x_2)f(x_0, x_1, x_2, x_3)$$
$$+ \ldots + (x - x_0)(x - x_1)(x - x_2) \ldots (x - x_{n-1})f(x_0, x_1, x_2, \ldots, x_n)$$
$$+ (x - x_0)(x - x_1)(x - x_2) \ldots (x - x_n)f(x, x_0, x_1, \ldots, x_n) \quad \ldots(6)$$

If $f(x)$ is a polynomial of degree n, then

$$f(x, x_0, x_1, \ldots, x_n) = 0 \qquad [\because (n+1)^{th} \text{ difference}]$$

Hence the equation (6) becomes,

$$f(x) = f(x_0) + (x - x_0) f(x_0, x_1)$$
$$+ (x - x_0)(x - x_1) f(x_0, x_1, x_2) + \ldots + \ldots + (x - x_0)(x - x_1) \ldots$$
$$(x - x_{n-1}) f(x_0, x_1, x_2, \ldots x_n)$$

This is called Newton's divided difference interpolation formula.

ILLUSTRATIVE EXAMPLES

Example 1. Using Newton's divided difference formula, find the value of $f(2)$, $f(8)$ and $f(15)$ for the following table:

x:	4	5	7	10	11	13
$f(x)$:	48	100	294	900	1210	2028

Solution. Since the intervals are not equal. So we form the divided difference table below:

x	$f(x)$	$\Delta f(x)$	$\Delta^2 f(x)$	$\Delta^3 f(x)$	$\Delta^4 f(x)$
4	48				
		$\dfrac{100-48}{5-4} = 52$			
5	100		$\dfrac{97-52}{7-4} = 15$		
		$\dfrac{294-100}{7-5} = 97$		$\dfrac{21-15}{10-4} = 1$	
7	294		$\dfrac{202-97}{10-5} = 21$		0
		$\dfrac{900-294}{10-7} = 202$		$\dfrac{27-21}{11-5} = 1$	
10	900		$\dfrac{310-202}{11-7} = 27$		0
		$\dfrac{1210-900}{10-7} = 310$		$\dfrac{33-27}{11-7} = 1$	
11	1210		$\dfrac{409-310}{13-10} = 33$		
		$\dfrac{2028-1210}{10-7} = 409$			
13	2028				

By Newton's divided difference interpolation formula

$$f(x) = f(x_0) + (x - x_0)f(x_0, x_1) + (x - x_0)(x - x_1)f(x_0, x_1, x_2)$$
$$+ (x - x_0)(x - x_1)(x - x_2)f(x_0, x_1, x_2, x_3) + \ldots \quad \ldots(1)$$

In our problem, $x_0 = 4, x_1 = 5, x_2 = 7, x_2 = 7, x_3 = 10, x_4 = 11, x_5 = 13$ and $f(x_0) = 48, f(x_0, x_1) = 52, f(x_0, x_1, x_2) = 15, f(x_0, x_1, x_2, x_3) = 1$

Putting these values in equation (1), we get

$$f(x) = 48 + (x - 4)52 + (x - 4)(x - 5)15 + (x - 4)(x - 5)(x - 7)1$$

$\therefore \quad f(2) = 4$

$f(8) = 448$

$f(15) = 3150$

Example 2. From the following table, find $f(x)$ and hence $f(6)$ by using Newton's interpolation formula:

x:	1	2	7	8
$f(x)$:	1	5	5	4

Solution. Since the intervals are not equal, so we form the divided difference table below:

x	$f(x)$	$\Delta f(x)$	$\Delta^2 f(x)$	$\Delta^3 f(x)$
1	1			
		$\dfrac{5-1}{2-1} = 4$		
2	5		$\dfrac{0-4}{7-1} = \dfrac{-2}{3}$	
		$\dfrac{5-5}{7-2} = 0$		$\dfrac{-1/6 + 2/3}{8-1} = \dfrac{1}{14}$
7	5		$\dfrac{-1-0}{8-2} = -1/6$	
		$\dfrac{4-5}{8-7} = -1$		
8	4			

By Newton's divided difference formula,

$$f(x) = f(x_0) + (x - x_0)f(x_0, x_1) + (x - x_0)(x - x_1)f(x_0, x_1, x_2) + \ldots$$

$$= 1 + (x - 1)(4) + (x - 1)(x - 2)(-2/3) + (x - 1)(x - 2)(x - 7)\left(\dfrac{1}{14}\right)$$

$f(x) = \dfrac{1}{42}[3x^3 - 58x^2 + 321x - 224]$

$f(6) = \dfrac{1}{42}[3 \times 216 - 58 \times 36 + 321 \times 6 - 224] = 6.23809524$

Example 3. Find the equation $y = f(x)$ of least degree and passing through the points $(-1, -21), (1, 15), (2, 12), (3, 3)$.

Solution. Since the intervals are not equal, so we form divided difference table below:

184 Numerical and Statistical Techniques

x	$f(x)$	$\Delta f(x)$	$\Delta^2 f(x)$	$\Delta^3 f(x)$
-1	-21			
		18		
1	15		-7	
		-3		1
2	12		-3	
		-9		
3	3			

By Newton's divided difference formula,
$$y = f(x) = -21 + (x+1)(18) + (x+1)(x-1)(-7) + (x+1)(x-1)(x-2)\cdot 1$$
$$y = f(x) = x^3 - 9x^2 + 17x + 6$$

Example 4. Find the third divided difference with arguments 2, 4, 9, 10 of the function $f(x) = x^3 - 2x$.

Solution. $x = 2, f(x) = f(2) = 4$
$$x = 4, f(x) = f(4) = 64 - 8 = 56$$
$$x = 9, f(x) = f(9) = 729 - 18 = 711$$
$$x = 10, f(x) = f(10) = 1000 - 20 = 980$$

The divided difference table is

x	$f(x)$	$\Delta f(x)$	$\Delta^2 f(x)$	$\Delta^3 f(x)$
2	4			
		26		
4	56		15	
		131		1
9	711		23	
		269		
10	980			

Hence third divided difference is 1.

Example 5. Prove that $\Delta^3_{bcd}\left(\dfrac{1}{a}\right) = -\dfrac{1}{abcd}$

Solution. If $f(x) = \dfrac{1}{x}$, then $f(a) = \dfrac{1}{a}$

By definition $f(a, b) = \underset{b}{\Delta} f(a) = \underset{b}{\Delta}\left(\dfrac{1}{a}\right) = \dfrac{\dfrac{1}{b} - \dfrac{1}{a}}{b - a} = -\dfrac{1}{ab}$

$$f(a, b, c) = \underset{bc}{\Delta^2} f(a)$$

$$= \frac{f(b,c)-f(a,b)}{c-a} = \frac{-\frac{1}{bc}+\frac{1}{ab}}{c-a}$$

$$= \frac{1}{abc}\left(\frac{c-a}{c-a}\right) = \frac{1}{abc}$$

$$f(a,b,c,d) = \Delta^3_{bcd} f(a) = \frac{f(b,c,d)-f(a,b,c)}{d-a}$$

$$= \frac{\frac{1}{bcd}-\frac{1}{abc}}{d-a} = \frac{1}{abcd}\left(\frac{a-d}{d-a}\right) = -\frac{1}{abcd}$$

Hence $\Delta^3_{bcd} f(a) = \Delta^3_{bcd}\left(\frac{1}{a}\right) = -\frac{1}{abcd}$

Example 6. Find the polynomial equation $y = f(x)$ passing through (5, 1335), (2, 9), (0, 5), (−1, 33), and (−4, 1245).

Solution. The divided difference table is

x	$f(x)$	$\Delta f(x)$	$\Delta^2 f(x)$	$\Delta^3 f(x)$	$\Delta^4 f(x)$
−4	1245				
		−404			
−1	33		94		
		−28		−14	
0	5		10		3
		2		13	
2	9		88		
		442			
5	1335				

By Newton's divided difference formula

$f(x) = 1245 + (x + 4)(-404) + (x + 4)(x + 1)94$
$\quad + (x + 4)(x + 1)(x - 0)(-14) + (x + 4)(x + 1)x(x - 2)(3)$
$= 1245 - 404 x - 1616 + 94x^2 + 470 x + 376 - 14x^3$
$\quad - 70x^2 - 56x + 3x^4 + 9x^3 - 18x^2 - 24 x$

$f(x) = 3x^4 - 5x^3 + 6x^2 - 14x + 5$

Example 7. Find $f'(10)$ from the following data:

x:	3	5	11	27	34
$f(x)$:	−13	23	899	17315	35606

Solution. The divided difference table is

x	f(x)	Δf(x)	Δ²f(x)	Δ³f(x)	Δ⁴f(x)
3	−13				
		18			
5	23		16		
		146		1	
11	899		40		0
		1026		1	
27	17315		69		
		2613			
34	35606				

By Newton's divided difference formula,
$$f(x) = -13 + (x-3)(18) + (x-3)(x-5)(16) + (x-3)(x-5)(x-11)1$$
$$f(x) = x^3 - 3x^2 - 7x + 8$$
$$\therefore \quad f'(x) = 3x^2 - 6x - 7$$
$$\therefore \quad f'(10) = 3(10)^2 - 6(10) - 7 = 233$$

Example 8. The following table gives same relation between steam pressure and temperature. find the pressure at temperature 372.1°.

T	361°	367°	378°	387°	399°
P	154.9	167.9	191.0	212.5	244.2

Solution. The divided difference table is

T	P	Δp	Δ²p	Δ³p	Δ⁴p
361°	154.9				
		2.01666666			
367°	167.0		0.009714795		
		2.18181818		0.000024566	
378°	191.0		0.010353535		0.000000739
		2.3888888		0.000052609	
387°	212.5		0.012037037		
		2.64166666			
399°	244.2				

By Newton's divided difference formula,
$$P(T = 372.1°) = 154.9 + (11.1)(2.01666666) + (11.1)(5.1)(0.009714795)$$
$$+ (11.1)(5.1)(-5.9)(0.000024566)$$
$$+ (11.1)(5.1)(-5.9)(-14.9)(0.000000739)$$
$$= 177.8394819$$

Example 9. Find the polynomial equation of degree four passing through the points (8, 1515), (7, 778), (5, 138), (4, 43) and (2, 3).

Solution. The divided difference table is

x	$f(x)$	$\Delta f(x)$	$\Delta^2 f(x)$	$\Delta^3 f(x)$	$\Delta^4 f(x)$
8	1515				
		737			
			139		
7	778			16	
		320			
			75		1
5	138				
		95		10	
4	43		25		
		20			
2	3				

By Newton's divided difference formula

$f(x) = 1515 + (x-8)(737) + (x-8)(x-7)(139)$
$\quad + (x-8)(x-7)(x-5)(16) + (x-8)(x-7)(x-5)(x-4)(1)$
$f(x) = x^4 - 10x^3 + 36x^2 - 36x - 5$

EXERCISE 4.4
---

1. Find a cubic polynomial of x for the following data:

x:	0	1	2	5
$f(x)$:	2	3	12	147

2. Using divided difference table, find $f(x)$ which takes the values 1, 4, 40, 85 as $x = 0, 1, 3, 4$.

3. Using the following table, find $f(x)$ as a polynomial by using Newton's formula:

x:	−1	0	3	6	7
$f(x)$:	3	−6	39	822	1611

4. Find a polynomial $f(x)$ of lowest degree which takes the values 3, 7, 9 and 19 when $x = 2, 4, 5, 10$.

5. Obtain the value of $\log_{10} 656$, from the following data:

   $\log_{10} 654 = 2.8156$, $\log_{10} 658 = 2.8182$, $\log_{10} 659 = 2.8189$ and $\log_{10} 666 = 2.8202$.

6. Find the polynomial equation of degree four passing through the points (8, 1515), (7, 778), (5, 138), (4, 43) and (2, 3).

188 Numerical and Statistical Techniques

7. If $f(x) = \dfrac{1}{x^2}$, find the divided differences $f(a, b)$, $f(a, b, c)$ and $f(a, b, c, d)$.

8. Find $f(x)$ as a polynomial of degree 2 for the following data:

x:	1	2	-4
$f(x)$:	3	-5	4

9. Given the values:

x:	5	7	11	13	17
$f(x)$:	150	392	1452	2366	5202

   Find $f(9)$ using Newton's divided difference formula.

10. Apply Newton's divided difference formula to find the value of $f(8)$ if, $f(1) = 3$, $f(3) = 31$, $f(6) = 223$, $f(10) = 1011$, $f(11) = 1343$.

11. Using Newton's divided difference formula to find $f(7)$, if $f(3) = 24$, $f(5) = 120$, $f(8) = 504$, $f(9) = 720$ and $f(12) = 1716$.

12. Find the polynomial equation for the following data by using Newton's divided difference formula:

x:	0.5	1.5	3.0	5.0	6.5	8.0
$f(x)$:	1.625	5.875	31.0	131.0	282.125	521.0

13. Given that $\log_{10} 2 = 0.3010$, $\log_{10} 3 = 0.4771$, $\log_{10} 7 = 0.8451$, find the value of $\log_{10} 33$.

ANSWERS
---

1. $x^3 + x^2 - x + 2$
2. $x^3 + x^2 + x + 1$
3. $x^4 - 3x^3 + 5x^2 - 6$
4. $2x - 1$
5. 2.8169
6. $y = x^4 - 10x^3 + 36x^2 - 36x - 5$
7. $-\left(\dfrac{a+b}{a^2 b^2}\right)$, $\dfrac{ab+bc+ca}{a^2 b^2 c^2}$, $-\left(\dfrac{abc+bcd+acd+abd}{a^2 b^2 c^2}\right)$
8. $-\dfrac{1}{10}(13x^2 + 41x - 84)$ 9. 810 10. 521
11. 328 12. $x^3 + x + 1$ 13. 1.5184

4.20 LAGRANGE'S INTERPOLATION FORMULA (FOR UNEQUAL INTERVALS)

The forward and backward interpolation formulae of Newton can be used only when the values of independent variable x are equally spaced. In cases, where the values of independent variable are not equally spaced and in cases when the differences of dependent variable are not small, ultimately we will use Lagrange's interpolation formula.

Let $y = f(x)$ be a function such that $f(x)$ takes the values $y_0, y_1, y_2, ... y_n$, corresponding to $x = x_0, x_1, x_2, ... x_n$. i.e.
$$y_i = f(x_i), i = 0, 1, 2, ..., n.$$

Given the $(n + 1)$ points $(x_0, y_0), ... (x_n, y_n)$, where the values of x need not be equally spaced, we wish to find a polynomial of degree n in x.

The polynomial $y = f(x)$ may be written as
$$y = f(x) = A_0(x - x_1)(x - x_2) ... (x - x_n) + A_1(x - x_0)(x - x_2) + ...(x - x_n)$$
$$+ ... + A_n(x - x_0)(x - x_1) ... (x - x_{n-1}) \quad ...(1)$$

where $A_0, A_1, A_2, ... A_n$ are constants to be determined.

Substituting $x = x_0, x_1, ... x_n$ in equation (1), we get
$$f(x_0) = A_0(x_0 - x_1)(x_0 - x_2) ...(x_0 - x_n)$$

$$\therefore A_0 = \frac{f(x_0)}{(x_0 - x_1)(x - x_2) \cdots (x_0 - x_n)}$$

Similarly, putting $x = x_1$ in equation (1), we get
$$A_1 = \frac{f(x_1)}{(x_1 - x_0)(x_1 - x_2) \cdots (x_1 - x_n)}$$

In the same way, we get
$$A_n = \frac{f(x_n)}{(x_n - x_0)(x_n - x_1) \cdots (x_n - x_{n-1})}$$

Substituting these values of $A_0, A_1, ..., A_n$ in equation (1), weget
$$y = f(x) = \frac{(x - x_1)(x - x_2) \cdots (x - x_n)}{(x_0 - x_1)(x_0 - x_2) \cdots (x_0 - x_n)} f(x_0)$$
$$+ \frac{(x - x_0)(x - x_2) \cdots (x - x_n)}{(x_1 - x_0)(x_1 - x_2) \cdots (x_1 - x_n)} f(x_1)$$
$$+ \cdots + \frac{(x - x_0)(x - x_1) \cdots (x - x_{n-1})}{(x_n - x_0)(x_n - x_1) \cdots (x_n - x_{n-1})} f(x_n)$$

This is called Lagrange's interpolation formula.
Error is Lagrange's interpolation formula
Remainder,
$$y(x) - L_n(x) = R_n(x) = \frac{\pi_{n+1}(x)}{(n+1)!} y^{(n+1)}(\xi), a < \xi < b$$

where Lagrange's formula is for the class of functions having continuous derivatives of order upto $(n + 1)$ on $[a, b]$.

Quantity $E_L = \underset{[a, b]}{\text{Max.}} |R_n(x)|$ may be taken as an estimate of error.

190 *Numerical and Statistical Techniques*

Let us assume that
$$|y^{(n+1)}(\xi)| \leq M_{n+1}, a \leq \xi \leq b$$
then
$$E_L \leq \frac{M_{n+1}}{(n+1)!} \underset{[a,b]}{\text{Max}} |\pi_{n+1}(x)|$$

ILLUSTRATIVE EXAMPLES

Example 1. Using Lagrange's interpolation formula to find $f(9.5)$ for the following data:

x:	7	8	9	10
$f(x)$:	3	1	1	9

Solution. By Lagrange's interpolation formula, we have

$$y = f(x) = \frac{(x-x_1)(x-x_2)(x-x_3)}{(x_0-x_1)(x_0-x_2)(x_0-x_3)} f(x_0)$$
$$+ \frac{(x-x_0)(x-x_2)(x-x_3)}{(x_1-x_0)(x_1-x_2)(x_1-x_3)} f(x_1) + \frac{(x-x_0)(x-x_1)(x-x_3)}{(x_2-x_0)(x_2-x_1)(x_2-x_3)} f(x_2)$$
$$+ \frac{(x-x_0)(x-x_1)(x-x_2)}{(x_3-x_0)(x_3-x_1)(x_3-x_2)} f(x_3)$$

Here, $x_0 = 7, x_1 = 8, x_2 = 9, x_3 = 10, f(x_0) = 3,$
$f(x_1) = 1, f(x_2) = 1, f(x_3) = 9$

$$f(x) = \frac{(x-8)(x-9)(x-10)}{(7-8)(7-9)(7-10)} (3) + \frac{(x-7)(x-9)(x-10)}{(8-7)(8-9)(8-10)} (1)$$
$$+ \frac{(x-7)(x-8)(x-10)}{(9-7)(9-8)(9-10)} (1) + \frac{(x-7)(x-8)(x-9)}{(10-7)(10-8)(10-9)} (9)$$
$$= 0.1875 - 0.3125 + 0.9375 + 2.8125$$
$$f(x) = 3.625$$

Example 2. Find the parabola of the form $= ax^2 + bx + c$, passing through the points (0, 0), (1, 1) and (2, 20).

Solution. Here, $x_0 = 0, x_1 = 1, x_2 = 2, f(x_0) = 0, f(x_1) = 1, f(x_2) = 20$

By Lagrange's interpolation formula, we have

$$y = f(x) = \frac{(x-1)(x-2)}{(0-1)(0-2)} (0) + \frac{(x-0)(x-2)}{(1-0)(1-2)} (1) + \frac{(x-0)(x-1)}{(2-0)(2-1)} (20)$$
$$= 0 - x(x-2) + 10x(x-1)$$
$$y = 9x^2 - 8x$$

Example 3. Using Lagrange's interpolation formula, find $f(10)$ from the following data:

x:	5	6	9	11
$f(x)$:	12	13	14	16

Solution. By Lagrange's interpolation formula, we have

$$f(x) = \frac{(x-x_1)(x-x_2)(x-x_3)}{(x_0-x_1)(x_0-x_2)(x_0-x_3)} f(x_0)$$

Here, $x_0 = 5, x_1 = 6, x_2 = 9, x_3 = 11, f(x_0) = 12, f(x_1) = 13,$
$f(x_2) = 14, f(x_3) = 16$

$$+ \frac{(x-x_0)(x-x_2)(x-x_3)}{(x_1-x_0)(x_1-x_2)(x_1-x_3)} f(x_1) + \frac{(x-x_0)(x-x_1)(x-x_3)}{(x_2-x_0)(x_2-x_1)(x_2-x_3)} f(x_2)$$

$$+ \frac{(x-x_0)(x-x_1)(x-x_2)}{(x_3-x_0)(x_3-x_1)(x_3-x_2)} f(x_3)$$

$$f(x) = \frac{(x-6)(x-9)(x-11)}{(5-6)(5-9)(5-11)} (12) + \frac{(x-5)(x-9)(x-11)}{(6-5)(6-9)(6-11)} (13)$$

$$+ \frac{(x-5)(x-6)(x-11)}{(9-5)(9-6)(9-11)} (14) + \frac{(x-5)(x-6)(x-9)}{(11-5)(11-6)(11-9)} (16)$$

Putting $x = 10$, we get

$$f(10) = \frac{4\times1\times-1}{-1\times-4\times-6} (12) + \frac{5\times1\times-1}{1\times-3\times-5} (13)$$

$$+ \frac{5\times4\times-1}{4\times3\times-2} (14) + \frac{5\times4\times1}{6\times5\times2} (16)$$

$$f(10) = 14.6666$$

Example 4. Using Lagrange's interpolation formula, find the value of $\sin\left(\frac{\pi}{6}\right)$ from the following data:

x:	0	$\pi/4$	$\pi/2$
$y = \sin x$:	0	0.70711	1.0

Solution. By Lagrange's interpolation formula, we have

$$\sin\left(\frac{\pi}{6}\right) = \frac{\left(\frac{\pi}{6}-0\right)\left(\frac{\pi}{6}-\frac{\pi}{2}\right)}{\left(\frac{\pi}{4}-0\right)\left(\frac{\pi}{4}-\frac{\pi}{2}\right)} (0.70711) + \frac{\left(\frac{\pi}{6}-0\right)\left(\frac{\pi}{6}-\frac{\pi}{4}\right)}{\left(\frac{\pi}{2}-0\right)\left(\frac{\pi}{2}-\frac{\pi}{4}\right)} \quad (1)$$

$$= \frac{8}{9}(0.70711) - \frac{1}{9} = \frac{4.65688}{9} = 0.51743$$

Example 5. Compute the value of $f(x)$ for $x = 2.5$ from the following table by using Lagrange's interpolation formula.

Solution. By Lagrange's interpolation formula, we have

$$f(x) = \frac{(x-2)(x-3)(x-4)}{(1-2)(1-3)(1-4)}(1) + \frac{(x-1)(x-3)(x-4)}{(2-1)(2-3)(2-4)}(8)$$

$$+ \frac{(x-1)(x-2)(x-4)}{(3-1)(3-2)(3-4)}(27) + \frac{(x-1)(x-2)(x-3)}{(4-1)(4-2)(4-3)}(64)$$

$$= -\frac{1}{6}(x-2)(x-3)(x-4) + 4(x-1)(x-3)(x-4)$$

$$-\frac{27}{2}(x-1)(x-2)(x-4) + \frac{32}{3}(x-1)(x-2)(x-3)$$

Putting $x = 2.5$, we get
$f(2.5) = -1/6(2.5-2)(2.5-3)(2.5-4)$

$$+ 4(2.5-1)(2.5-3)(2.5-4) - \frac{27}{2}(2.5-1)(2.5-2)(2.5-4)$$

$$+ \frac{32}{3}(2.5-1)(2.5-2)(2.5-3)$$

$f(2.5) = 15.625$

Example 6. Find the cubic polynomial from the following data:

x:	0	1	2	5
$f(x)$:	2	3	12	47

Solution. By Lagrange's formula, we have

$$f(x) = \frac{(x-x_1)(x-x_2)(x-x_3)}{(x_0-x_1)(x_0-x_2)(x_0-x_3)} f(x_0)$$

$$+ \frac{(x-x_0)(x-x_2)(x-x_3)}{(x_1-x_0)(x_1-x_2)(x_1-x_3)} f(x_1) + \frac{(x-x_0)(x-x_1)(x-x_3)}{(x_2-x_0)(x_2-x_1)(x_2-x_3)} f(x_2)$$

$$+ \frac{(x-x_0)(x-x_1)(x-x_2)}{(x_3-x_0)(x_3-x_1)(x_3-x_2)} f(x_3)$$

Here $x_0 = 0$, $x_1 = 1$, $x_2 = 2$, $x_3 = 5$, $f(x_0) = 2$,
$f(x_1) = 3$, $f(x_2) = 12$, $f(x_3) = 47$

$\therefore \quad f(x) = -\frac{1}{5}(x-1)(x-2)(x-5) + \frac{3}{4}x(x-2)(x-5)$

$$- 2x(x-1)(x-5) + \frac{49}{20}x(x-1)(x-2)$$

$$= -\frac{1}{5}(x^3 - 8x^2 + 17x - 10) + \frac{3}{4}(x^3 - 7x^2 + 10x)$$

$$- 2(x^3 - 6x^2 + 5x) + \frac{49}{20}(x^3 - 3x^2 + 2x)$$

$f(x) = x^3 + x^2 - x + 2$

Example 7. Find the polynomial $f(x)$ of degree 2 such that:
$f(1) = 1, f(3) = 27, f(4) = 64$, by using Lagrange's interpolation formula.
Solution. By Lagrange's formula, we have

$$f(x) = \frac{(x-x_1)(x-x_2)}{(x_0-x_1)(x_0-x_2)} f(x_0) + \frac{(x-x_0)(x-x_2)}{(x_1-x_0)(x_1-x_2)} f(x_1) + \frac{(x-x_0)(x-x_1)}{(x_2-x_0)(x_2-x_1)} f(x_2)$$

Here $x_0 = 1, x_1 = 3, x_2 = 4, f(x_0) = 1, f(x_1) = 27, f(x_2) = 64$

$$f(x) = \frac{(x-3)(x-4)}{(1-3)(1-4)}(1) + \frac{(x-1)(x-4)}{(3-1)(3-4)}(27) + \frac{(x-1)(x-3)}{(4-1)(4-3)}(64)$$

$$= \frac{1}{6}(x^2 - 7x + 12) - \frac{27}{2}(x^2 - 5x + 4) + \frac{64}{3}(x^2 - 4x + 3)$$

$f(x) = 8x^2 - 19x + 12$

Example 8. Find the cubic polynomial from the following data by using Lagrange's interpolation formula.

x:	0	1	2	5
$f(x)$:	2	3	12	147

Solution. By Lagrange's interpolation formula, we have

$$f(x) = \frac{(x-x_1)(x-x_2)(x-x_3)}{(x_0-x_1)(x_0-x_2)(x_0-x_3)} f(x_0)$$
$$+ \frac{(x-x_0)(x-x_2)(x-x_3)}{(x_1-x_0)(x_1-x_2)(x_1-x_3)} f(x_1) + \frac{(x-x_0)(x-x_1)(x-x_3)}{(x_2-x_0)(x_2-x_1)(x_2-x_3)} f(x_2)$$
$$+ \frac{(x-x_0)(x-x_1)(x-x_2)}{(x_3-x_0)(x_3-x_1)(x_3-x_2)} f(x_3)$$

$$f(x) = \frac{(x-1)(x-2)(x-5)}{(0-1)(0-2)(0-5)}(2) + \frac{(x-0)(x-2)(x-5)}{(1-0)(1-2)(1-5)}(3)$$
$$+ \frac{(x-0)(x-1)(x-5)}{(2-0)(2-1)(2-5))}(12) + \frac{(x-0)(x-1)(x-2)}{(5-0)(5-1)(5-2)}(147)$$

$$= -\frac{1}{5}(x-1)(x-2)(x-5) + \frac{3}{4}x(x-2)(x-5) - 2x(x-1)(x-5)$$
$$+ \frac{49}{20}x(x-1)(x-2)$$

$$= -\frac{1}{5}(x^3 - 8x^2 + 17x - 10) + \frac{3}{4}(x^3 - 7x^2 + 10x) - 2(x^3 - 6x^2 + 5x)$$
$$+ \frac{49}{20}(x^3 - 3x^2 + 2x)$$

$f(x) = x^3 + x^2 - x + 2$

EXERCISE 4.5

1. Use Lagrange's interpolation formula to fit a polynomial for the following data:

x:	-1	0	2	3
u_x:	-8	3	1	12

   Hence or otherwise find the value of u_1.

2. Applying Lagrange's formula, find a polynomial for the following data:

x:	-2	-1	2	3
$f(x)$:	-10	-8	3	5

3. Use Lagrange's formula to find $f(6)$ from the following data:

x:	2	5	7	10	12
$f(x)$:	18	180	448	1210	2028

4. Find the value of u_5, if $u_0 = 1$, $u_3 = 19$, $u_4 = 49$, $u_6 = 181$, by using Lagrange's formula.

5. Apply Lagrange's formula to find $f(5)$ and $f(6)$ given that $f(2) = 4$, $f(1) = 2$, $f(3) = 8$, $f(7) = 128$.

6. Find the value of $\tan 33°$ by Lagrange's formula, if $\tan 30° = 0.5774$, $\tan 32° = 0.6249$, $\tan 35° = 0.7002$, $\tan 38° = 0.7813$.

7. Find $f(x)$ for the following data by using Lagrange's interpolation formula:
   $f(0) = -18, f(1) = 0, f(3) = 0, f(5) = -248, f(6) = 0, f(9) = 13104$

8. Find $f(x)$ for the following data by using Lagrange's interpolation formula:

x:	0	1	4	5
$f(x)$:	4	3	24	39

9. If $y_3 = 16$, $y_5 = 36$, $y_7 = 64$, $y_8 = 81$ and $y_9 = 100$, find y_4.

10. Using Lagrange's interpolation formula, fit a polynomial for the following data:

x:	0	1	3	4
y:	-12	0	6	12

    Also find y at $x = 2$.

ANSWERS

1. $u_x = 2x^3 - 6x^2 + 3x + 3$; $u_1 = 2$
2. $-\dfrac{1}{15}x^3 - \dfrac{3}{20}x^2 + \dfrac{241}{60}x - 3.9$
3. 293.99856
4. 100.99999
5. 37.8; 73
6. 0.64942084
7. $x^5 - 9x^4 + 18x^3 - x^2 + 9x - 18$
8. $2x^2 - 3x + 4$
9. 25
10. $y = x^3 - 7x^2 + 18x - 12$; 4

4.21 INVERSE INTERPOLATION

So far, we treat y as a function of x. For a given set of values of x and y, we were finding the values of y corresponding to some $x = x_k$ (which is not given in the table). Now the problem is, given some $y = y_r$, we should find the corresponding x. This process of finding x, for given y is called inverse interpolation. When the value of x are unequal, Lagrange's method is used by interchanging x and y. i – e.

$$x = \frac{(y-y_1)(y-y_2)\cdots(y-y_n)}{(y_0-y_1)(y_0-y_2)\cdots(y_0-y_n)} x_0 + \frac{(y-y_0)(y-y_2)\cdots(y-y_n)}{(y_1-y_0)(y_1-y_2)\cdots(y_1-y_n)} x_1 + \cdots + \frac{(y-y_0)(y-y_1)\cdots(y-y_{n-1})}{(y_n-y_0)(y_n-y_1)\cdots(y_n-y_{n-1})} x_n$$

This formula is called formula of inverse interpolation.

ILLUSTRATIVE EXAMPLES

Example 1. From the following table:

x:	20	25	30	35
$y(x)$:	0.342	0.423	0.5	0.65

Find the value of x for $y(x) = 0.390$

Solution. By inverse interpolation formula, we have

$$x = \frac{(y-y_1)(y-y_2)(y-y_3)}{(y_0-y_1)(y_0-y_2)(y_0-y_3)} x_0$$

$$+ \frac{(y-y_0)(y-y_2)(y-y_3)}{(y_1-y_0)(y_1-y_2)(y_1-y_3)} x_1 + \frac{(y-y_0)(y-y_1)(y-y_3)}{(y_2-y_0)(y_2-y_1)(y_2-y_3)} x_2$$

$$+ \frac{(y-y_0)(y-y_1)(y-y_2)}{(y_3-y_0)(y_3-y_1)(y_3-y_2)} x_3$$

$$x = \frac{(0.390-0.423)(0.390-0.5)(0.390-0.65)}{(0.342-0.423)(0.342-0.5)(0.342-0.65)} (20)$$

$$+ \frac{(0.390-0.342)(0.390-0.5)(0.390-0.65)}{(0.423-0.342)(0.423-0.5)(0.423-0.65)} (25)$$

$$+ \frac{(0.390-0.342)(0.390-0.423)(0.390-0.65)}{(0.5-0.342)(0.5-0.423)(0.5-0.65)} (30)$$

$$+ \frac{(0.390-0.342)(0.390-0.423)(0.390-0.5)}{(0.65-0.342)(0.65-0.423)(0.65-0.5)} (35)$$

$x = 22.84057797$.

Example 2. Find the age corresponding to the annuity value 13.6 given the table

Age (x)	30	35	40	45	50
Annuity value(y)	15.9	14.9	14.1	13.3	12.5

Solution. $x = \dfrac{(13.6-14.9)(13.6-14.1)(13.6-13.3)(13.6-12.5)}{(15.9-14.9)(15.9-14.1)(15.9-13.3)(15.9-12.5)} \times 30$

$+ \dfrac{(13.6-15.9)(13.6-14.1)(13.6-13.3)(13.6-12.5)}{(14.9-15.9)(14.9-14.1)(14.9-13.3)(14.9-12.5)} \times 35$

$+ \dfrac{(13.6-15.9)(13.6-14.9)(13.6-13.3)(13.6-12.5)}{(14.1-15.9)(14.1-14.9)(14.1-13.3)(14.1-12.5)} \times 40$

$+ \dfrac{(13.6-15.9)(13.6-14.9)(13.6-14.1)(13.6-12.5)}{(13.3-15.9)(13.3-14.9)(13.3-14.1)(13.3-12.5)} \times 45$

$+ \dfrac{(13.6-15.9)(13.6-14.9)(13.6-14.1)(13.6-13.3)}{(12.5-15.9)(12.5-14.9)(12.5-14.1)(12.5-13.3)} \times 50$

$x \approx 43$

EXERCISE 4.6
---

1. Find x corresponding to $y = 85$ for the following data:

x:	2	5	8	14
y:	94.8	87.9	81.3	68.7

2. Find x corresponding to $y = 100$ for the following data:

x:	3	5	7	9	11
y:	6	24	58	108	174

3. Find x for $y = 0.3$ from the following data:

x:	0.4	0.6	0.8
y:	0.3683	0.3332	0.2897

4. The values of elliptic integral $f(\theta) = \sqrt{2}\int_0^\theta \dfrac{d\theta}{\sqrt{1+\cos^2\theta}}$ are given below:

θ:	21°	23°	25°
f(θ):	0.3706	0.4068	0.4433

   find θ for which $f(\theta) = 0.3887$

ANSWERS
---

1. 6.5928   2. 8.656   3. 0.07575
4. 22.0020

4.22 CUBIC SPLINE

The fundamental idea behind cubic spline interpolation is based on the engineer's tool used to draw smooth curvfes through a number of points. This spline consists of weights attached to a flat surface at the points to be connected. A flexible strip is then bent across each of these weights, resulting in a pleasingly smooth curve. The mathematical spline is similar in principle. The points, in this case are numerical data. The weights are the coefficients on the cubic polynomials used to interpolate the data. These coefficients 'bend' the line so that it passes through each of the data points without any erratic behavior or breaks in continuity.

Real world numerical data is usually difficult to analyze. Any fuction which would effectively correlate the data would be difficult to obtain and highly unwieldy. To this end, the idea of the cubic spline was developed. Using this process, a series of unique cubic polynomials are fitted between each of the data points, with the stipulation that the curve obtained be continuous and appear smooth. These cubic splines can then be used to determine rates of change and cumulative change over an interval.

Suppose we want to interpolate between the given $(n + 1)$ data points (x_i, y_i), $i = 0, 1, 2,..., n$. where $a = x_0 < x_1 < x_2 ... < x_n = b$, by means of a smooth polynomial curve. Here x_i are called knots. A cubic spline $S(x)$, which is the commonly used, is defined by the following properties:

1. $S(x)$ interpolate all data points *i.e.*, $S(x_i) = y_i$, $i = 0, 1, 2,...n$.
2. $S(x)$, $S'(x)$, $S''(x)$ are continuous on the interval $[x_0, x_n]$
3. $S(x)$ is a cubic polynomial, say $s_i(x)$, in each subinterval (x_{i-1}, x_i), $i = 1, 2,...,n$.

From property (3) we have

$$S(x) = \begin{cases} s_1(x), x_0 \leq x \leq x_1 \\ s_2(x), x_1 \leq x \leq x_2 \\ \dots\dots\dots\dots\dots \\ s_n(x), x_{n-1} \leq x \leq x_n \end{cases} \qquad ...(1)$$

From interpolation property (1) we have

$$\left. \begin{array}{l} S(x_0) = s_1(x_0) = y_0 \\ S(x_1) = s_2(x_1) = y_1 \\ \dots\dots\dots\dots\dots \\ S(x_n) = s_n(x_n) = y_n \end{array} \right\} \qquad ...(2)$$

The continuity property (2) gives the following relations

$$\left. \begin{array}{l} s_1(x_1) = s_2(x_1) \\ s'_1(x_1) = s'_2(x_1) \\ s''_1(x_1) = s''_2(x_1) \\ \quad etc. \end{array} \right\} \qquad ...(3)$$

Now, assuming that the knots are arbitrary and using the properties (1), (2), (3) above, equations of the cubic spline can be obtained as follows:

According to property (3) $s_i(x)$ is a cubic polynomial in the subinterval (x_{i-1}, x_i). Hence, we can write

$$s_i''(x) = \frac{1}{h_i}[(x_i - x)M_{i-1} + (x - x_{i-1})M_i] \qquad (4)$$

where $h_i = x_i - x_{i-1}$, $M_{i-1} = S''(x_{i-1})$ and $M_i = S''(x_i)$.

Now, on intergrating equation (4) twice, we have

$$s_i(x) = \frac{1}{h_i}\left[\frac{(x_i - x)^3}{6} M_{i-1} + \frac{(x - x_{i-1})^3}{6} M_i\right]$$
$$+ c_i(x_i - x) + d_i(x - x_{i-1}) \qquad (5)$$

where c_i and d_i are constants to be determined.

Using conditions $s_i(x_{i-1}) = y_{i-1}$ and $s_i(x_i) = y_i$, we have

$$c_i = \frac{1}{h_i}\left[y_{i-1} - \frac{h_i^2}{6} M_{i-1}\right]$$

and
$$d_i = \frac{1}{h_i}\left[y_i - \frac{h_i^2}{6} M_i\right]$$
$\qquad (6)$

Substituting for c_i and d_i in equation (5), we have

$$s_i(x) = \frac{1}{h_i}\left[\frac{(x_i - x)^3}{6} M_{i-1} + \frac{(x - x_{i-1})^3}{6} M_i + \left(y_{i-1} - \frac{h_i^2}{6} M_{i-1}\right)(x_i - x) + \left(y_i - \frac{h_i^2}{6} M_i\right)(x - x_{i-1})\right] \qquad (7)$$

In equation (7), the spline second derivatives, M_i are not known. To find M_i, we use the property of continuity of $s_i'(x)$. Now on differentiating equation (7) w.r.t. x, we have

$$s_i'(x) = \frac{1}{h_i}\left[-\frac{(x_i - x)^2}{2} M_{i-1} + \frac{(x - x_{i-1})^2}{2} M_i\right.$$
$$\left. - \left(y_{i-1} - \frac{h_i^2}{6} M_{i-1}\right) + \left(y_i - \frac{h_i^2}{6} M_i\right)\right] \qquad (8)$$

Putting $x = x_i$ in equation (8), we get the left hand derivative

$$s_i'(x_i^-) = \frac{h_i}{2} M_i - \frac{1}{h_i}\left(y_{i-1} - \frac{h_i^2}{6} M_{i-1}\right) + \frac{1}{h_i}\left(y_i - \frac{h_i^2}{6} M_i\right)$$
$$= \frac{1}{h_i}(y_i - y_{i-1}) + \frac{h_i}{6} M_{i-1} + \frac{h_i}{3} M_i \qquad (9)$$

To obtain the right hand derivative, put $i = i + 1$ in equation (7), we have

$$s_{i+1}(x) = \frac{1}{h_{i+1}} \left[\frac{(x_{i+1}-x)^3}{6} M_i + \frac{(x-x_i)^3}{6} M_{i+1} + \left(y_i - \frac{h_{i+1}^2}{6} M_i \right) \right. \\ \left. (x_{i+1}-x) + \left(y_{i+1} - \frac{h_{i+1}^2}{6} M_{i+1} \right)(x-x_i) \right] \quad (10)$$

where $h_{i+1} = x_{i+1} - x_i$

Now differentiating equation (10) and putting $x = x_i$, we get the right hand derivative

$$s'_{i+1}(x_i^+) = \frac{1}{h_{i+1}}(y_{i+1} - y_i) - \frac{h_{i+1}}{6} M_{i+1} - \frac{h_{i+1}}{3} M_i \quad (11)$$

$$(i = 0, 1, 2,..., n-1)$$

On equating the equations (9) and (11), we get the following recurrence relation.

$$\frac{h_i}{6} M_{i-1} + \frac{1}{3}(h_i + h_{i+1}) M_i + \frac{h_{i+1}}{6} M_{i+1} = \frac{y_{i+1} - y_i}{h_{i+1}} - \frac{y_i - y_{i-1}}{h_i} \quad (12)$$

$$(i = 1, 2,..., n-1)$$

In case of equal intervals, we have $h_i = h_{i+1} = h$ and equation (12) becomes

$$h^2[M_{i-1} + 4M_i + M_{i+1}] = 6(y_{i+1} - 2y_i + y_{i-1}), (i = 1, 2,..., n-1) \quad (13)$$

Equation (13) gives a system of $(n-1)$ equations with $(n+1)$ unkowns $M_0, M_1, ..., M_n$.

Note: In natural spline $M_0 = M_n = 0$.

ILLUSTRATIVE EXAMPLES

Example 1. Obtain the natural cubic spline for the following data:

x	4	9	16
y	2	3	4

Solution. Here, $n = 2$, $M_0 = M_2 = 0$ (natural cubic spline),
$h_1 = x_1 - x_0 = 9 - 4 = 5$; $h_2 = x_2 - x_1 = 16 - 9 = 7$
$y_0 = 2, y_1 = 3, y_2 = 4$
Since the $x_i's$ are given at unequal interval, so we use equation (12) i.e.,

$$\frac{h_i}{6} M_{i-1} + \frac{1}{3}(h_i + h_{i+1}) M_i + \frac{h_{i+1}}{6} M_{i+1} = \frac{y_{i+1} - y_i}{h_{i+1}} - \frac{y_i - y_{i-1}}{h_i}$$

Now, put $i = 1$ in the above equation, we have

$$\frac{h_1}{6} M_0 + \frac{1}{3}(h_1 + h_2) M_1 + \frac{h_2}{6} M_2 = \frac{y_2 - y_1}{h_2} - \frac{y_1 - y_0}{h_1}$$

$$\frac{1}{3}(5+7)M_1 = \frac{4-3}{7} - \frac{3-2}{5} \quad (\because M_0 = M_2 = 0)$$

On solving we get
$$M_1 = -0.0143$$

Since number of intervals are two, so there will be two cubic splines as

For $4 \leq x \leq 9$

$$s_1(x) = \frac{1}{h_1}\left[\frac{(x_1-x)^3}{6}M_0 + \frac{(x-x_0)^3}{6}M_1 + \left(y_0 - \frac{h_1^2}{6}M_0\right)\right.$$

$$\left.(x_1-x) + \left(y_1 - \frac{h_1^2}{6}M_1\right)(x-x_0)\right]$$

$$= \frac{1}{5}\left[\frac{(x-4)^3}{6}(-0.0143) + (2)(9-x) + \left(3 - \frac{25}{6}(-0.0143)\right)(x-4)\right]$$

$$= \frac{1}{5}\left[-0.002383(x-4)^3 + 1.06x + 5.76\right].$$

For $9 \leq x \leq 16$

$$s_2(x) = \frac{1}{h_2}\left[\frac{(x_2-x)^3}{6}M_1 + \frac{(x-x_1)^3}{6}M_2 + \left(y_1 - \frac{h_2^2}{6}M_1\right)\right.$$

$$\left.(x_2-x) + \left(y_2 - \frac{h_2^2}{6}M_2\right)(x-x_1)\right]$$

$$= \frac{1}{7}\left[\frac{(16-x)^3}{6}(-0.0143) + \left(3 - \frac{49}{6}(-0.0143)\right)(16-x) + (4)(x-9)\right]$$

$$= \frac{1}{7}\left[-0.002383(16-x)^3 + 0.8832x + 13.8688\right]$$

Example 2. Obtain cubic spline for every subinterval from the given data:

x	0	1	2	3
$y = f(x)$	1	2	33	244

with the end conditions $M_0 = M_3 = 0$. Hence find $f(2.5)$.

Solution. Here, $n = 3$, $M_0 = M_3 = 0$
$h_1 = x_1 - x_0 = 1 - 0 = 1; h_2 = x_2 - x_1 = 2 - 1 = 1; h_3 = x_3 - x_2 = 3 - 2 = 1$
i.e.,
$$h_1 = h_2 = h_3 = h = 1$$
$$y_0 = 1, y_1 = 2, y_2 = 33, y_3 = 244$$

Since the $x_i's$ are given at equal interval, so we use equation (13) i.e.,
$$h^2[M_{i-1} + 4M_i + M_{i+1}] = 6(y_{i+1} - 2y_i + y_{i-1})$$

Now, put $i = 1$ and $i = 2$ in the above equation respectively, we get
$$[M_0 + 4M_1 + M_2] = 6(y_2 - 2y_1 + y_0)$$

$$4M_1 + M_2 = 180 \quad (\because M_0 = 0)$$

and
$$[M_1 + 4M_2 + M_3] = 6(y_3 - 2y_2 + y_1)$$
$$M_1 + 4M_2 = 1080 \ (\because M_3 = 0)$$

On solving these equations for M_1 and M_2, we get
$$M_1 = -24 \text{ and } M_2 = 276$$

Since number of intervals are three, so there will be three cubic splines as

For $0 \le x \le 1$

$$s_1(x) = \frac{1}{h_1}\left[\frac{(x_1-x)^3}{6}M_0 + \frac{(x-x_0)^3}{6}M_1 + \left(y_0 - \frac{h_1^2}{6}M_0\right)(x_1-x) + \left(y_1 - \frac{h_1^2}{6}M_1\right)(x-x_0)\right]$$

$$= \left[\frac{(x-0)^3}{6}(-24) + (1)(1-x) + \left(2 - \frac{1}{6}(-24)\right)(x-0)\right]$$

$$= -4x^3 + 5x + 1$$

For $1 \le x \le 2$

$$s_2 = \frac{1}{h_2}\left[\frac{(x_2-x)^3}{6}M_1 + \frac{(x-x_1)^3}{6}M_2 + \left(y_1 - \frac{h_2^2}{6}M_1\right)(x_2-x) + \left(y_2 - \frac{h_2^2}{6}M_2\right)(x-x_1)\right]$$

$$= \left[\frac{(2-x)^3}{6}(-24) + \frac{(x-1)^3}{6}(276) + \left(2 - \frac{1}{6}(-24)\right)(2-x) + \left(33 - \frac{1}{6}(276)\right)(x-1)\right]$$

$$= -4(2-x)^3 + 46(x-1)^3 + 6(2-x) - 13(x-1)$$
$$= 50x^3 - 162x^2 + 167x - 53$$

For $2 \le x \le 3$

$$s_3(x) = \frac{1}{h^3}\left[\frac{x_3-x)^3}{6}M_2 + \frac{(x-x_2)^3}{6}M_3 + \left(y_2 - \frac{h_3^2}{6}M_2\right)(x_3-x) + \left(y_3 - \frac{h_3^2}{6}M_3\right)(x-x_2)\right]$$

$$= \left[\frac{(3-x)^3}{6}(276) + \left(33 - \frac{1}{6}(276)\right)(3-x) + (244)(x-2)\right]$$

$$= 46(3-x)^3 - 13(3-x) + 244(x-2)$$
$$= -46x^3 + 414x^2 - 985x + 715$$

Now, y at $x = 2.5$
$$y(2.5) = f(2.5) = -46(2.5)^3 + 414(2.5)^2 - 985(2.5) + 715$$
$$= 121.25$$

EXERCISE 4.6

1. Obtain cubic spline for the interval (3, 4) from the given data:

x	1	2	3	4
f(x)	3	10	29	65

   with the end conditions $M_0 = M_4 = 0$.

2. Obtain the natural cubic spline from the following data:

x	1	2	3	4
f(x)	0	1	0	0

3. Fit a cubic spline for the following data:

x	0	1	2	3
f(x)	1	2	5	11

   with the end conditions $M_0 = M_3 = 0$. Also evaluate $f(2.5)$.

ANSWERS

1. $M_1 = 12.4$, $M_2 = 22.4$; $f(x) = \dfrac{-56x^3 + 672x^2 - 2092x + 2175}{15}$

2. $M_1 = -3.6$, $M_2 = 2.4$; For $1 \leq x \leq 2$, $f(x) = \dfrac{-3x^3 + 9x^2 - x - 5}{5}$

   For $2 \leq x \leq 3$, $f(x) = \dfrac{5x^3 - 39x^2 + 95x - 69}{5}$

   For $3 \leq x \leq 4$, $f(x) = \dfrac{-2x^3 + 24x^2 - 94x + 120}{5}$

3. For $0 \leq x \leq 1$, $f(x) = \dfrac{x^3 + 2x + 3}{3}$; For $1 \leq x \leq 2$, $f(x) = \dfrac{x^3 + 2x + 3}{3}$

   For $2 \leq x \leq 3$, $f(x) = \dfrac{-2x^3 + 18x^2 - 34x + 27}{3}$; $f(2.5) = 7.66$

Chapter 5
Numerical Differentiation and Integration

NUMERICAL DIFFERENTIATION

Numerical differentiation is concerned with the method of finding the successive derivatives of a function at a given argument, using the given table of entries corresponding to a set of arguments, equally or unequally spaced. In case of equidistant values of x, if the derivative is required at a point nearer to the starting value in the table, we use newton's forward interpolation formula. If the derivative is required at the end of the table, we use newton's backward interpolation formula. If the value of derivative is required near the middle of the table value, then we use one of the central difference interpolation formulae. However, in case of unequal intervals, we can use newton's divided difference formula or Lagrange's interpolation formula to get the derivative value. In this chapter we shall be concerned with the problems of numerical differentiation and integration. That is to say, given the set of values of x and y, we shall derive formulae to compute

(i) $\dfrac{dy}{dx}, \dfrac{d^2y}{dx^2}, \ldots$, for any value of x in $[x_0, x_n]$ and

(ii) $\int_{x_0}^{x_n} y\,dx$ i.e., $\int_{x_0}^{x_n} f(x)\,dx$

5.1 NEWTON'S FORWARD DIFFERENCE FORMULA TO GET THE DERIVATIVE

$$y = y(x) = y(x_0 + uh)$$
$$= y_0 + u\,\Delta y_0 + \frac{u(u-1)}{2!}\Delta^2 y_0 + \frac{u(u-1)(u-2)}{3!}\Delta^3 y_0$$
$$+ \frac{u(u-1)(u-2)(u-3)}{4!}\Delta^4 y_0 + \ldots \qquad \ldots(1)$$

where $\quad x = x_0 + uh \Rightarrow u = \dfrac{x - x_0}{h} \qquad \ldots(2)$

we know that
$$\frac{dy}{dx} = \frac{dy}{du}\frac{du}{dx} \qquad ...(3)$$
Differentiating equation (2) w.r.t. x, we get
$$\frac{du}{dx} = \frac{1}{h}$$
∴ From equation (3), we have
$$\frac{dy}{dx} = \frac{1}{h}\frac{dy}{du} \qquad ...(4)$$
Differentiating equation (1) w.r.t. u, we get
$$\frac{dy}{du} = \Delta y_0 + \frac{2u-1}{2}\Delta^2 y_0 + \frac{3u^2 - 6u + 2}{6}\Delta^3 y_0$$
$$+ \frac{2u^3 - 9u^2 + 11u - 3}{12}\Delta^4 y_0 + \ldots \qquad ...(5)$$
From equation (4) and (5), we get
$$\frac{dy}{dx} = \frac{1}{h}\left[\Delta y_0 + \frac{2u-1}{2}\Delta^2 y_0 + \frac{3u^2 - 6u + 2}{6}\Delta^3 y_0\right.$$
$$\left. + \frac{2u^3 - 9u^2 + 11u - 3}{12}\Delta^4 y_0 + \ldots\right] \qquad ...(6)$$
Differentiating equation (6) w.r.t. x, we get
$$\frac{d^2y}{dx^2} = \frac{d}{dx}\left(\frac{dy}{dx}\right) = \frac{d}{du}\left(\frac{dy}{dx}\right)\frac{du}{dx}$$
$$= \frac{1}{h}\left[\Delta^2 y_0 + (u-1)\Delta^3 y_0 + \left(\frac{6u^2 - 18u + 11}{12}\right)\Delta^4 y_0 + \ldots\right]\frac{1}{h}$$
$$= \frac{1}{h^2}\left[\Delta^2 y_0 + (u-1)\Delta^3 y_0 + \left(\frac{6u^2 - 18u + 11}{12}\right)\Delta^4 y_0 + \ldots\right] \qquad ...(7)$$
similarly
$$\frac{d^3y}{dx^3} = \frac{1}{h^3}\left[\Delta^3 y_0 + \left(\frac{2u-3}{2}\right)\Delta^4 y_0 + \ldots\right] \qquad ...(8)$$
and so on.

Equation (6), (7) and (8) gives the first, second and third derivative respectively at any general x.

Now putting $x = x_0$ i.e., $u = 0$ in equation (6), (7) and (8), we get
$$\left(\frac{dy}{dx}\right)_{x=x_0} = \left(\frac{dy}{dx}\right)_{u=0}$$
$$= \frac{1}{h}\left[\Delta y_0 - \frac{1}{2}\Delta^2 y_0 + \frac{1}{3}\Delta^3 y_0 - \frac{1}{4}\Delta^4 y_0 + \ldots\right] \qquad ...(9)$$

$$\left(\frac{d^2y}{dx^2}\right)_{x=x_0} = \left(\frac{d^2y}{dx^2}\right)_{u=0} = \frac{1}{h^2}\left[\Delta^2 y_0 - \Delta^3 y_0 + \frac{11}{12}\Delta^4 y_0 - ....\right] \quad ...(10)$$

$$\left(\frac{d^3y}{dx^3}\right)_{x=x_0} = \left(\frac{d^3y}{dx^3}\right)_{u=0} = \frac{1}{h^3}\left[\Delta^3 y_0 - \frac{3}{2}\Delta^4 y_0 + ...\right] \quad ...(11)$$

Equation (9), (10) and (11) gives the value of first, second and third derivative at $x = x_0$ i.e., $u = 0$.

Aliter: we know that
$$E = e^{hD}$$
$$\Rightarrow 1 + \Delta = e^{hD}$$
$$\therefore \quad hD = \log(1+\Delta) = \Delta - \frac{\Delta^2}{2} + \frac{\Delta^3}{3} - \frac{\Delta^4}{4} + ...$$

$$\Rightarrow \quad D = \frac{1}{h}\left[\Delta - \frac{\Delta^2}{2} + \frac{1}{3}\Delta^3 - \frac{1}{4}\Delta^4 + ...\right]$$

similarly
$$D^2 = \frac{1}{h^2}\left[\Delta - \frac{\Delta^2}{2} + \frac{1}{3}\Delta^3 - \frac{1}{4}\Delta^4 + ...\right]^2$$

$$= \frac{1}{h^2}\left[\Delta^2 - \Delta^3 + \frac{11}{12}\Delta^4 - \frac{5}{6}\Delta^5 + ...\right]$$

and
$$D^3 = \frac{1}{h^3}\left[\Delta^3 - \frac{3}{2}\Delta^4 + ...\right] \text{ and so on.}$$

5.2 NEWTON'S BACKWARD DIFFERENCE FORMULA TO GET THE DERIVATIVE

$$y = y(x) = y(x_n + uh)$$
$$= y_n + u\nabla y_n + \frac{u(u+1)}{2!}\nabla^2 y_n + \frac{u(u+1)(u+2)}{3!}\nabla^3 y_n$$
$$+ \frac{u(u+1)(u+2)(u+3)}{4!}\nabla^4 y_n + ... \quad ...(1)$$

where
$$x = x_n + uh \Rightarrow u = \frac{x - x_n}{h} \quad ...(2)$$

we know that
$$\frac{dy}{dx} = \frac{dy}{du}\frac{du}{dx} \quad ...(3)$$

Differentiating equation (2) w.r.t. x, we get
$$\frac{du}{dx} = \frac{1}{h}$$

∴ From equation (3), we have

$$\frac{dy}{dx} = \frac{1}{h}\frac{dy}{du} \qquad ...(4)$$

Differentiating equation (1) w.r.t. u, we get

$$\frac{dy}{du} = \nabla y_n + \left(\frac{2u+1}{2}\right)\nabla^2 y_n + \left(\frac{3u^2+6u+2}{6}\right)\nabla^3 y_n$$
$$+ \left(\frac{4u^3+18u^2+22u+6}{24}\right)\nabla^4 y_n + ... \qquad ...(5)$$

From equation (4) and (5), we get

$$\frac{dy}{dx} = \frac{1}{h}\left[\nabla y_n + \left(\frac{2u+1}{2}\right)\nabla^2 y_n + \left(\frac{3u^2+6u+2}{6}\right)\nabla^3 y_n + \left(\frac{4u^3+18u^2+22u+6}{24}\right)\nabla^4 y_n + ...\right] \qquad ...(6)$$

Again differentiating equation (6) w.r.t. x, we get

$$\frac{d^2y}{dx^2} = \frac{d}{dx}\left(\frac{dy}{dx}\right) = \frac{d}{du}\left(\frac{dy}{dx}\right)\frac{du}{dx}$$
$$= \frac{1}{h}\left[\nabla^2 y_n + (u+1)\nabla^3 y_n + \left(\frac{6u^2+18u+11}{12}\right)\nabla^4 y_n + ...\right]\frac{1}{h}$$

$$\frac{d^2y}{dx^2} = \frac{1}{h^2}\left[\nabla^2 y_n + (u+1)\nabla^3 y_n + \left(\frac{6u^2+18u+11}{12}\right)\nabla^4 y_n + ...\right] \qquad ...(7)$$

similarly

$$\frac{d^3y}{dx^3} = \frac{1}{h^3}\left[\nabla^3 y_n + \left(\frac{u+3}{2}\right)\nabla^4 y_n + ...\right] \qquad ...(8)$$

and so on.

Equation (6), (7) and (8) gives the first, second and third derivative respectively at any general x.

Now putting $x = x_n$ i.e., $u = 0$ in equation (6), (7) and (8) we get

$$\left(\frac{dy}{dx}\right)_{x=x_n} = \left(\frac{dy}{dx}\right)_{u=0}$$
$$= \frac{1}{h}\left[\nabla y_n + \frac{1}{2}\nabla^2 y_n + \frac{1}{3}\nabla^3 y_n + \frac{1}{4}\nabla^4 y_n + ...\right] \qquad ...(9)$$

$$\left(\frac{d^2y}{dx^2}\right)_{x=x_n} = \left(\frac{d^2y}{dx^2}\right)_{u=0} = \frac{1}{h^2}\left[\nabla^2 y_n + \nabla^3 y_n + \frac{11}{12}\nabla^4 y_n + ...\right] \qquad ...(10)$$

$$\left(\frac{d^3y}{dx^3}\right)_{x=x_n} = \left(\frac{d^3y}{dx^3}\right)_{u=0} = \frac{1}{h^3}\left[\nabla^3 y_n + \frac{3}{2}\nabla^4 y_n + ...\right] \qquad ...(11)$$

Equation (9), (10) and (11) gives the value of first, second and third derivative at $x = x_n$ i.e., $u = 0$.

Aliter: We know that

$$E^{-1} = 1 - \nabla$$

$$e^{-hD} = 1 - \nabla$$

$$\therefore \quad -hD = \log(1-\nabla) = -\left(\nabla + \frac{1}{2}\nabla^2 + \frac{1}{3}\nabla^3 + \frac{1}{4}\nabla^4 + ...\right)$$

$$\Rightarrow \quad D = \frac{1}{h}\left(\nabla + \frac{1}{2}\nabla^2 + \frac{1}{3}\nabla^3 + \frac{1}{4}\nabla^4 + ...\right)$$

similarly

$$D^2 = \frac{1}{h^2}\left(\nabla + \frac{1}{2}\nabla^2 + \frac{1}{3}\nabla^3 + ...\right)^2$$

$$D^2 = \frac{1}{h^2}\left(\nabla^2 + \nabla^3 + \frac{11}{12}\nabla^4 + ...\right)$$

and

$$D^3 = \frac{1}{h^3}\left(\nabla^3 + \frac{3}{2}\nabla^4 + ...\right) \text{ and so on.}$$

5.3. STIRLING'S INTERPOLATION FORMULA TO GET THE DERIVATIVE

$$y = y(x) = y(x_0 + uh)$$

$$= y_0 + \frac{u}{1!}\left(\frac{\Delta y_0 + \Delta y_{-1}}{2}\right) + \frac{u^2}{2!}\Delta^2 y_{-1} + \frac{u(u^2-1^2)}{3!}\left(\frac{\Delta^3 y_{-1} + \Delta^3 y_{-2}}{2}\right)$$

$$+ \frac{u(u^2-1^2)}{4!}\Delta^4 y_{-2} + \frac{u(u^2-1^2)(u^2-2^2)}{5!}\left(\frac{\Delta^5 y_{-2} + \Delta^5 y_{-3}}{2}\right) + ... \quad (1)$$

where $\qquad x = x_0 + uh \Rightarrow u = \frac{x-x_0}{h} \qquad ...(2)$

we know that

$$\frac{dy}{dx} = \frac{dy}{du}\frac{du}{dx} \qquad ...(3)$$

Differentiating equation (2) w.r.t. x, we get

$$\frac{du}{dx} = \frac{1}{h}$$

∴ From equation (3), we have
$$\frac{dy}{dx} = \frac{1}{h}\frac{dy}{du} \qquad \ldots(4)$$

Differentiating equation (1) w.r.t. u, we get

$$\frac{dy}{du} = \frac{\Delta y_0 + \Delta y_{-1}}{2} + u\Delta^2 y_{-1} + \left(\frac{3u^2-1}{6}\right)\left(\frac{\Delta^3 y_{-1} + \Delta^3 y_{-2}}{2}\right)$$
$$+ \left(\frac{4u^3 - 2u}{4!}\right)\Delta^4 y_{-2} + \left(\frac{5u^4 - 15u^2 + 4}{5!}\right)\left(\frac{\Delta^5 y_{-2} + \Delta^5 y_{-3}}{2}\right) + \ldots \qquad (5)$$

From equation (4) and (5), we get

$$\frac{dy}{dx} = \frac{1}{h}\left[\frac{\Delta y_0 + \Delta y_{-1}}{2} + u\Delta^2 y_{-1} + \left(\frac{3u^2-1}{6}\right)\left(\frac{\Delta^3 y_{-1} + \Delta^3 y_{-2}}{2}\right)\right.$$
$$\left. + \left(\frac{4u^3 - 2u}{4!}\right)\Delta^4 y_{-2} + \left(\frac{5u^4 - 15u^2 + 4}{5!}\right)\left(\frac{\Delta^5 y_{-2} + \Delta^5 y_{-3}}{2}\right) + \ldots\right] \qquad \ldots(6)$$

Differentiating equation (6) w.r.t. x, we get

$$\frac{d^2y}{dx^2} = \frac{d}{dx}\left(\frac{dy}{dx}\right) = \frac{d}{du}\left(\frac{dy}{dx}\right)\frac{du}{dx}$$

$$\frac{d^2y}{dx^2} = \frac{1}{h^2}\left[\Delta^2 y_{-1} + u\left(\frac{\Delta^3 y_{-1} + \Delta^3 y_{-2}}{2}\right) + \left(\frac{6u^2-1}{12}\right)\Delta^4 y_{-2}\right.$$
$$\left. + \left(\frac{2u^3 - 3u}{12}\right)\left(\frac{\Delta^5 y_{-2} + \Delta^5 y_{-3}}{2}\right) + \ldots\right] \qquad \ldots(7)$$

similarly

$$\frac{d^3y}{dx^3} = \frac{1}{h^3}\left[\left(\frac{\Delta^3 y_{-1} + \Delta^3 y_{-2}}{2}\right) + u\Delta^4 y_{-2}\right.$$
$$\left. + \left(\frac{2u^2-1}{4}\right)\left(\frac{\Delta^5 y_{-2} + \Delta^5 y_{-3}}{2}\right) + \ldots\right] \qquad \ldots(8)$$

and so on.

Equation (6), (7) and (8) gives the first, second and third derivative at any general x.

Now putting $x = x_0$ i.e., $u = 0$, in equation (6), (7) and (8), we get

$$\left(\frac{dy}{dx}\right)_{x=x_0} = \left(\frac{dy}{dx}\right)_{u=0}$$
$$= \frac{1}{h}\left[\left(\frac{\Delta y_0 + \Delta y_{-1}}{2}\right) - \frac{1}{6}\left(\frac{\Delta^3 y_{-1} + \Delta^3 y_{-2}}{2}\right) + \frac{1}{30}\left(\frac{\Delta^5 y_{-2} + \Delta^5 y_{-3}}{2}\right) - \ldots\right] \qquad \ldots(9)$$

$$\left(\frac{d^2y}{dx^2}\right)_{x=x_0} = \left(\frac{d^2y}{dx^2}\right)_{u=0} = \frac{1}{h^2}\left[\Delta^2 y_{-1} - \frac{1}{12}\Delta^4 y_{-2} + \frac{1}{90}\Delta^6 y_{-3} - \ldots\right] \ldots(10)$$

$$\left(\frac{d^3y}{dx^3}\right)_{x=x_0} = \left(\frac{d^3y}{dx^3}\right)_{u=0}$$

$$= \frac{1}{h^3}\left[\left(\frac{\Delta^3 y_{-1} + \Delta^3 y_{-2}}{2}\right) - \frac{1}{4}\left(\frac{\Delta^5 y_{-2} + \Delta^5 y_{-3}}{2}\right) + \ldots\right] \ldots(11)$$

Equation (9), (10) and (11) gives the value of first, second and third derivative at $x = x_0$ i.e., $u = 0$.

5.4 BESSEL'S FORMULA TO GET THE DERIVATIVE

$$y = \frac{y_0 + y_1}{2} + \left(u - \frac{1}{2}\right)\Delta y_0 + \frac{u(u-1)}{2!}\left(\frac{\Delta^2 y_{-1} + \Delta^2 y_0}{2}\right)$$

$$+ \frac{\left(u - \frac{1}{2}\right)u(u-1)}{3!}\Delta^3 y_{-1} + \frac{(u+1)u(u-1)(u-2)}{4!}\left(\frac{\Delta^4 y_{-2} + \Delta^4 y_{-1}}{2}\right)$$

$$+ \frac{(u+1)u(u-1)(u-2)\left(u - \frac{1}{2}\right)}{5!}\Delta^5 y_{-2} + \ldots \quad \ldots(1)$$

where $\quad u = \dfrac{x - x_0}{h} \quad \ldots(2)$

we know that $\quad \dfrac{dy}{dx} = \dfrac{dy}{du}\dfrac{du}{dx} \quad \ldots(3)$

Differentiating equation (2) w.r.t. x, we get

$$\frac{du}{dx} = \frac{1}{h}$$

$\therefore$ From equation (3), we have

$$\frac{dy}{dx} = \frac{1}{h}\frac{dy}{du} \quad \ldots(4)$$

Differentiating equation (1) w.r.t. u, we get

$$\frac{dy}{du} = \Delta y_0 + \left(\frac{2u-1}{2!}\right)\left(\frac{\Delta^2 y_{-1} + \Delta^2 y_0}{2}\right) + \left(\frac{3u^2 - 3u + \frac{1}{2}}{3!}\right)\Delta^3 y_{-1}$$

$$+ \left(\frac{4u^3 - 6u^2 - 2u + 2}{4!}\right)\left(\frac{\Delta^4 y_{-2} + \Delta^4 y_{-1}}{2}\right)$$

$$+ \left(\frac{5u^4 - 10u^3 + 5u - 1}{5!}\right)\Delta^5 y_{-2} + \ldots \quad \ldots(5)$$

From equation (4) and (5), we get

$$\frac{dy}{dx} = \frac{1}{h}\left[\Delta y_0 + \left(\frac{2u-1}{2!}\right)\left(\frac{\Delta^2 y_{-1} + \Delta^2 y_0}{2}\right) + \left(\frac{3u^2 - 3u + \frac{1}{2}}{3!}\right)\Delta^3 y_{-1}\right.$$

$$+ \left(\frac{4u^3 - 6u^2 - 2u + 2}{4!}\right)\left(\frac{\Delta^4 y_{-2} + \Delta^4 y_{-1}}{2}\right)$$

$$\left. + \left(\frac{5u^4 - 10u^3 + 5u - 1}{5!}\right)\Delta^5 y_{-2} + \ldots\right] \qquad \ldots(6)$$

Differentiating equation (6) w.r.t. x, we get

$$\frac{d^2 y}{dx^2} = \frac{d}{dx}\left(\frac{dy}{dx}\right) = \frac{d}{du}\left(\frac{dy}{dx}\right)\frac{du}{dx}$$

$$\frac{d^2 y}{dx^2} = \frac{1}{h^2}\left[\left(\frac{\Delta^2 y_{-1} + \Delta^2 y_0}{2}\right) + \left(\frac{2u-1}{2}\right)\Delta^3 y_{-1} + \left(\frac{6u^2 - 6u - 1}{12}\right)\right. \qquad \ldots(7)$$

$$\left.\left(\frac{\Delta^4 y_{-2} + \Delta^4 y_{-1}}{2}\right)\left(\frac{4u^3 - 6u^2 + 1}{24}\right)\Delta^5 y_{-2} + \ldots\right]$$

similarly

$$\frac{d^3 y}{dx^3} = \frac{1}{h^3}\left[\Delta^3 y_{-1} + \left(u - \frac{1}{2}\right)\left(\frac{\Delta^4 y_{-2} + \Delta^4 y_{-1}}{2}\right) + \left(\frac{u^2 - u}{2}\right)\Delta^5 y_{-2} + \ldots\right]$$

$$\ldots(8)$$

Equation (6), (7) and (8) gives the first, second and third derivative at any general x.

Now putting $x = x_0$ i.e., $u = 0$ in equation (6), (7) and (8) we get

$$\left(\frac{dy}{dx}\right)_{x = x_0} = \left(\frac{dy}{dx}\right)_{u = 0}$$

$$= \frac{1}{h}\left[\Delta y_0 - \frac{1}{2}\left(\frac{\Delta^2 y_{-1} + \Delta^2 y_0}{2}\right) + \frac{1}{12}\Delta^3 y_{-1} + \frac{1}{12}\left(\frac{\Delta^4 y_{-2} + \Delta^4 y_{-1}}{2}\right) - \frac{1}{120}\Delta^5 y_{-2} - \ldots\right]$$

$$\ldots(9)$$

$$\left(\frac{d^2 y}{dx^2}\right)_{x = x_0} = \left(\frac{d^2 y}{dx^2}\right)_{u = 0}$$

$$= \frac{1}{h^2}\left[\left(\frac{\Delta^2 y_{-1} + \Delta^2 y_0}{2}\right) - \frac{1}{2}\Delta^3 y_{-1} - \frac{1}{12}\left(\frac{\Delta^4 y_{-2} + \Delta^4 y_{-1}}{2}\right)\right.$$

$$\left. + \frac{1}{24}\Delta^5 y_{-2} + \ldots\right] \qquad \ldots(10)$$

$$\left(\frac{d^3y}{dx^3}\right)_{x=x_0} = \left(\frac{d^3y}{dx^3}\right)_{u=0} = \frac{1}{h^3}\left[\Delta^3 y_{-1} - \frac{1}{2}\left(\frac{\Delta^4 y_{-2} + \Delta^4 y_{-1}}{2}\right) + ...\right] \quad ...(11)$$

Equation (9), (10) and (11) gives the value of first, second and third derivative at $x = x_0$, i.e., $u = 0$.

Note: If the interval of differencing is not constant (i.e., x's are not equally spaced), we get newton's divided difference formula or lagrange's interpolation formula, and then differentiating it w.r.t. x, we get the derivatives at any x in the range.

(i) Newton's divided difference formula is

$$f(x) = f(x_0) + (x - x_0) f(x_0, x_1) + (x - x_0)(x - x_1) f(x_0, x_1, x_2)$$
$$+ (x - x_0)(x - x_1)(x - x_2) f(x_0, x_1, x_2, x_3) + ...$$
$$+ (x-x_0)(x-x_1)...(x-x_{n-1}) f(x_0, x_1, ..x_n)$$

or $f(x) = f(x_0) + (x - x_0) \Delta f(x_0) + (x - x_0)(x - x_1) \Delta^2 f(x_0) + (x - x_0)(x - x_1)(x - x_2) \Delta^3 f(x_0) + ... + (x - x_0)(x - x_1)...(x - x_{n-1}) \Delta^n f(x_0)$

$f'(x)$ is given by

$$f'(x) = \Delta f(x_0) + [2x - (x_0 + x_1)] \Delta^2 f(x_0) + [3x^2 - 2x(x_0 + x_1 + x_2) + (x_0 x_1 + x_1 x_2 + x_2 x_0)] \Delta^3 f(x_0) + ...$$

(ii) Lagrange's interpolation formula is

$$f(x) = \frac{(x-x_1)(x-x_2)...(x-x_n)}{(x_0-x_1)(x_0-x_2)...(x_0-x_n)} f(x_0)$$
$$+ \frac{(x-x_0)(x-x_2)...(x-x_n)}{(x_1-x_0)(x_1-x_2)...(x_1-x_n)} f(x_1) + .. \quad ...(1)$$

or
$$y = f(x) = \frac{(x-x_1)(x-x_2)...(x-x_n)}{(x_0-x_1)(x_0-x_2)...(x_0-x_n)} y_0$$
$$+ \frac{(x-x_0)(x-x_2)...(x-x_n)}{(x_1-x_0)(x_1-x_2)...(x_1-x_n)} y_1 + ..$$

$f'(x)$ can be obtained by differentiating $f(x)$ is equation (1).

ILLUSTRATIVE EXAMPLES

Example 1. From the following table, find the first two derivatives of $(x)^{1/3}$ at $x = 50$ and $x = 56$.

x:	50	51	52	53	54	55	56
$y = x^{1/3}$:	3.6840	3.7084	3.7325	3.7563	3.7798	3.8030	3.8259

Solution Since we require y', y'' at $x = 50$, we use newton's forward formula and at $x = 56$, we use newton's backward formula.

x	y	Δy	$\Delta^2 y$	$\Delta^3 y$
50	3.6840	0.0244	−0.0003	0
51	3.7084	0.0241	−0.0003	0
52	3.7325	0.0238	−0.0003	0
53	3.7563	0.0235	−0.0003	0
54	3.7798	0.0232	−0.0003	
55	3.8030	0.0229		
56	3.8259			

$$u = \frac{x - x_0}{h} = \frac{50 - 50}{1} = 0$$

$$\left(\frac{dy}{dx}\right)_{x=50} = \left(\frac{dy}{dx}\right)_{u=0} = \frac{1}{h}\left[\Delta y_0 - \frac{1}{2}\Delta^2 y_0 + \frac{1}{3}\Delta^3 y_0\right]$$

$$= \frac{1}{1}\left[0.0244 - \frac{1}{2}(-0.0003) + \frac{1}{3}(0)\right] = 0.02455$$

$$\left(\frac{d^2 y}{dx^2}\right)_{x=50} = \frac{1}{h^2}\left[\Delta^2 y_0 - \Delta^3 y_0 + ...\right] = -0.0003$$

Now $$u = \frac{x - x_n}{h} = \frac{56 - 56}{1} = 0$$

By newton's backward difference formula

$$\left(\frac{dy}{dx}\right)_{x=56} = \frac{1}{h}\left[\nabla y_n + \frac{1}{2}\nabla^2 y_n + \frac{1}{3}\nabla^3 y_n + ...\right]$$

$$= \frac{1}{1}\left[0.0229 + \frac{1}{2}(-0.0003) + 0\right] = 0.02275$$

$$\left(\frac{d^2 y}{dx^2}\right)_{x=56} = \frac{1}{h^2}\left[\nabla^2 y_n + \nabla^3 y_n + ...\right]$$

$$= \frac{1}{1^2}[-0.0003] = -0.0003$$

Example 2. Find $\frac{dy}{dx}$ at $x = 0.1$ from the following table :

x :	0.1	0.2	0.3	0.4
y :	0.9975	0.9900	0.9776	0.9604

Solution: The difference table is

x	y	Δy	$\Delta^2 y$	$\Delta^3 y$
0.1	0.9975	−0.0075	−0.0049	0.0001
0.2	0.9900	−0.0124	−0.0048	
0.3	0.9776	−0.0172		
0.4	0.9604			

Here $h = 0.1$, $x_0 = 0.1$, $u = \dfrac{x - x_0}{h} = \dfrac{0.1 - 0.1}{0.1} = 0$

$$\left(\dfrac{dy}{dx}\right)_{x=0.1} = \dfrac{1}{h}\left[\Delta y_0 - \dfrac{1}{2}\Delta^2 y_0 + \dfrac{1}{3}\Delta^3 y_0 + \ldots\right]$$

$$= \dfrac{1}{0.1}\left[-0.0075 - \dfrac{1}{2}(-0.0049) + \dfrac{1}{3}(0.0001)\right]$$

$$= -0.050167$$

Example 3. The population of a certain town is given below. Find the rate of growth of the population in 1931.

year x :	1931	1941	1951	1961	1971
population (thousands) y :	40.62	60.80	79.95	103.56	132.65

Solution: The differeance table is

x	y	Δy	$\Delta^2 y$	$\Delta^3 y$	$\Delta^4 y$
1931	40.62	20.18	−1.03	5.49	−4.47
1941	60.80	19.15	4.46	1.02	
1951	79.95	23.61	5.48		
1961	103.56	29.09			
1971	132.65				

Here $x_0 = 1931$, $h = 10$ and $u = \dfrac{x - x_0}{h} = \dfrac{1931 - 1931}{10} = 0$

$$\left(\dfrac{dy}{dx}\right)_{x=1931} = \left(\dfrac{dy}{dx}\right)_{u=0} = \dfrac{1}{h}\left[\Delta y_0 - \dfrac{1}{2}\Delta^2 y_0 + \dfrac{1}{3}\Delta^3 y_0 - \dfrac{1}{4}\Delta^4 y_0 + \ldots\right]$$

$$= \dfrac{1}{10}\left[20.18 - \dfrac{1}{2}(-1.03) + \dfrac{1}{3}(5.49) - \dfrac{1}{4}(-4.47)\right]$$

$$= \dfrac{1}{10}[20.18 + 0.515 + 1.83 + 1.1175] = 2.36425$$

Example 4. The table given below reveals the velocity 'v' of a body during the time 't' specified. Find its acceleration at $t = 1.1$.

t :	1.0	1.1	1.2	1.3	1.4
v :	43.1	47.7	52.1	56.4	60.8

Solution: The difference table is

t	v	Δv	$\Delta^2 v$	$\Delta^3 v$	$\Delta^4 v$
1.0	43.1	4.6	−0.2	0.1	0.1
1.1	47.7	4.4	−0.1	0.2	
1.2	52.1	4.3	0.1		
1.3	56.4	4.4			
1.4	60.8				

Here $h = 0.1$, $u = \dfrac{t-t_0}{h} = \dfrac{1.1-1.0}{0.1} = 1$

Acceleration is given by $\left(\dfrac{dv}{dt}\right)$

$$= \dfrac{1}{h}\left[\Delta v_0 + \dfrac{2u-1}{2}\Delta^2 v_0 + \dfrac{3u^2-6u+2}{6}\Delta^3 v_0 \right.$$

$$\left. + \dfrac{(4u^3-18u^2+22u-6)}{24}\Delta^4 v_0 + ...\right]$$

$\left(\dfrac{dv}{dt}\right)_{t=1.1} = \left(\dfrac{dv}{dt}\right)_{u=1} = \dfrac{1}{0.1}\left[4.6 + \dfrac{1}{2}(-0.2) - \dfrac{1}{6}(0.1) + \dfrac{1}{12}(0.1)\right] = 44.917$

Example 5. Find the gradient of the road at the middle point of the elevation above a datum line of seven points of road which are given below:

| x: | 0 | 300 | 600 | 900 | 1200 | 1500 | 1800 |
| y: | 135 | 149 | 157 | 183 | 201 | 205 | 193 |

Solution: The difference table is

x	y	Δy	$\Delta^2 y$	$\Delta^3 y$	$\Delta^4 y$	$\Delta^5 y$	$\Delta^6 y$
0	135	14	−6	24	−50	70	−86
300	149	8	18	−26	20	−16	
600	157	26	−8	−6	4		
900	183	18	−14	−2			
1200	201	4	−16				
1500	205	−12					
1800	193						

$$u = \dfrac{x-x_0}{h} = \dfrac{900-900}{300} = 0$$

Here we require $\left(\dfrac{dy}{dx}\right)_{x=900}$. Since $x = 900$ lies in middle of the table, we use one of the central difference formula, in particular stirling's formula.

$\left(\dfrac{dy}{dx}\right)_{x=x_0}$

$= \dfrac{1}{h}\left[\dfrac{1}{2}(\Delta y_0 + \Delta y_{-1}) - \dfrac{1}{12}(\Delta^3 y_{-1} + \Delta^3 y_{-2}) + \dfrac{1}{60}(\Delta^5 y_{-2} + \Delta^5 y_{-3}) + ...\right]$

$= \dfrac{1}{300}\left[\dfrac{1}{2}(18+26) - \dfrac{1}{12}(-6-26) + \dfrac{1}{60}(70-16)\right]$

$$= \frac{1}{300}[22 + 2.6666 + 0.9] = 0.085222$$

Hence, the gradient of the road at the middle point is 0.085222.

Example 6. Find the first and second derivative of the function tabulated below at $x = 0.6$.

| x : | 0.4 | 0.5 | 0.6 | 0.7 | 0.8 |
| y : | 1.5836 | 1.7974 | 2.0442 | 2.3275 | 2.6511 |

Solution: The difference table is

x	y	Δy	$\Delta^2 y$	$\Delta^3 y$	$\Delta^4 y$
0.4	1.5836	0.2138	0.0330	0.0035	0.0003
0.5	1.7974	0.2468	0.0365	0.0038	
0.6	2.0442	0.2833	0.0403		
0.7	2.3275	0.3236			
0.8	2.6511				

Here $h = 0.1$, $u = \dfrac{0.6 - 0.6}{0.1} = 0$

Since $x = 0.6$ is in the middle of the table of the table, so we use stirling's formula.

$$\left(\frac{dy}{dx}\right)_{x=x_0} = \frac{1}{h}\left[\frac{1}{2}(\Delta y_0 + \Delta y_{-1}) - \frac{1}{12}(\Delta^3 y_{-1} + \Delta^3 y_{-2})\right.$$

$$\left. + \frac{1}{60}(\Delta^5 y_{-2} + \Delta^5 y_{-3}) + ...\right]$$

$$\left(\frac{dy}{dx}\right)_{x=0.6} = \frac{1}{0.1}\left[\frac{1}{2}(0.2833 + 0.2468) - \frac{1}{12}(0.0038 + 0.0035)\right]$$

$$= 2.64442$$

$$\left(\frac{d^2 y}{dx^2}\right)_{x=x_0} = \frac{1}{h^2}\left[\Delta^2 y_{-1} - \frac{1}{12}\Delta^4 y_{-2} + ...\right]$$

$$\left(\frac{d^2 y}{dx^2}\right)_{x=0.6} = \frac{1}{(0.1)^2}\left[0.0365 - \frac{1}{12}(0.0003)\right] = 3.6475$$

Example 7. Find the first and second derivatives of the function at $x = 1.5$ from the table given below:

| x : | 1.5 | 2.0 | 2.5 | 3.0 | 3.5 | 4.0 |
| y : | 3.375 | 7.0 | 13.625 | 24.0 | 38.875 | 59.0 |

Solution: Here $h = 0.5$ and $u = \dfrac{x - x_0}{h} = \dfrac{1.5 - 1.5}{0.5} = 0$

The difference table is

x	y	Δy	$\Delta^2 y$	$\Delta^3 y$	$\Delta^4 y$
1.5	3.375	3.625	3.0	0.75	0
2.0	7.0	6.625	3.75	0.75	0
2.5	13.625	10.375	4.50	0.75	
3.0	24.0	14.875	5.25		
3.5	38.875	20.125			
4.0	59.0				

$$\left(\frac{dy}{dx}\right)_{x=x_0} = \left(\frac{dy}{dx}\right)_{u=0} = \frac{1}{h}\left[\Delta y_0 - \frac{1}{2}\Delta^2 y_0 + \frac{1}{3}\Delta^3 y_0 + ...\right]$$

$$\left(\frac{dy}{dx}\right)_{x=1.5} = \frac{1}{0.5}\left[3.625 - \frac{1}{2}(3) + \frac{1}{3}(0.75) + ...\right]$$

$$= \frac{1}{0.5}[3.625 - 1.5 + 0.25]$$

$$= \frac{1}{0.5}[2.375] = 4.75$$

$$\left(\frac{d^2 y}{dx^2}\right)_{x=x_0} = \frac{1}{h^2}\left[\Delta^2 y_0 - \Delta^3 y_0 + \frac{11}{12}\Delta^4 y_0 + ...\right]$$

$$\left(\frac{d^2 y}{dx^2}\right)_{x=1.5} = \frac{1}{(0.5)^2}\left[3 - 0.75 + \frac{11}{12} \times 0\right] = \frac{2.25}{0.25} = 9$$

Example 8. Find $y'(6)$ for the following data:

x:	0	2	3	4	7	9
y:	4	26	58	112	466	922

Solution: Since the arguments are not equally spaced, so we use newton's divided difference formula or lagrange's formula.

Divided difference table

x	$y = f(x)$	$\Delta f(x)$	$\Delta^2 f(x)$	$\Delta^3 f(x)$	$\Delta^4 f(x)$
0	4	11	7	1	0
2	26	32	11	1	0
3	58	54	16	1	
4	112	118	22		
7	466	228			
9	922				

By Newton's divided difference folmula,

$$y = f(x) = f(x_0) + (x - x_0) f(x_0, x_1) + (x - x_0)(x - x_1) f(x_0, x_1, x_2)$$
$$+ (x - x_0)(x - x_1)(x - x_2) f(x_0, x_1, x_2, x_3) + \ldots$$
$$= 4 + (x - 0) \times 11 + (x - 0)(x - 2) \times 7 + (x - 0)(x - 2)(x - 3) \times 1 + \ldots$$
$$y = f(x) = x^3 + 2x^2 + 3x + 4$$
$$y'(x) = 3x^2 + 4x + 3$$
$$y'(6) = 3(6)^2 + 4(6) + 3 = 135.$$

Example 9. Find $f'(4)$ from the following data:

x:	0	2	5	1
$f(x)$:	0	8	125	1

Solution: Since the arguments are not equally spaced, so we use lagrange's formula or newton's divided difference formula.

$$f(x) = \frac{(x-2)(x-5)(x-1)}{(0-2)(0-5)(0-1)} 0 + \frac{(x-0)(x-5)(x-1)}{(2-0)(2-5)(2-1)} (8)$$
$$+ \frac{(x-0)(x-2)(x-1)}{(5-0)(5-2)(5-1)} (125) + \frac{(x-0)(x-2)(x-5)}{(1-0)(1-2)(1-5)} (1)$$
$$= \frac{-4}{3}(x^3 - 6x^2 + 5x) + \frac{25}{12}(x^3 - 3x^2 + 2x) + \frac{1}{4}(x^3 - 7x^2 + 10x)$$
$$f(x) = x^3$$
$$f'(x) = 3x^2$$
$$\therefore \qquad f'(4) = 3(4)^2 = 48$$

Example 10. The table below gives the results of an observation; θ is the observed temperature in degrees centigrade of a vessel of cooling water; t is the time in minutes from the begining of observation.

t:	1	3	5	7	9
θ:	85.3	74.5	67.0	60.5	54.3

find the approximate rate of cooling at $t = 3$ and 3.5.

Solution: Here $\dfrac{d\theta}{dt}$ represents the rate of cooling and $h = 2$.

$$\therefore \qquad u = \frac{t - t_0}{h} = \frac{t-1}{2}$$

At $t = 3, u = 1$ and at $t = 3.5, u = 1.25$

$$\frac{d\theta}{dt} = \frac{1}{h} \left[\Delta \theta_0 + \frac{2u-1}{2} \Delta^2 \theta_0 + \frac{3u^2 - 6u + 2}{6} \Delta^3 \theta \right.$$
$$\left. + \frac{(4u^3 - 18u^2 + 22u - 6)}{24} \Delta^4 \theta_0 + \ldots \right] \quad (1)$$

The difference table is

t	θ	Δθ	Δ²θ	Δ³θ	Δ⁴θ
1	85.3	−10.8	3.3	−2.3	1.6
3	74.5	−7.5	1.0	−0.7	
5	67.0	−6.5	0.3		
7	60.5	−6.2			
9	54.3				

(a) Putting $u = 1$ in equation (1), we get

$$\left(\frac{d\theta}{dt}\right)_{u=1} = \frac{1}{2}\left[-10.8 + \frac{1}{2}(3.3) - 1/6(-2.3) + \frac{1}{12}(1.6)\right] = -4.31667$$

(b) Putting $u = 1.25$ in equation (1), we get

$$\left(\frac{d\theta}{dt}\right)_{u=1.25} = \frac{1}{2}[-10.8 + 0.75(3.3) - (0.1354)(-2.3)$$
$$+ (0.04948)(1.6)]$$

$$= \frac{1}{2}[-10.8 + 2.475 + 0.31142 + 0.079168]$$

$$\left(\frac{d\theta}{dt}\right)_{u=1.25} = \frac{1}{2}[-7.934412] = -3.967206$$

EXERCISE 5.1

1. Find the first and second derivative of $\sqrt{x}$ at $x = 15$ from the table given below:

x :	15	17	19	21	23	25
$\sqrt{x}$:	3.873	4.123	4.359	4.583	4.796	5.000

2. Obtain the second derivative of y at $x = 0.96$ from the following data:

x :	0.96	0.98	1.00	1.02	1.04
y :	0.7825	0.7739	0.7651	0.7563	0.7473

3. Find the first three derivatives of the function at $x = 1.5$ from the table given below:

x :	1.5	2.0	2.5	3.0	3.5	4.0
y :	3.375	7.0	13.625	24.0	38.875	59.0

4. Find $y'(0)$ and $y''(0)$ from the given table:

x :	0	1	2	3	4	5
y :	4	8	15	7	6	2

5. Find $y'(4)$ from the table given below:

x :	1	2	4	8	10
y :	0	1	5	21	27

6. Find $f'(8)$, given $f(6) = 1.556$, $f(7) = 1.690$, $f(9) = 1.908$, $f(12) = 2.158$.

7. A curve passes through the points $(0, 18)$, $(1, 10)$, $(3, -18)$, and $(6, 90)$. Find the slope of the curve at $x = 2$.

8. Evalvate y' and y'' at $x = 2$, for the table given below :

x :	0	1	3	6
y :	18	10	-18	40

9. From the following data find $f'(5)$:

x :	0	2	3	4	7	9
$f(x)$:	4	26	58	112	486	922

10. Given the following table of values, find $f'(8)$:

x :	6	7	9	12
$f(x)$:	1.556	1.690	1.908	2.158

11. Find $f'(6)$ from the following table:

x :	0	1	3	4	5	7	9
$f(x)$:	150	108	0	-54	-100	-144	-84

12. The following table gives values of pressure P and specific volume V of saturated steam:

P :	105	42.7	25.3	16.7	13
V :	2	4	6	8	10

find

(i) the rate of change of pressure w.r.t. volume at $V = 2$.

(ii) the rate of change of volume w.r.t. pressure at $P = 105$.

ANSWERS
---

1. $0.1289, -0.004$
2. -1.91666
3. $4.75, 9.0, 6.0$
4. $-27.9, 117.67$
5. 2.8326
6. 0.109
7. -16
8. -16.4
9. 98
10. 0.10848
11. -23
12. (i) -52.4 (ii) -0.01908

5.5 NUMERICAL INTEGRATION

If a set of tabulated values of the integrand $f(x)$ is given, then determining the value of the integral $\int_{x_0}^{x_n} f(x)dx$ is called numerical integration.

We subdivide the given interval into a large number of sub intervals of equal width h and replace the function tabulated at the points of sub division by any one of the interpolating polynomial, over each of the sub intervals and evaluate the integrals.

5.6 NEWTON-COTE'S QUADRATURE FORMULA

Consider the function $y = f(x)$, which assumes the values $y_0, y_1, y_2, ..., y_n$ corresponding to the values of the arguments $x = x_0, x_1, x_2, ..., x_n$.

Let $$I = \int_a^b f(x)dx$$

Divide the interval of integration (a, b) into n sub intervals of equal width h, where $h = \dfrac{b-a}{n}$, such that $x_0 = a$, $x_1 = x_0 + h$, $x_2 = x_0 + 2h$, ..., $x_n = x_0 + nh = b$.

$$\therefore \quad I = \int_a^b f(x)dx = \int_{x_0}^{x_0+h} f(x)dx \qquad ...(1)$$

Now, Newton's forward interpolation formula is given by,

$$y = f(x) = y_0 + u\Delta y_0 + \frac{u(u-1)}{2!}\Delta^2 y_0 + \frac{u(u-1)(u-2)}{3!}\Delta^3 y_0 + ...$$

where $u = \dfrac{x - x_0}{h} \Rightarrow du = \dfrac{1}{h}dx \Rightarrow dx = h\,du$

∴ Equation (1) becomes

$$I = h\int_0^n \left[y_0 + u\Delta y_0 + \frac{u(u-1)}{2!}\Delta^2 y_0 + \frac{u(u-1)(u-2)}{3!}\Delta^3 y_0 + ... \right]du$$

$$= h\left[uy_0 + \frac{u^2}{2}\Delta y_0 + \frac{1}{2}\left(\frac{u^3}{3} - \frac{u^2}{2}\right)\Delta^2 y_0 + \frac{1}{6}\left(\frac{u^4}{4} - u^3 + u^2\right)\Delta^3 y_0 + ... \right]_0^n$$

$$= h\left[ny_0 + \frac{n^2}{2}\Delta y_0 + \frac{1}{2}\left(\frac{n^3}{3} - \frac{n^2}{2}\right)\Delta^2 y_0 + \frac{1}{6}\left(\frac{n^4}{4} - n^3 + n^2\right)\Delta^3 y_0 + ... \right]$$

$$= nh\left[y_0 + \frac{n}{2}\Delta y_0 + \frac{1}{2}\left(\frac{n^2}{3} - \frac{n}{2}\right)\Delta^2 y_0 + \frac{1}{6}\left(\frac{n^3}{4} - n^2 + n\right)\Delta^3 y_0 + ... \right]$$

$$\int_{x_0}^{x_n} f(x)dx = nh\left[y_0 + \frac{n}{2}\Delta y_0 + \frac{n(2n-3)}{12}\Delta^2 y_0 + \frac{n(n-2)^2}{24}\Delta^3 y_0 + ... \right]$$

This is called general quadrature formula or Newton- Cote's quadrature formula.

5.7 TRAPEZOIDAL RULE (FOR n = 1)

Put $n = 1$ in the quadrature formula, *i.e.* The curve is approximated by a polynomial of degree one (linear polynomial) so that the differences of second and higher orders are all zero, we get

$$\int_{x_0}^{x_0+h} f(x)\,dx = h\left[y_0 + \frac{1}{2}\Delta y_0\right] = h\left[y_0 + \frac{1}{2}(y_1 - y_0)\right] = \frac{h}{2}[y_0 + y_1]$$

$$\int_{x_0}^{x_0+h} f(x)\,dx = \frac{h}{2}[y_0 + y_1]$$

Similarly,
$$\int_{x_0+h}^{x_0+2h} f(x)\,dx = \frac{h}{2}[y_1 + y_2]$$

....

$$\int_{x_0+(n-1)h}^{x_0+nh} f(x)\,dx = \frac{h}{2}[y_{n-1} + y_n]$$

Adding the above integrals, we get

$$\int_{x_0}^{x_0+nh} f(x)\,dx = \frac{h}{2}\left[(y_0 + y_1) + (y_1 + y_2) + \ldots + (y_{n-1} + y_n)\right]$$

$$\int_{x_0}^{x_0+nh} f(x)\,dx = \frac{h}{2}\left[(y_0 + y_n) + 2(y_1 + y_2 + \ldots + y_{n-1})\right]$$

$$\Rightarrow \int_{x_0}^{x_0+nh} f(x)\,dx = \frac{h}{2}\left[\begin{array}{l}(\text{sum of the first and last ordinate}) \\ + 2(\text{sum of the remaining ordinates})\end{array}\right]$$

This is known as Trapezoidal Rule.

Here, n must be a multiple of 1.

Note: In Trapezoidal rule, the polynomial is linear. This rule is least accurate.

5.8 SIMPSON'S 1/3rd RULE (FOR n = 2)

Put $n = 2$ in the quadrature formula, i.e. The curve is approximated by a polynomial of degree two so that the differences of third and higher orders vanish, we get

$$\int_{x_0}^{x_0+2h} f(x)\,dx = 2h\left[y_0 + \Delta y_0 + \frac{1}{6}\Delta^2 y_0\right]$$

$$= 2h\left[y_0 + (y_1 - y_0) + \frac{1}{6}(y_2 - 2y_1 + y_0)\right]$$

$$\int_{x_0}^{x_0+2h} f(x)\,dx = \frac{h}{3}[y_0 + 4y_1 + y_2]$$

Similarly, $\int_{x_0+2h}^{x_0+4h} f(x)\,dx = \frac{h}{3}[y_2 + 4y_3 + y_4]$

....

$\int_{x_0+(n-2)h}^{x_0+nh} f(x)\,dx = \frac{h}{3}[y_{n-2} + 4y_{n-1} + y_n]$

Adding the above integrals, we get

$$\int_{x_0}^{x_0+nh} f(x)\,dx = \frac{h}{3}[(y_0 + 4y_1 + y_2) + (y_2 + 4y_3 + y_4) + (y_{n-2} + 4y_{n-1} + y_n)]$$

$$\int_{x_0}^{x_0+nh} f(x)\,dx = \frac{h}{3}\Big[(y_0 + y_n) + 4(y_1 + y_3 + \dots + y_{n-1}) + 2(y_2 + y_4 + \dots + y_{n-2})\Big]$$

$$\int_{x_0}^{x_0+nh} f(x)\,dx = \frac{h}{3}\begin{bmatrix}(\text{sum of the first and last ordinate})\\ + 4(\text{sum of remaining even ordinates})\\ + 2(\text{sum of remaining odd ordinates})\end{bmatrix}$$

This is known as **Simpson's $1/3^{rd}$ Rule**. In this rule, n must be a multiple of 2.

Note. 1. Here, though y_2 has even suffix, it is third (odd) ordinate and y_1 has odd suffix, it is second (even) ordinate.

2. In Simpson's $1/3^{rd}$ rule, the polynomial is of degree two. For this rule, the number of intervals n must be even, i.e. the number of ordinates must be odd.

5.9 SIMPSON'S $3/8^{th}$ RULE (FOR $n = 3$)

Put $n = 3$ in the quadrature formula, i.e. The curve is approximated by a polynomial of degree three so that the differences of fourth and higher orders vanish, we get

$$\int_{x_0}^{x_0+3h} f(x)\,dx = 3h\left[y_0 + \frac{3}{2}\Delta y_0 + \frac{3}{4}\Delta^2 y_0 + \frac{1}{8}\Delta^3 y_0\right]$$

$$= \frac{3h}{8}[8y_0 + 12(y_1 - y_0) + 6(y_2 - 2y_1 + y_0) + (y_3 - 3y_2 + 3y_1 - y_0)]$$

$$\int_{x_0}^{x_0+3h} f(x)\,dx = \frac{3h}{8}[y_0 + 3y_1 + 3y_2 + y_3]$$

Similarly, $\int_{x_0+3h}^{x_0+6h} f(x)\,dx = \dfrac{3h}{8}[y_3 + 3y_4 + 3y_5 + y_6]$

....

$\int_{x_0+(n-3)h}^{x_0+nh} f(x)\,dx = \dfrac{3h}{8}[y_{n-3} + 3y_{n-2} + 3y_{n-1} + y_n]$

Adding the above integrals, we get

$\int_{x_0}^{x_0+nh} f(x)\,dx = \dfrac{3h}{8}\begin{bmatrix}(y_0 + 3y_1 + 3y_2 + y_3) + (y_3 + 3y_4 + 3y_5 + y_6) + \ldots \\ \ldots\ldots + (y_{n-3} + 3y_{n-2} + 3y_{n-1} + y_n)\end{bmatrix}$

$\int_{x_0}^{x_0+nh} f(x)\,dx = \dfrac{3h}{8}\begin{bmatrix}(y_0 + y_n) + 3(y_1 + y_2 + y_4 + y_5 + \ldots + y_{n-2} + y_{n-1}) \\ + 2(y_3 + y_6 + y_9 + \ldots + y_{n-3})\end{bmatrix}$

This is known as **Simpson's 3/8th Rule**. This rule is applicable only when n is a multiple of 3.

Note: In Simpson's 3/8th rule, the polynomial is of degree three.

5.10 BOOLE'S RULE (FOR $n = 4$)

Putting $n = 4$ in the Newton Cote's formula, we get Boole's rule. Here all the differences of order greater than four are zero.

Now,

$\int_{x_0}^{x_0+4h} f(x)\,dx = 4h\left[y_0 + 2\Delta y_0 + \dfrac{5}{3}\Delta^2 y_0 + \dfrac{2}{3}\Delta^3 y_0 + \dfrac{7}{90}\Delta^4 y_0\right]$

$= \dfrac{2h}{45}[7y_0 + 32y_1 + 12y_2 + 32y_3 + 7y_4]$

$\int_{x_0}^{x_0+4h} f(x)\,dx = \dfrac{2h}{45}[7y_0 + 32y_1 + 12y_2 + 32y_3 + 7y_4]$

Similarly, $\int_{x_0+4h}^{x_0+8h} f(x)\,dx = \dfrac{2h}{45}[7y_4 + 32y_5 + 12y_6 + 32y_7 + 7y_8]$

....

$\int_{x_0+(n-4)h}^{x_0+nh} f(x)\,dx = \dfrac{2h}{45}[7y_{n-4} + 32y_{n-3} + 12y_{n-2} + 32y_{n-1} + 7y_n]$

Adding the above integrals, we get

$$\int_{x_0}^{x_0+nh} f(x)\,dx = \frac{2h}{45}\left[\begin{array}{l}7(y_0+y_n)+32(y_1+y_3+y_5+\ldots)\\ +12(y_2+y_6+y_{10}+\ldots)+14(y_4+y_8+y_{12}+\ldots)\end{array}\right]$$

This is known as **Boole's Rule**. This rule is applicable only when the number of sub intervals, n is taken as a multiple of 4.

Note: In Boole's rule, the polynomial is of degree four.

5.11 WEDDLE'S RULE (FOR $n = 6$)

Putting $n = 6$ in the Newton Cote's formula and neglecting all the differences of order greater than six, we get Weddle's rule.

Now,

$$\int_{x_0}^{x_0+6h} f(x)\,dx = 6h\left[y_0+3\Delta y_0+\frac{9}{2}\Delta^2 y_0+4\Delta^3 y_0+\frac{123}{60}\Delta^4 y_0+\frac{11}{20}\Delta^5 y_0+\frac{41}{840}\Delta^6 y_0\right]$$

$$= \frac{3h}{10}[y_0+5y_1+y_2+6y_3+y_4+5y_5+y_6]$$

Similarly, $\displaystyle\int_{x_0+6h}^{x_0+12h} f(x)\,dx = \frac{3h}{10}[y_6+5y_7+y_8+6y_9+y_{10}+5y_{11}+y_{12}]$

....

$$\int_{x_0+(n-6)h}^{x_0+nh} f(x)\,dx = \frac{3h}{10}[y_{n-6}+5y_{n-5}+y_{n-4}+6y_{n-3}+y_{n-2}+5y_{n-1}+y_n]$$

Adding the above integrals, we get

$$\int_{x_0}^{x_0+nh} f(x)\,dx = \frac{3h}{10}\left[\begin{array}{l}(y_0+5y_1+y_2+6y_3+y_4+5y_5)\\ +(2y_6+5y_7+y_8+6y_9+y_{10}+5y_{11})\\ +\ldots+(2y_{n-6}+5y_{n-5}+y_{n-4}+6y_{n-3}+y_{n-2}+5y_{n-1}+y_n)\end{array}\right]$$

This is known as **Weddle's Rule**. In this rule n must be a multiple of 6.

Note.1. In the above formula, the coefficients may be remembered in groups as

    First group - coefficients : 1, 5, 1, 6, 1, 5

    All interior groups - coefficients : 2, 5, 1, 6, 1, 5

    Last group - coefficients : 2, 5, 1, 6, 1, 5, 1

2. If there are only seven ordinates, the coefficients are 1, 5, 1, 6, 1, 5, 1
3. In Weddle's rule, the polynomial is of degree six.
4. In Weddle's rule, a minimum of 7 ordinates are necessary.

ILLUSTRATIVE EXAMPLES

Example 1. Evaluate $\int_0^6 \dfrac{1}{1+x^2}\,dx$ by (i) Trapezoidal rule (ii) Simpson's $1/3^{rd}$ rule (iii) Simpson's $3/8^{th}$ rule (iv) Weddle's rule. Also check the result by actual integration.

Solution: We take $n = 6$, $h = \dfrac{b-a}{n} = \dfrac{6-0}{6} = 1$, which is a multiple of 1,2,3 and 6 and hence all the above rules will be applied.

∴ The above interval will be divided into six equal parts with $h = 1$, given as

x	0	1	2	3	4	5	6
$f(x) = \dfrac{1}{1+x^2}$	1.00	0.500	0.200	0.100	0.0588	0.0385	0.027
	y_0	y_1	y_2	y_3	y_4	y_5	y_6

(i) Trapezoidal rule:

$$\int_{x_0}^{x_0+nh} f(x)\,dx = \frac{h}{2}\left[(y_0+y_n) + 2(y_1+y_2+\ldots+y_{n-1})\right]$$

$$\int_0^6 \frac{1}{1+x^2}\,dx = \frac{1}{2}\left[(1.00+0.027)+2(0.5+0.2+0.1+0.0588+0.0385)\right]$$

$$= \frac{1}{2}[1.027+2\times 0.8973] = \frac{1}{2}\times 2.8216 = 1.4108$$

(ii) Simpson's $1/3^{rd}$ rule:

$$\int_{x_0}^{x_0+nh} f(x)\,dx = \frac{h}{3}\left[\begin{array}{l}(y_0+y_n)+4(y_1+y_3+\ldots+y_{n-1})\\ +2(y_2+y_4+\ldots+y_{n-2})\end{array}\right]$$

$$\int_0^6 \frac{1}{1+x^2}\,dx = \frac{1}{3}\left[\begin{array}{l}(1.00+0.027)+4(0.5+0.1+0.0385)\\ +2(0.2+0.0588)\end{array}\right]$$

$$= \frac{1}{3}[1.027+2.554+0.5176] = \frac{1}{3}\times 4.0986 = 1.3662.$$

(iii) Simpson's $3/8^{th}$ rule:

$$\int_{x_0}^{x_0+nh} f(x)\,dx = \frac{3h}{8}\left[\begin{array}{l}(y_0+y_n)+3(y_1+y_2+y_4+y_5+\ldots+y_{n-2}+y_{n-1})\\ +2(y_3+y_6+3y_9+\ldots+y_{n-3n})\end{array}\right]$$

$$\int_0^6 \frac{1}{1+x^2} dx = \frac{3}{8}\left[\begin{array}{c}(1.00+0.027)+3(0.5+0.2+0.0588+0.0385)\\+2(0.1)\end{array}\right]$$

$$= \frac{3}{8}[1.027+2.3919+0.2] = \frac{3}{8} \times 3.6189 = 1.35708$$

(iv) Weddle's rule:

$$\int_{x_0}^{x_0+nh} f(x)\,dx = \frac{3h}{10}[y_0+5y_1+y_2+6y_3+y_4+5y_5+y_6]$$

$$\int_0^6 \frac{1}{1+x^2} dx = \frac{3}{10}[1.00+5(0.5)+0.2+6(0.1)+0.0588+5(0.0385)+0.027]$$

$$= \frac{3}{10}[1+2.5+0.2+0.6+0.0588+0.1925+0.027]$$

$$= \frac{3}{10} \times 4.5783 = 1.37349$$

(v) Actual integration:

$$\int_0^6 \frac{1}{1+x^2} dx = \left[\tan^{-1} x\right]_0^6 = \tan^{-1}(6) - \tan^{-1}(0) = 1.405647$$

Note: Here, the value by trapezoidal rule is most accurate.

Example 2. Calculate the value of $\int_1^2 \frac{1}{x} dx$ by Trapezoidal rule and Simpson's rule, taking $h = \frac{1}{4}$.

Solution: Given, $h = \frac{1}{4} \Rightarrow n = \frac{b-a}{h} = \frac{2-1}{1/4} = 4$, which is multiple of 1 and 2 and hence Trapezoidal rule and Simpson's 1/3rd rule will be applied.

The required table is:

x	1	5/4	3/2	7/4	2
$f(x) = \frac{1}{x}$	1.00	0.800	0.6666	0.5714	0.5
	y_0	y_1	y_2	y_3	y_4

(i) Trapezoidal rule:

$$\int_{x_0}^{x_0+nh} f(x)\,dx = \frac{h}{2}\left[\begin{array}{l}(\text{sum of the first and last ordinate})\\+2(\text{sum of the remaining ordinates})\end{array}\right]$$

$$\Rightarrow \int_1^2 \frac{1}{x} dx = \frac{1}{4\times 2}\Big[(1.00+0.5)+2(0.800+0.6666+0.5714)\Big]$$

$$= \frac{1}{8}[1.5+2\times 2.038] = \frac{1}{8}\times 5.576 = 0.697$$

(*ii*) **Simpson's 1/3rd rule:**

$$\int_{x_0}^{x_0+nh} f(x)dx = \frac{h}{3}\begin{bmatrix}(\text{sum of the first and last ordinate})\\ +4(\text{sum of remaining even ordinates})\\ +2(\text{sum of remaining odd ordinates})\end{bmatrix}$$

$$\int_1^2 \frac{1}{1+x^2} dx = \frac{1}{4\times 3}\Big[(1.00+0.5)+4(0.800+0.5714)+2(0.6666)\Big]$$

$$= \frac{1}{12}[1.5+5.4856+1.3332] = 0.6932.$$

Example 3. Compute the value of $\int_0^\pi \sin x\, dx$ by Trapezoidal rule and Simpson's rule dividing the range into ten equal parts.

Solution: Given, $n = 10 \Rightarrow h = \frac{b-a}{n} = \frac{\pi-0}{10} = \frac{\pi}{10}$. Here Trapezoidal and Simpson's $1/3^{\text{rd}}$ rules will be applied.

The values of $f(x) = \sin x$ for different values of x are:

x	0	$\pi/10$	$2\pi/10$	$3\pi/10$	$4\pi/10$	$5\pi/10$
$f(x) = \sin x$	0.0	0.3090	0.5878	0.8090	0.9511	1.0000

$6\pi/10$	$7\pi/10$	$8\pi/10$	$9\pi/10$	π
0.9511	0.8090	0.5878	0.3090	0.0

(i) **Trapezoidal rule:**

$$\int_0^\pi \sin x\, dx = \frac{\pi}{10\times 2}\begin{bmatrix}(0+0)+2(0.3090+0.5878+0.8090+0.9511+1.0\\ +0.9511+0.8090+0.5878+0.3090)\end{bmatrix}$$

$= 1.9843$ approx.

(ii) **Simpson's 1/3rd rule:**

$$\int_0^\pi \sin x\, dx = \frac{\pi}{10\times 3}\begin{bmatrix}(0+0)+4(0.3090+0.8090+1.0+0.8090+0.3090)\\ +2(0.5878+0.9511+0.9511+0.5878)\end{bmatrix}$$

$= 2.00091$ approx.

Example 4. Evaluate $\int_{4}^{5.2} \log x \, dx$ by using Weddle's rule.

Solution: For Weddle's rule take $n = 6 \Rightarrow h = \dfrac{5.2 - 4}{6} = 0.2$ and the values of $f(x) = \log x$ for different values of x are:

x	4.0	4.2	4.4	4.6	4.8	5.0	5.2
$f(x) = \log x$	1.3862	1.4350	1.4816	1.5261	1.5686	1.6094	1.6486

By Weddle's rule:

$$\int_{x_0}^{x_0+nh} f(x)\,dx = \frac{3h}{10}\left[y_0 + 5y_1 + y_2 + 6y_3 + y_4 + 5y_5 + y_6\right]$$

$$\Rightarrow \int_{4}^{5.2} \log x \, dx = \frac{3 \times 0.2}{10}\begin{bmatrix} 1.3862 + 5(1.4350) + 1.4816 + 6(1.5261) \\ + 1.5686 + 5(1.6094) + 1.6486 \end{bmatrix}$$

$$= 0.06\left[6.085 + 7.175 + 9.1566 + 8.047\right]$$

$$= 0.06 \times 30.4636 = 1.82782$$

Example 5. Evaluate $\int_{0}^{4} \dfrac{dx}{4x + 5}$ by using Boole's rule.

Solution: For Boole's rule take $n = 4$, $h = 1$ and the values of $f(x) = \dfrac{1}{4x + 5}$ for different values of x are:

x	0	1	2	3	4
$f(x) = \dfrac{1}{4x+5}$	1/5	1/9	1/13	1/17	1/21

By Boole's rule:

$$\int_{x_0}^{x_0+nh} f(x)\,dx = \frac{2h}{45}\begin{bmatrix} 7(y_0 + y_n) + 32(y_1 + y_3 + y_5 + \ldots) \\ + 12(y_2 + y_6 + y_{10} + \ldots) + 14(y_4 + y_8 + y_{12} + \ldots) \end{bmatrix}$$

$$\Rightarrow \int_{0}^{4} \frac{1}{4x+5} = \frac{2 \times 1}{45}\left[7\left(\frac{1}{5}\right) + 32\left(\frac{1}{9} + \frac{1}{17}\right) + 12\left(\frac{1}{13}\right) + 14\left(\frac{1}{21}\right)\right]$$

$$= \frac{2}{45}\left[1.4 + 5.43786 + 0.92307 + 0.66666\right] = 0.374559.$$

Example 6. Evaluate $\int_{1}^{10} x^2 \, dx$ by taking 10 ordinates.

Solution: For 10 ordinates, and $h = 1$ hence Simpson's 3/8th rule is applicable and the values of $f(x) = x^2$ for different values of x are as follows:

x	1	2	3	4	5	6	7	8	9	10
$f(x) = x^2$	1	4	9	16	25	36	49	64	81	100

By Simpson's 3/8th rule:

$$\int_{x_0}^{x_0+nh} f(x)dx = \frac{3h}{8}\left[(y_0+y_n)+3(y_1+y_2+y_4+y_5+\ldots+y_{n-2}+y_{n-1}) + 2(y_3+y_6+y_9+\ldots+y_{n-3})\right]$$

$$\int_1^{10} x^2 dx = \frac{3 \times 1}{8}\left[(1+100)+3(4+9+25+36+64+81)+2(16+49)\right]$$

$$= \frac{3}{8}[101+657+130] = \frac{3}{8} \times 888 = 333.$$

Example 7. A curve passes through the points (1,3), (2,5), (3,7), (4,9), (5,11), (6,13) and (7,15). Find the area bounded by the curve, the x-axis and the lines $x = 1$ and $x = 4$. Also find the volume of solid of revolution got by revolving this area about the x-axis.

Solution: Given:

x	1	2	3	4	5	6	7
y	3	5	7	9	11	13	15

Since, there are seven ordinates, i.e. six intervals; we can use Simpson's 1/3rd rule.

$$\therefore \quad \text{Area} = \int_1^7 y\,dx \; ; \; h = 1$$

$$= \frac{1}{3}\left[(3+15)+4(5+9+13)+2(7+11)\right]$$

$$= \frac{1}{3}[18+108+36] = \frac{1}{3} \times 162 = 54 \text{ square units.}$$

x	1	2	3	4	5	6	7
y^2	9	25	49	81	121	169	225

$$\therefore \text{Volume} = \pi \int_1^7 y^2 dx$$

$$= \frac{\pi}{3}\left[(9+225)+4(25+81+169)+2(49+121)\right]$$

$$= \frac{3.14}{3}[234+1100+340] = \frac{3.14}{3} \times 1674 = 1752.12 \text{ cubic units.}$$

Example 8. A river is 80 meters wide. The depth 'd' in meters at a distance x meters from one bank is given by the following table. Find the area of cross-section of the river using Simpson's rule:

x	0	10	20	30	40	50	60	70	80
y	0	4	7	9	12	15	14	8	3

Solution: Here, $h = 10$ and $n = 8$; we use Simpson's $1/3^{rd}$ rule.

$$\therefore \text{Area of cross-section} = \int_0^{80} y\, dx$$

$$\therefore A = \frac{10}{3}\left[(0+3) + 2(7+12+14) + 4(4+9+15+8)\right]$$

$$A = \frac{10}{3}[3 + 66 + 144] = \frac{10}{3} \times 213 = 710 \text{ square meters.}$$

Example 9. A rocket is launched from the ground. Its acceleration is registered during the first 60 seconds and is given in the following table. Find the velocity of the rocket at $t = 60$ seconds by Simpson's 3/8th rule.

t(sec.)	0	10	20	30	40	50	60
a(cm/sec²)	29	30.85	34.17	36.67	41.23	42.55	44.05

Solution: Here, $h = 10$ and $n = 6$; we can use Simpson's $3/8^{th}$ rule.

Now, since acceleration; $a = \dfrac{dv}{dt} \Rightarrow v = \int_0^{60} a\, dt$

$$\therefore \text{velocity} \quad v = \int_0^{60} a\, dt$$

$$= \frac{3 \times 10}{8}\left[\begin{array}{l}(29+44.05) + 3(30.85+34.17+41.23+42.55) \\ +2(36.67)\end{array}\right]$$

$$= \frac{15}{4}[73.05 + 3 \times 148.80 + 73.34]$$

$$= \frac{15}{4} \times 592.79 = 2222.9625\, cm/\sec.$$

$$= 22.23\, m/\sec.$$

Example 10. The velocity (v km/min.) of a scooter, starting from rest, is given at fixed interval of time (t min.) as follows:

t	2	4	6	8	10	12	14	16	18	20
v	10	18	25	29	31	20	11	5	2	0

Calculate the approximate distance covered in 20 minutes.

Solution: Here, $h = 2$ and $n = 9$; we can use Trapezoidal and Simpson's $3/8^{th}$ rules.

Now, since velocity; $v = \dfrac{ds}{dt} \Rightarrow s = \int_0^{20} v\, dt$

∴ Distance by Trapezoidal rule,

$$s = \int_0^{20} v\, dt = \dfrac{2}{2}\left[(10+0) + 2(18+25+29+31+20+11+5+2)\right]$$

$= 10 + 282 = 283\, km.$

Distance by Simpson's $3/8^{th}$ rule,

$$s = \int_0^{20} v\, dt = \dfrac{3 \times 2}{8}\left[\begin{array}{l}(10+0) + 3(18+25+31+20+5+2) \\ + 2(29+11)\end{array}\right]$$

$= \dfrac{3}{4}[10 + 303 + 80] = \dfrac{3}{4} \times 393 = 294.75\, km.$

EXERCISE 5.2

1. Evaluate $\int_0^1 \dfrac{dx}{1+x^2}$ by using

   (i) Simpson's $1/3^{rd}$ rule taking $h = \dfrac{1}{4}$

   (ii) Simpson's $3/8^{th}$ rule taking $h = \dfrac{1}{6}$

   (iii) Weddle's rule taking $h = \dfrac{1}{6}$.

   (iv) Also check the result by actual integration.

2. Evaluate $\int_0^2 \dfrac{dx}{x^2 + x + 1}$ to three decimal places, dividing the range into 8 equal parts using Simpson's rule.

3. Find the area bounded by the curve and the x-axis from $x = 7.47$ to $x = 7.52$, for the following table:

x	0	1	2	3	4	5	6
$y = f(x)$	1.00 y_0	0.500 y_1	0.200 y_2	0.100 y_3	0.0588 y_4	0.0385 y_5	0.027 y_6

by using Trapezoidal rule.

4. Evaluate $\int_0^2 e^x dx$ taking 6 intervals.

5. Evaluate $\int_0^1 \dfrac{1}{1+x} dx$ by (i) Trapezoidal rule (ii) Simpson's 1/3rd rule (iii) Simpson's 3/8th rule (iv) Weddle's rule.

6. A curve passes through the points (1,0.2), (2,0.7), (3,1), (4,1.3), (5,1.5), (6,1.7), (7,1.9), (8,2.1) and (9,2.3). Find the volume of the solid of revolution got by revolving the area between the curve, the x-axis and $x = 1, x = 9$ about the x-axis.

7. The velocity 'v' of a car, running on a straight road at interval of time $t = 2$ minutes are given below

t(min).	0	2	4	6	8	10	12
v(km/hr)	0	22	33	27	18	7	0

Calculate the approximate distance covered by the car in 12 minutes. (Use Simpson's rule).

8. The velocity 'v' of a train at a distance 'S' from a point on its track is given by the table below:

S(meter)	0	10	20	30	40	50	60
v(m/sec.)	47	58	64	65	61	52	38

Estimate the time taken to travel 60 meters by using Simpson's 1/3rd rule, Simpson's 3/8th rule and Weddle's rule.

Hint. $v = \dfrac{ds}{dt}$, $t = \int_0^{60} \dfrac{1}{v} ds$; take $y = \dfrac{1}{v}$.

9. Using Boole's rule, find the value of the integral $\int_{0.4}^{1.6} \dfrac{x}{\sinh x} dx$ (take $n = 12$).

10. Evaluate $\int_{\pi/4}^{\pi/2} \dfrac{\sin x}{x} dx$ using Simpson's 1/3rd rule.

11. The velocity 'v' of a particle moving in a straight line covers a distance 'x' in time 't' are related as follows:

x	0	10	20	30	40
v	45	60	65	54	42

Find the time taken to travel the distance of 40 units.

12. Evaluate $\int_0^1 \cos x \, dx$, taking $h = 0.2$.

ANSWERS

1. (i) 0.7854 (ii) 0.7854 (iii) 0.7854
   (iv) 0.785
2. 0.815
3. 0.09965
4. 6.4481, 6.3894
5. (i) 0.6949 (ii) 0.6931 (iii) 0.6932
   (iv) 0.6932
6. 59.68
7. 3.5555 km.
8. 1.06351, 1.06406, 1.06242
9. 1.01078
10. 0.6118
11. 0.725 units of time
12. 0.83865

5.12 ROMBERG'S METHOD

In trapezoidal rule, the error for an interval of size h is

$$= -\frac{(b-a)h^2}{12} y''(\xi), \quad a < \xi < b$$

$$= kh^2$$

where $k = -\frac{(b-a)}{12} y''(\xi)$ is a constant.

Suppose we want to evaluate $I = \int_{x_0}^{x_n} y\,dx$ by the trapezoidal rule with two different values of h say h_1 and h_2, then

$$I = I_1 + E_1 = I_1 + kh_1^2 \qquad \ldots(1)$$

$$I = I_2 + E_2 = I_2 + kh_2^2 \qquad \ldots(2)$$

where I_1 and I_2 are the approximations with errors E_1 and E_2 respectively.

$\therefore \quad I_1 + kh_1^2 = I_2 + kh_2^2$

$\Rightarrow \quad k = \dfrac{I_1 - I_2}{h_2^2 - h_1^2}$

From equation (1), we have

$$I = I_1 + \left(\frac{I_1 - I_2}{h_2^2 - h_1^2}\right) h_1^2 = \frac{I_1 h_1^2 - I_2 h_1^2}{h_2^2 - h_1^2}$$

This I gives a better approximations than I_1 or I_2.

If, we take $h_1 = h$ and $h_2 = \dfrac{h}{2}$, then we get

$$I = \frac{I_1\left(\frac{h^2}{4}\right) - I_2 h^2}{\frac{h^2}{4} - h^2} = \frac{4I_2 - I_1}{3}$$

$$I = I_2 + \frac{1}{3}(I_2 - I_1) \qquad \ldots(3)$$

we obtained this result by applying Trapezoidal rule twice. By applying the rule several times, every time halving h, we get a sequence of results $A_1, A_2, A_3, A_4 \ldots$ in which the error is reduced by $\frac{1}{4}$ every time.

Again, apply the formula (3) to each pair of A's i.e., A_1, A_2; A_2, A_3; etc., to get improved result say $B_1, B_2, B_3, B_4 \ldots$ Now again, applying the formula (3) to the pairs B_1, B_2; B_2, B_3; etc, we get another sequence of better results $C_1, C_2, C_3, C_4 \ldots$ Continuing this process until two successive values are very close to each other. This process is called Romberg's method or Romberg's integration.

ILLUSTRATIVE EXAMPLE

Example. Evaluate $\int_0^1 \frac{dx}{1+x^2}$ using Romberg's method and hence obtain an approximate value of π.

Solution: By taking $h = 0.5, 0.25, 0.125$ respectively, let us evaluate the given integral using trapezoidal rule.

(i) when h = 0.5, the values of $y = \frac{1}{1+x^2}$ are as follows:

x :	0	0.5	1
y :	1	0.8	0.5

$$\therefore \quad I = \frac{h}{2}\left[\begin{array}{l}(\text{sum of the first and last ordinate})\\ +2(\text{sum of the remaining ordinates})\end{array}\right]$$

$$I = \frac{0.5}{2}\bigl[(1+0.5)+2(0.8)\bigr] = 0.775$$

(ii) when $h = 0.25$, we have

x :	0	0.25	0.5	0.75	1
y :	1	0.9412	0.8	0.64	0.5

$$\therefore \quad I = \frac{0.25}{2}\bigl[(1+0.5)+2(0.9412+0.8+0.64)\bigr] = 0.78280$$

(iii) when h = 0.125, we have

x:	0	0.125	0.25	0.375	0.5	0.625	0.75	0.875	1
y:	1	0.9846	0.9412	0.8767	0.8	0.7191	0.64	0.5664	0.5

$$\therefore \quad I = \frac{0.125}{2}\big[(1+0.5) + 2(0.9846 + 0.9412 + 0.8767 + 0.8$$
$$+ 0.7191 + 0.64 + 0.5664)\big]$$
$$= 0.784750$$

Hence the different values obtained by trapezoidal rule for various $h's$ are
$$I_1 = 0.775,\ I_2 = 0.78280,\ I_3 = 0.784750.$$

Applying the formula $I = I_2 + \frac{1}{3}(I_2 - I_1)$, we will get two improved values, say

$$I_1^* = 0.78280 + \frac{1}{3}(0.78280 - 0.77500) = 0.7854$$

and $\quad I_2^* = 0.78475 + \frac{1}{3}(0.78475 - 0.78280) = 0.7854$

Since these two values obtained above are the same, hence $I = 0.7854$.

By actual integration, we have

$$\int_0^1 \frac{dx}{1+x^2} = \big[\tan^{-1} x\big]_0^1 = \tan^{-1} 1 - \tan^{-1} 0 = \frac{\pi}{4} - 0 = \frac{\pi}{4}$$

$$\Rightarrow \qquad \frac{\pi}{4} \cong 0.7854$$
$$\Rightarrow \qquad \pi \cong 3.1416$$

5.13 EULER-MACLAURIN'S FORMULA

Let $f(x)$ be a function such that

$$\Delta F(x) = f(x) \qquad \ldots(1)$$

i.e., $\quad F(x) = \Delta^{-1} f(x)$

Let $x_0, x_1, x_2, \ldots, x_n$ are the equidistant values of x with width h.
Then from (1), we have

$$\Delta F(x_0) = f(x_0) \text{ i.e., } F(x+h) - F(x) = f(x_0)$$
$$F(x_1) - F(x_0) = f(x_0)$$
$$F(x_2) - F(x_1) = f(x_1)$$
$$\ldots \qquad \ldots \qquad \ldots$$
$$\ldots \qquad \ldots \qquad \ldots$$
$$F(x_n) - F(x_{n-1}) = f(x_{n-1})$$

236 *Numerical and Statistical Techniques*

Adding all these equations, we have

$$F(x_n) - F(x_0) = \sum_{i=0}^{n-1} f(x_i) \qquad \ldots(2)$$

Also, we have $F(x) = \Delta^{-1} f(x) = (E-1)^{-1} f(x) \quad [\because E - 1 = \Delta]$

$$= \left(e^{hD} - 1\right)^{-1} f(x) \quad [\because E = e^{hD}]$$

Expanding by Binomial theorem

$$F(x) = \left[\left(1 + \frac{hD}{1!} + \frac{h^2 D^2}{2!} + \frac{h^3 D^3}{3!} + \ldots\right) - 1\right]^{-1} f(x)$$

Where $D = \dfrac{d}{dx}$

$$= \left\{\frac{hD}{1!} + \frac{h^2 D^2}{2!} + \frac{h^3 D^3}{3!} + \ldots\right\}^{-1} f(x)$$

$$= (hD)^{-1} \left\{1 + \left(\frac{hD}{2!} + \frac{h^2 D^2}{3!} + \ldots\right)\right\}^{-1} f(x)$$

$$= \frac{1}{h} D^{-1} \left\{1 - \left(\frac{hD}{2!} + \frac{h^2 D^2}{3!} + \ldots\right) + \left(\frac{hD}{2!} + \frac{h^2 D^2}{3!} + \ldots\right)^2 + \ldots\right\} f(x)$$

$$= \frac{1}{h} D^{-1} \left\{1 - \frac{hD}{2!} + \frac{h^2 D^2}{12} - \frac{h^4 D^4}{720} + \ldots\right\} f(x)$$

$$= \frac{1}{h} \left\{D^{-1} - \frac{h}{2!} + \frac{h^2 D}{12} - \frac{h^4 D^3}{720} + \ldots\right\} f(x)$$

$$= \frac{1}{h} \left\{D^{-1} f(x) - \frac{h}{2!} f(x) + \frac{h^2}{12} Df(x) - \frac{h^4}{720} D^3 f(x) + \ldots\right\}$$

$$\Rightarrow F(x) = \frac{1}{h} \int f(x) dx - \frac{1}{2!} f(x) + \frac{h}{12} f'(x) - \frac{h^3}{720} f'''(x) + \ldots \qquad \ldots(3)$$

Putting $x = x_n$ and $x = x_0$ in equation (2) and then subtracting, we have

$$F(x_n) - F(x_0) = \frac{1}{h} \int_{x_0}^{x_n} f(x) dx - \frac{1}{2} [f(x_n) - f(x_0)]$$

$$+ \frac{h}{12} [f'(x_n) - f'(x_0)] - \frac{h^3}{720} [f'''(x_n) - f'''(x_0)] + \ldots$$

From (2), we have

$$\sum_{i=0}^{n-1} f(x_i) = \frac{1}{h}\int_{x_0}^{x_n} f(x)dx - \frac{1}{2}[f(x_n)-f(x_0)]$$
$$+ \frac{h}{12}[f'(x_n)-f'(x_0)] - \frac{h^3}{720}[f'''(x_n)-f'''(x_0)] + \ldots$$

$$\frac{1}{h}\int_{x_0}^{x_n} f(x)dx = \sum_{i=0}^{n-1} f(x_i) + \frac{1}{2}[f(x_n)-f(x_0)]$$
$$- \frac{h}{12}[f'(x_n)-f'(x_0)] + \frac{h^3}{720}[f'''(x_n)-f'''(x_0)] + \ldots$$

$$\Rightarrow \int_{x_0}^{x_n} ydx = h\sum_{i=0}^{n-1} y_i + \frac{h}{2}[y_n - y_0] - \frac{h^2}{12}[y_n' - y_0'] + \frac{h^4}{720}[y_n''' - y_0'''] + \ldots$$

$$\Rightarrow \int_{x_0}^{x_n} ydx = h[y_0 + y_1 + y_2 + \ldots + y_{n-1}] + \frac{h}{2}[y_n - y_0]$$
$$- \frac{h^2}{12}[y_n' - y_0'] + \frac{h^4}{720}[y_n''' - y_0'''] + \ldots$$

which gives

$$\int_{x_0}^{x_n} ydx = \frac{h}{2}[y_0 + 2y_1 + 2y_2 + \ldots + 2y_{n-1} + y_n]$$
$$- \frac{h^2}{12}[y_n' - y_0'] + \frac{h^4}{720}[y_n''' - y_0'''] + \ldots$$

This is called **Euler-Maclaurin's Formula**.

Note. This formula is often used to find the sum of a series of the form

$$y(x_0) + y(x_0 + h) + y(x_0 + 2h) + \ldots + y(x_0 + nh).$$

---- ILLUSTRATIVE EXAMPLES ----

Example 1. Evaluate $\int_{1}^{5} \frac{dx}{x}$ using Euler-Maclaurin's Formula.

Solution: Here $y = \frac{1}{x}$ and $x_0 = 1$, $x_n = 5$.

Taking $n = 8$, $h = 0.5$

Also, $y' = -\dfrac{1}{x^2}$, $y'' = \dfrac{2}{x^3}$, $y''' = -\dfrac{6}{x^4}$

From Euler-Maclaurin's Formula

$$\int_{x_0}^{x_n} y\,dx = \dfrac{h}{2}[y_0 + 2y_1 + 2y_2 + \ldots + 2y_{n-1} + y_n]$$

$$-\dfrac{h^2}{12}[y'_n - y'_0] + \dfrac{h^4}{720}[y'''_n - y'''_0] + \ldots$$

$$\int_1^5 \dfrac{1}{x}dx = \dfrac{0.5}{2}\left[\dfrac{1}{1} + \dfrac{2}{1.5} + \dfrac{2}{2.0} + \dfrac{2}{2.5} + \dfrac{2}{3.0} + \dfrac{2}{3.5} + \dfrac{2}{4.0} + \dfrac{2}{4.5} + \dfrac{1}{5.0}\right]$$

$$-\dfrac{(0.5)^2}{12}\left[\dfrac{-1}{5^2} - \left(\dfrac{-1}{1^2}\right)\right] + \dfrac{(0.5)^4}{720}\left[\dfrac{-6}{5^4} - \left(\dfrac{-6}{1^4}\right)\right] + \ldots$$

$$= \dfrac{0.5}{2}[1 + 2(0.66667 + 0.5 + 0.4 + 0.33333 + 0.28571 + 0.25 + 0.22222) + 0.2]$$

$$-\dfrac{0.25}{12}[-0.004 + 1] + \dfrac{0.0625}{720}[-0.0096 + 6] + \ldots$$

$$= 1.48611 - 0.02075 + 0.00052 + \ldots$$

$$= 1.46484$$

Example 2. Using Euler-Maclaurin's Formula find the value $\int_0^{10} \dfrac{1}{1+x}dx$.

Solution: Here $y = \dfrac{1}{1+x}$ and $x_0 = 0$, $x_n = 10$.

Taking $n = 10$, $h = 1$

Also, $y' = -\dfrac{1}{(1+x)^2}$, $y'' = \dfrac{2}{(1+x)^3}$, $y''' = -\dfrac{6}{(1+x)^4}$

From Euler-Maclaurin's Formula

$$\int_{x_0}^{x_n} y\,dx = \dfrac{h}{2}[y_0 + 2y_1 + 2y_2 + \ldots + 2y_{n-1} + y_n]$$

$$-\dfrac{h^2}{12}[y'_n - y'_0] + \dfrac{h^4}{720}[y'''_n - y'''_0] + \ldots$$

$$\int_0^{10} \frac{1}{x} dx = \frac{1}{2}\left[\frac{1}{1} + \frac{2}{2} + \frac{2}{3} + \frac{2}{4} + \frac{2}{5} + \frac{2}{6} + \frac{2}{7} + \frac{2}{8} + \frac{2}{9} + \frac{2}{10} + \frac{1}{11}\right]$$

$$-\frac{(1)^2}{12}\left[\frac{-1}{11^2} - \left(\frac{-1}{1^2}\right)\right] + \frac{(1)^4}{720}\left[\frac{-6}{11^4} - \left(\frac{-6}{1^4}\right)\right] + \ldots$$

$$= \frac{1}{2}[1 + 1 + 0.66667 + 0.5 + 0.4 + 0.33333 + 0.28571 + 0.25 + 0.22222$$

$$+ 0.2 + 0.09091] - \frac{1}{12}[-0.008264 + 1] + \frac{1}{720}[-0.00041 + 6] + \ldots$$

$$= 2.47442 - 0.08265 + 0.008333$$

$$= 2.4001$$

Example 3. Find the sum of the series

$$\frac{1}{(201)^2} + \frac{1}{(203)^2} + \frac{1}{(205)^2} + \ldots\ldots + \frac{1}{(299)^2}$$

using Euler-Maclaurin's Formula.

Solution: Here $y = \frac{1}{x^2}$ and $x_0 = 201$, $x_n = 299$, $h = 2$, $n = 49$.

Also, $y' = -\frac{2}{x^3}$, $y'' = \frac{6}{x^4}$, $y''' = -\frac{24}{x^5}$

From Euler-Maclaurin's Formula

$$\int_{x_0}^{x_n} y\,dx = \frac{h}{2}[y_0 + 2y_1 + 2y_2 + \ldots + 2y_{n-1} + y_n]$$

$$-\frac{h^2}{12}[y_n' - y_0'] + \frac{h^4}{720}[y_n''' - y_0'''] + \ldots$$

$$\int_{201}^{200} \frac{1}{x^2} dx = \frac{2}{2}\left[\frac{1}{(201)^2} + \frac{2}{(203)^2} + \frac{2}{(205)^2} + \ldots + \frac{2}{(297)^2} + \frac{1}{(299)^2}\right]$$

$$-\frac{(2)^2}{12}\left[\frac{-2}{(299)^3} - \left(\frac{-2}{(201)^3}\right)\right] + \frac{(2)^4}{720}\left[\frac{-24}{(299)^5} - \left(\frac{-24}{(201)^5}\right)\right] + \ldots$$

$$\Rightarrow \left[\frac{1}{(201)^2} + \frac{2}{(203)^2} + \frac{2}{(205)^2} + \ldots + \frac{2}{(297)^2} + \frac{1}{(299)^2}\right]$$

$$= \int_{201}^{200} \frac{1}{x^2} dx + \frac{(2)^2}{12}\left[\frac{-2}{(299)^3} - \left(\frac{-2}{(201)^3}\right)\right] - \frac{(2)^4}{720}\left[\frac{-24}{(299)^5} - \left(\frac{-24}{(201)^5}\right)\right] + \ldots$$

or $\left[\dfrac{2}{(201)^2}+\dfrac{2}{(203)^2}+\dfrac{2}{(205)^2}+....+\dfrac{2}{(297)^2}+\dfrac{2}{(299)^2}\right]$

$$=\int_{201}^{200}\dfrac{1}{x^2}dx+\left(\dfrac{1}{(201)^2}+\dfrac{1}{(299)^2}\right)+\dfrac{(2)^2}{12}\left[\dfrac{-2}{(299)^3}-\left(\dfrac{-2}{(201)^3}\right)\right]$$

$$-\dfrac{(2)^4}{720}\left[\dfrac{-24}{(299)^5}-\left(\dfrac{-24}{(201)^5}\right)\right]+...$$

$$2\left[\dfrac{1}{(201)^2}+\dfrac{1}{(203)^2}+......+\dfrac{1}{(299)^2}\right]=\left(-\dfrac{1}{x}\right)_{201}^{299}+\left(\dfrac{1}{(201)^2}+\dfrac{1}{(299)^2}\right)$$

$$+\dfrac{2}{3}\left[\dfrac{-1}{(299)^3}+\dfrac{1}{(201)^3}\right]-\dfrac{8}{15}\left[\dfrac{-1}{(299)^5}+\dfrac{1}{(201)^5}\right]+...$$

$$2\left[\dfrac{1}{(201)^2}+\dfrac{1}{(203)^2}+......+\dfrac{1}{(299)^2}\right]=\left(-\dfrac{1}{299}+\dfrac{1}{201}\right)+\left(\dfrac{1}{(201)^2}+\dfrac{1}{(299)^2}\right)$$

$$+\dfrac{2}{3}\left[\dfrac{-1}{(299)^3}+\dfrac{1}{(201)^3}\right]-\dfrac{8}{15}\left[\dfrac{-1}{(299)^5}+\dfrac{1}{(201)^5}\right]+...$$

$$2\left[\dfrac{1}{(201)^2}+\dfrac{1}{(203)^2}+......+\dfrac{1}{(299)^2}\right]=$$

$$(0.001631)+(0.000036)+\dfrac{2}{3}[0.000000085]-\dfrac{8}{15}[0.000000....]+...$$

$$2\left[\dfrac{1}{(201)^2}+\dfrac{1}{(203)^2}+......+\dfrac{1}{(299)^2}\right]=$$

$$(0.001631)+(0.000036)+0.000000056-0.0000...+...$$

$$2\left[\dfrac{1}{(201)^2}+\dfrac{1}{(203)^2}+......+\dfrac{1}{(299)^2}\right]=0.001667$$

$$\therefore\quad \dfrac{1}{(201)^2}+\dfrac{1}{(203)^2}+......+\dfrac{1}{(299)^2}=0.000834$$

Example 4. Prove that: $\sum_{1}^{n}x^3=\left\{\dfrac{n(n+1)}{2}\right\}^2$

using Euler-Maclaurin's Formula.

Solution: Here $y = x^3$ and $x_0 = 1$, $x_n = n$, $h = 1$.

Also, $y' = 3x^2$, $y'' = 6x$, $y''' = 6$

From Euler-Maclaurin's formula, we have

$$\int_{x_0}^{x_n} y\,dx = \frac{h}{2}[y_0 + 2y_1 + 2y_2 + \ldots + 2y_{n-1} + y_n]$$

$$-\frac{h^2}{12}[y'_n - y'_0] + \frac{h^4}{720}[y'''_n - y'''_0] + \ldots$$

$$\int_1^n x^3\,dx = \frac{1}{2}\left[1^3 + 2\left(2^3 + 3^3 + \ldots + (n-1)^3\right) + n^3\right]$$

$$-\frac{(1)^2}{12}\left[3n^2 - 3(1)^2\right] + \frac{(1)^4}{720}[6 - 6]$$

$$\frac{1}{2}\left[1^3 + 2\left(2^3 + 3^3 + \ldots + (n-1)^3\right) + n^3\right] = \int_1^n x^3\,dx + \frac{1}{12}\left[3n^2 - 3\right]$$

$$\frac{2}{2}\left[1^3 + 2^3 + 3^3 + \ldots + (n-1)^3 + n^3\right] = \left(\frac{x^4}{4}\right)_1^n + \frac{1}{2}\left(1^3 + n^3\right) + \frac{1}{12}\left[3n^2 - 3\right]$$

$$\sum_1^n x^3 = \left(\frac{n^4}{4} - \frac{1}{4}\right) + \frac{1}{2}\left(n^3 + 1\right) + \frac{1}{4}\left[n^2 - 1\right]$$

$$\sum_1^n x^3 = \frac{n^4}{4} + \frac{n^3}{2} + \frac{n^2}{4}$$

$$\sum_1^n x^3 = \frac{n^4 + 2n^3 + n^2}{4} = \frac{n^2\left(n^2 + 2n + 1\right)}{4}$$

$$\sum_1^n x^3 = \frac{n^2(n+1)^2}{4} = \left\{\frac{n(n+1)}{2}\right\}^2$$

5.14 GAUSSIAN QUADRATURE FORMULA

The numerical integration formulas, discussed so far (i.e., the trapezoidal rule and Simpson's Rule) involve equally spaced argument values. Gauss gave a formula which uses the same number of argument values but with different spacing with more accuracy called Gaussian quadrature. In the evaluation of an integral on the interval a to b, it is not necessary to evaluate $f(x)$ at the endpoints, ie. at a or b of the interval.

The simplest form of Gaussian integration is based on the use of an optimally chosen polynomial to approximate the integrand $f(x)$ over the interval $[-1, +1]$. The simplest form uses a uniform weighting over the interval, and the particular points at which to evaluate $f(x)$ are the roots of a particular class of polynomials, the Legendre polynomials, over the interval. It can be shown that the best estimate of the integral is then:

$$\int_{-1}^{+1} f(x)\,dx = \sum_{i=1}^{n} w_i\, f(x_i)$$

where x_i and w_i are the abscissae and weights respectively. If the number of points at which the function $f(x)$ is evaluated is n, the resulting value of the integral is of the same accuracy as a simple polynomial method (such as Simpson's rule) of about twice the degree (i.e., of degree $2n$). Thus the carefully designed choice of function evaluation points in the Gauss-Legendre form results in the same accuracy for about half the number of function evaluations, and thus at about half the computing effort.

Gaussian quadrature formulae are evaluating using abscissae and weights from the following table. The choice of value of n is not always clear, and experimentation is useful to see the influence of choosing a different number of points. When choosing to use n points, we call the method an n-point Gaussian method.

n	Values of x_i	Weights w_i
2	$x_1 = -0.577350269$	$w_1 = 1$
	$x_2 = -x_1$	$w_2 = 1$
3	$x_1 = -0.77459667$	$w_1 = 0.55555555$
	$x_2 = 0.0$	$w_2 = 0.88888889$
	$x_3 = -x_1$	$w_3 = w_1$
4	$x_1 = -0.86113631$	$w_1 = 0.34785485$
	$x_2 = 0.33998104$	$w_2 = 0.65214515$
	$x_3 = -x_2$	$w_3 = w_2$
	$x_4 = -x_1$	$w_4 = w_1$
5	$x_1 = -0.90617985$	$w_1 = 0.23692689$
	$x_2 = 0.53846931$	$w_2 = 0.47862867$
	$x_3 = 0.0$	$w_3 = 0.56888889$
	$x_4 = -x_2$	$w_4 = w_2$
	$x_5 = -x_1$	$w_5 = w_1$

The Gauss-Legendre integration formula given here evaluates an estimate of the required integral on the interval for x of $[-1, +1]$. In most cases we will want to evaluate the integral on a more general interval, say $[a, b]$. We will use the variable x on this more general interval, and linearly map the interval $[a, b]$ for x onto the $[-1,+1]$ interval for t using the linear transformation:

$$x = \frac{(b+a)}{2} + \frac{(b-a)}{2}t \text{ then } dx = \frac{(b-a)}{2}dt$$

It is easily verified that substituting $t = -1$ gives $x = a$ and $t = 1$ gives $x = b$. We can now write the integral as:

$$I = \int_a^b f(x)\,dx = \frac{(b-a)}{2} \int_{-1}^{+1} f\left[\frac{(b+a)}{2} + \frac{(b-a)}{2}t\right]dt$$

The factor of $\frac{(b-a)}{2}$ in the second integral arises from the change of the variable of integration from x to t, which introduces the factor $\frac{dx}{dt}$. Finally, we can write the Gauss-Legendre estimate of the integral as:

$$I = \int_a^b f(x)\,dx = \frac{(b-a)}{2} \sum_{i=1}^{n} w_i f\left[\frac{(b+a)}{2} + \frac{(b-a)}{2}t_i\right]$$

ILLUSTRATIVE EXAMPLES

Example 1: Evaluate $\int_0^1 \frac{dx}{x^2+1}$ by using Gaussian quadrature 3 points formula.

Solution: Here, we have $a = 0, b = 1$.

$$x = \frac{(b+a)}{2} + \frac{(b-a)}{2}t$$

$$x = \frac{(1+0)}{2} + \frac{(1-0)}{2}t = \frac{1}{2} + \frac{1}{2}t \text{ and } dx = \frac{1}{2}dt$$

Now, $I = \int_0^1 \frac{dx}{x^2+1} = \frac{1}{2}\int_{-1}^1 \frac{dt}{\left(\frac{1}{2}+\frac{1}{2}t\right)^2 + 1} = 2\int_{-1}^1 \frac{1}{(t+1)^2 + 4}dt$

$$= 2\int_{-1}^1 f(t)\,dt \approx 2\sum_{i=1}^{3} w_i f(t_i)$$

$$= 2[w_1 f(t_1) + w_2 f(t_2) + w_3 f(t_3)]$$

Now the values of t_i $(i=1,2,3)$, coefficients w_i can be taken from the above table and the corresponding values of the $f(t_i)$ can be obtained as follows

i	1	2	3
t_i	−0.77459667	0	0.77459667
$f(t_i)$	0.2468644	0.2	0.1398759

$$I = 2[w_1 f(t_1) + w_2 f(t_2) + w_3 f(t_3)]$$
$$= 2[(0.55555555)(0.2468644)+(0.88888889)(0.2)$$
$$+ (0.55555555)(0.1398759)]$$
$$= 2(0.13714689+0.17777778+0.07770883) = 0.785267$$

Example 2. Evaluate $\int_{-2}^{2} e^{-\frac{x}{2}} dx$ by using Gaussian quadrature 3 points formula.

Solution: Here, we have $a = -2, b = 2$.

$$x = \frac{(b+a)}{2} + \frac{(b-a)}{2}t$$

$$x = \frac{(2-2)}{2} + \frac{(2+2)}{2}t = 2t \text{ and } dx = 2dt$$

Now, $I = \int_{-2}^{2} e^{-\frac{x}{2}} dx = 2\int_{-1}^{1} e^{-t} dt$

$$= 2\int_{-1}^{1} f(t)dt \approx 2\sum_{i=1}^{3} w_i f(t_i)$$

$$= 2[w_1 f(t_1) + w_2 f(t_2) + w_3 f(t_3)]$$

Now the values of t_i ($i = 1, 2, 3$) coefficients w_i can be taken from the above table and the corresponding values of the $f(t_i)$ can be obtained as follows

i	1	2	3
t_i	−0.77459667	0.0	0.77459667
$f(t_i)$	2.1697167	1.0	0.4608896

$$I = 2[w_1 f(t_1) + w_2 f(t_2) + w_3 f(t_3)]$$
$$= 2[(0.55555555)(2.1697167)+(0.88888889)(1.0)$$
$$+ (0.55555555)(0.4608896)]$$
$$= 2(1.20539815 + 0.88888889 + 0.25604978) = 4.7067364$$

Example 3. Evaluate $\int_{2}^{4} (x^4 + 1) dx$ by using Gaussian quadrature 2 point formula.

Solution: Here, we have $a = 2, b = 4$.

$$x = \frac{(b+a)}{2} + \frac{(b-a)}{2}t$$

$$x = \frac{(4+2)}{2} + \frac{(4-2)}{2}t = 3+t \text{ and } dx = dt$$

Now,
$$I = \int_2^4 (x^4 + 1)\, dx = \int_{-1}^1 \left[(t+3)^4 + 1\right] dt$$

$$= \int_{-1}^1 f(t)\,dt \approx \sum_{i=1}^2 w_i f(t_i)$$

$$= [w_1 f(t_1) + w_2 f(t_2)]$$

Now the values of t_i ($i = 1, 2$) coefficients w_i can be taken from the above table and the corresponding values of the $f(t_i)$ can be obtained as follows

i	1	2
t_i	−0.577350269	0.577350269
$f(t_i)$	35.447879	164.77433

$$I = [w_1 f(t_1) + w_2 f(t_2)]$$
$$= [(1.0)(35.447879) + (1.0)(164.77433)]$$
$$= 200.222209$$

EXERCISE 5.3
---

1. Evaluate $\int_0^1 \frac{dx}{1+x}$ using Romberg's method correct to three decimal places and hence evaluate $\log_e 2$.

2. Using Euler-Maclaurin's Formula find the value of $\int_0^1 \frac{dx}{1+x}$.

3. Evaluate $\int_0^{\frac{\pi}{2}} \sin x\, dx$ using the Euler-Maclaurin's Formula.

4. Using Euler-Maclaurin's Formula prove that $\sum_1^n x^2 = \frac{n(n+1)(2n+1)}{6}$.

5. Using Euler-Maclaurin's Formula prove that $\sum_1^n x^4 = \frac{n^5}{5} + \frac{n^4}{2} + \frac{n^3}{3} + \frac{n}{30}$.

6. Using Euler-Maclaurin's Formula, find the sum of the series
$$\frac{1}{400} + \frac{1}{402} + \frac{1}{404} + \ldots\ldots + \frac{1}{500}$$

7. Using Euler-Maclaurin's Formula, find the sum of the series
$$\frac{1}{(51)^2} + \frac{1}{(53)^2} + \frac{1}{(55)^2} + \ldots\ldots + \frac{1}{(99)^2}$$

8. Evaluate $\int_0^1 x\, dx$ using Gaussian quadrature four point formula.

9. Evaluate $\int_{-1}^1 \frac{1}{1+x^2}\, dx$ using Gaussian quadrature two point formula.

10. Evaluate $\int_0^{\pi/2} \sin x\, dx$ using Gaussian quadrature three point formula.

11. Evaluate $\int_1^2 e^x\, dx$ using Gaussian quadrature three point formula.

12. Evaluate $\int_5^{12} \frac{1}{x}\, dx$ using Gaussian quadrature three point formula.

ANSWERS

1. 0.6931
2. 0.69314
3. 1.000003
6. 0.01138
7. 0.00499
8. 0.49999
9. 1.50
10. 1.00002
11. 4.67077
12. 0.87534

Chapter – 6

NUMERICAL SOLUTION OF ORDINARY DIFFERENTIAL EQUATIONS

6.1 INTRODUCTION

In the fields of science and engineering, we come across physical and natural phenomenon which, when represented by mathematical models happen to differential equations. For example, simple harmonic motion, equation of motion, newtons law of cooling, deflection of a beam etc., are represented by differential equations. Many analytical methods exists for solving such equations. But these methods can be applied to solve only a selected class of differential equations. Those equations which govern physical systems do not possess, in general, closed form solutions. In such situations, we go for numerical solution of differential equations.

In solving a differential equation for approximate solution, we find numerical values of $y_1, y_2, y_3, \ldots$ corresponding to given numerical values of independent variables $x_1, x_2, x_3, \ldots$ so that the ordered pairs $(x_1, y_1), (x_2, y_2), \ldots$ satisfy a particular solution (though approximately). A solution of this type is called point wise solution.

Consider the first order ordinary differential equation,

$$\frac{dy}{dx} = f(x, y) \qquad \ldots(1)$$

with the initial condition $\quad y(x_0) = y_0 \qquad \ldots(2)$

By numerical solution of the differential equation, let $y(x_0) = y_0$, $y(x_1)$, $y(x_2), \ldots$ be the solutions of y at $x = x_0, x_1, x_2, \ldots$. Let $y = y(x)$ be the exact solution. If we draw the graph of $y = y(x)$ (exact value) and also draw the approximate curve by plotting $(x_0, y_0), (x_1, y_1), (x_2, y_2), \ldots$ we get two curves.

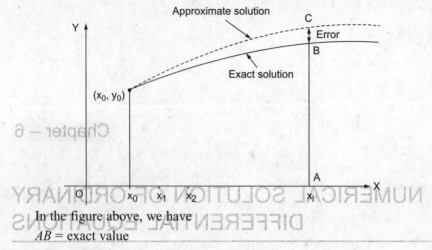

In the figure above, we have
AB = exact value

AC = Approximate value at $x = x_i$

then $BC = AC - AB = y_i - y(x_i) = \epsilon_i$ is called the **truncation error** at $x = x_i$.

6.2 INITIAL AND BOUNDARY VALUE PROBLEMS

It is well known that a differential equation of the nth order will have n arbitrary constants in its general solution. In order to compute the numerical solution of such an equation, we therefore need n conditions. Problems in which all the initial conditions are specified at the initial point only are called **initial value problems**. On the other hand, problems involving second and higher order differential equations, in which the conditions at two or more points are specified, are called **boundary value problems**.

6.3 NUMERICAL METHODS OF SOLVING ORDINARY DIFFERENTIAL EQUATIONS:

In this chapter, we shall discuss various numerical methods of solving ordinary differential equations. We know that these methods will yields the solution in one of the two forms:

(i) A series for y in terms of powers of x from which the value of y can be obtained by direct substitution.

(ii) A set of tabulated values of x and y.

The methods of picards and Taylor belong to class (i) where as those of Euler, Runge-Kutta etc. belong to class (ii).

6.4 PICARD'S METHOD OF SUCCESSIVE APPROXIMATIONS

Aim: To solve the differential equation

$$\frac{dy}{dx} = f(x, y) \qquad \ldots(1)$$

with the initial condition $y(x_0) = y_0$ \qquad ...(2)

$$\therefore \quad dy = f(x, y)\, dx$$

Integrating equation (1) between the limits x_0 and x and corresponding limits y_0 and y, we get

$$\int_{y_0}^{y} dy = \int_{x_0}^{x} f(x, y)\, dx$$

$$y - y_0 = \int_{x_0}^{x} f(x, y)\, dx$$

$$\Rightarrow \quad y = y_0 + \int_{x_0}^{x} f(x, y)\, dx \qquad \ldots(3)$$

This equation is called an integral equation and solved by the method of successive approximations. To obtain the first approximation we replace y by y_0 in the R.H.S. of equation (3), hence the first approximation is

$$y^{(1)} = y_0 + \int_{x_0}^{x} f(x, y_0)\, dx$$

similarly, for a second approximation, replace y_0 by $y^{(1)}$ in $f(x, y_0)$ which gives

$$y^{(2)} = y_0 + \int_{x_0}^{x} f(x, y^{(1)})\, dx$$

Proceeding like this, we obtain $y^{(3)}, y^{(4)}, \ldots y^{(n-1)}, y^{(n)}$, where

$$y^{(n)} = y_0 + \int_{x_0}^{x} f(x, y^{(n-1)})\, dx, \text{ with } y(x_0) = y_0$$

As a matter of fact, the process is stopped when the two values of y eg. $y^{(n-1)}$ and $y^{(n)}$ are same to the desired degree of a accuracy.

Note: This method yields a sequence of approximations $y^{(1)}, y^{(2)}, \ldots, y^{(n)}$ and it can be proved that if the function $f(x, y)$ is bounded in some region about the point (x_0, y_0) and if $f(x, y)$ satisfies the Lipschitz condition, viz.,

$$|f(x, y) - f(x, \bar{y})| \le k |y - \bar{y}|, k \text{ being a constant, then the sequence}$$

$y^{(1)}, y^{(2)} \ldots$ converges to the solution of (1).

ILLUSTRATIVE EXAMPLES

Example 1. Solve $\dfrac{dy}{dx} = x^2 + y^2$ with $y(0) = 1$, by using picard's method of successive approximations.

Solution. Here $x_0 = 0$, $y_0 = 1$.

By picard's method, we have

$$y = y_0 + \int_{x_0}^{x} f(x, y)\, dx = 1 + \int_{0}^{x} (x^2 + y^2)\, dx \qquad \ldots(1)$$

First approximation,

$$y^{(1)} = 1 + \int_0^x (x^2 + 1) \, dx = +1 + x + \frac{x^3}{3} \qquad \ldots(2)$$

Second approximation,

$$y^{(2)} = 1 + \int_0^x \left[x^2 + \left(1 + x + \frac{x^3}{3}\right)^2 \right] dx$$

$$= 1 + \left[x + \frac{2x^3}{3} + \frac{x^7}{63} + x^2 + \frac{2}{15}x^5 + \frac{1}{6}x^4 \right]$$

$$y^{(2)} = 1 + x + x^2 + \frac{2}{3}x^3 + \frac{1}{6}x^4 + \frac{1}{63}x^7 + \frac{2}{15}x^5 \qquad \ldots(3)$$

It is obvious that the integration might become more and more difficult as we proceed to higher approximations. Hence approximate value is $y^{(2)}$ and is given by (3).

Example 2. Solve $y' + y = e^x$, $y(0) = 0$, by picard's method.

Solution. Here $x_0 = 0$, $y_0 = 0$

By picard's method, we have

$$y = y_0 + \int_{x_0}^x f(x, y) \, dx$$

$$y = 0 + \int_0^x (e^x - y) \, dx$$

$$y = \int_0^x (e^x - y) \, dx \qquad \ldots(1)$$

$$\therefore \quad y^{(1)} = \int_0^x (e^x - 0) \, dx = (e^x - 1)$$

$$y^{(2)} = \int_0^x (e^x - e^x + 1) \, dx = x$$

$$y^{(3)} = \int_0^x (e^x - x) \, dx = e^x - \frac{x^2}{2} - 1$$

$$y^{(4)} = \int_0^x \left[e^x - \left(e^x - \frac{x^2}{2} - 1 \right) \right] dx$$

$$y^{(4)} = \frac{x^3}{6} + x$$

$$y^{(5)} = \int_0^x \left(e^x - x - \frac{x^3}{6} \right) dx$$

$$y^{(5)} = e^x - \frac{x^2}{2} - \frac{x^4}{24} - 1$$

Hence approximate value of $y = e^x - \frac{x^2}{2} - \frac{x^4}{24} - 1$

Example 3. Solve $\dfrac{dy}{dx} = x + y$, given $y(0) = 1$. obtain the values of $y(0.1)$, $y(0.2)$ by using picard's method and check your answer with the exact solution.

Solution. Here $x_0 = 0$, $y_0 = 1$ and $f(x, y) = x + y$

By Picard's method

$$y = y_0 + \int_{x_0}^{x} f(x, y)\, dx \qquad \ldots(1)$$

First approximation

$$y^{(1)} = y_0 + \int_{x_0}^{x} f(x, y_0)\, dx = 1 + \int_{0}^{x} (x+1)\, dx$$

$$y^{(1)} = 1 + x + \frac{x^2}{2} \qquad \ldots(2)$$

Second approximation

$$y^{(2)} = 1 + \int_{0}^{x} \left(x + 1 + x + \frac{x^2}{2} \right) dx$$

$$y^{(2)} = 1 + x + x^2 + \frac{1}{6}x^3 \qquad \ldots(3)$$

$$y^{(3)} = 1 + \int_{0}^{x} \left(x + 1 + x + x^2 + \frac{1}{6}x^3 \right) dx = 1 + x + x^2 + \frac{x^3}{3} + \frac{1}{24}x^4$$

$$\therefore \quad y(x) = y = 1 + x + x^2 + \frac{x^3}{3} + \frac{x^4}{24} + \ldots$$

put $x = 0.1$, then

$$y(0.1) = 1 + 0.1 + 0.01 + \frac{1}{3}(0.001) + \frac{1}{24}(0.0001)$$

$$= 1 + 0.1 + 0.01 + 0.0003333 + 0.0000041$$

$$y(0.1) = 1.1103374$$

similarly

$$y(0.2) = 1 + 0.2 + (0.2)^2 + \frac{(0.2)^3}{3} + \frac{(0.2)^4}{24}$$

$$y(0.2) = 1.242733$$

Now we shall obtain exact value,

$$\frac{dy}{dx} = x + y$$

$$\frac{dy}{dx} - y = x$$

which is linear and general solution is given by

$$ye^{-x} = -e^{-x}(1 + x) + c \qquad [\text{I.F} = e^{-x}]$$

$$y = -x - 1 + ce^{x}$$

putting $y = 1, x = 0$, we get $c = 2$

$\therefore \quad y = -x - 1 + 2e^x$

$\therefore \quad y(0.1) = -0.1 - 1 + 2e^{0.1} = 1.11034184$

and $\quad y(0.2) = -0.2 - 1 + 2e^{0.2} = 1.24280555$

In both cases, $y(0.1)$ are same (correct to 4 decimal places) and the value of $y(0.2)$ differ only by 0.0001.

Example 4. Solve $\dfrac{dy}{dx} = 1 + xy$, with $x_0 = 2$, $y_0 = 0$, by using picard's method.

Solution. Here $x_0 = 2$, $y_0 = 0$ and $f(x, y) = 1 + xy$

By picard's method, we have

$$y = y_0 + \int_{x_0}^{x} f(x,y)\, dx \qquad \ldots(1)$$

$\therefore \quad y^{(1)} = y_0 + \int_{2}^{x} f(x, y_0)\, dx = 0 + \int_{2}^{x} f(x, 0)\, dx$

$y^{(1)} = \int_{2}^{x} (1 + x.0)\, dx = x - 2$

$y^{(2)} = \int_{2}^{x} [1 + x(x-2)]\, dx = \left(x - x^2 + \dfrac{x^3}{3} \right)_{2}^{x}$

$y^{(2)} = -\dfrac{2}{3} + x - x^2 + \dfrac{x^3}{3}$

$y^{(3)} = \int_{2}^{x} \left[1 + x \left(-\dfrac{2}{3} + x - x^2 + \dfrac{x^3}{3} \right) \right] dx$

$y^{(3)} = -\dfrac{22}{15} + x - \dfrac{x^2}{3} + \dfrac{x^3}{3} - \dfrac{x^4}{4} + \dfrac{x^5}{15}$

Example 5. If $\dfrac{dy}{dx} = \dfrac{y-x}{y+x}$ and $y(0) = 1$. Find the value of y at $x = 0.1$ by using picard's method.

Solution. Here $x_0 = 0$, $y_0 = 1$ and $f(x, y) = \dfrac{y-x}{y+x}$

By picard's method, we have

$$y = y_0 + \int_{x_0}^{x} f(x, y)\, dx \qquad \ldots(1)$$

$\therefore \quad y^{(1)} = y_0 + \int_{0}^{x} f(x, y_0)\, dx = 1 + \int_{0}^{x} f(x, 1)\, dx$

$= 1 + \int_{0}^{x} \left(\dfrac{1-x}{1+x} \right) dx = 1 + \int_{0}^{x} \left(\dfrac{2}{1+x} - 1 \right) dx$

similarly
$$y^{(1)} = 1 - x + 2\log(1+x)$$
$$y^{(2)} = 1 + x - 2\int_0^x \frac{x\,dx}{1+2\log(1+x)}$$

which is difficult to integrate.
Hence, when $x = 0.1$, $y^{(1)} = 1 - 0.1 + 2\log(1.1) = 0.9828$

In this example only first approximation can be obtained and so it gives the approximate value of y for $x = 0.1$.

Example 6. Using picard's method to obtain y for $x = 0.1$, given that $\frac{dy}{dx} = 3x + y^2$; $y = 1$ at $x = 0$.

Solution. Here $x_0 = 0$, $y_0 = 1$ and $f(x, y) = 3x + y^2$

By picard's method, we have

$$y = y_0 + \int_{x_0}^x f(x, y)\, dx$$

∴
$$y^{(1)} = y_0 + \int_0^x f(x, y_0)\, dx$$

$$= 1 + \int_0^x f(x, 1)\, dx = 1 + \int_0^x (3x + 1)\, dx$$

$$y^{(1)} = 1 + x + \frac{3}{2}x^2$$

similarly
$$y^{(2)} = 1 + x + \frac{5}{2}x^2 + \frac{4}{3}x^3 + \frac{3}{4}x^4 + \frac{9}{20}x^5$$

$$y^{(3)} = 1 + x + \frac{5}{2}x^2 + 2x^3 + \frac{23}{12}x^4 + \frac{25}{12}x^5 + \frac{68}{45}x^6$$

$$+ \frac{1157}{1260}x^7 + \frac{17}{32}x^8 + \frac{47}{240}x^9 + \frac{27}{400}x^{10} + \frac{81}{4400}x^{11}$$

when $x = 0.1$ we get
$y^{(1)} = 1.115$, $y^{(2)} = 1.1264$, $y^{(3)} = 1.12721$

Thus $y = 1.127$ when $x = 0.1$.

EXERCISE 6.1
---

1. Given $y' = \frac{x^2}{1+y^2}$ and $y(0) = 0$, find $y(0.25)$, $y(0.5)$ by using picard's method.

2. Solve $y' = x^2 + y^2$, given $y(0) = 0$, by using picard's method.

3. Find $y(0.2)$, if $\frac{dy}{dx} = \log(x+y)$; $y(0) = 1$ by using picard's method.

4. Apply picard's method to find $y(0.2)$ and $y(0.4)$ given that $\dfrac{dy}{dx} = 1 + y^2$ and $y(0) = 0$.

5. Solve $y' = y - x^2$, $y(0) = 1$, by picard's method upto the third approximation. Hence find the value of $y(0.1)$ and $y(0.2)$.

6. Solve $\dfrac{dy}{dx} = x + y^2 + 1$, given $y(0) = 0$.

7. Solve $y' = 2x - y$ with $y(1) = 3$. Also find $y(1.1)$.

ANSWERS

1. 0.005, 0.042
2. $y = \dfrac{1}{5}x^3 + \dfrac{1}{63}x^7 + \dfrac{2}{2079}x^{11} + \ldots$
3. 1.0082
4. 0.2027, 0.4227
5. $y = 1 + x + \dfrac{x^2}{2} - \dfrac{x^3}{6} - \dfrac{x^4}{12} - \dfrac{x^5}{60} + \ldots$; 1.1048249, 1.218528
6. $y = x + \dfrac{x^2}{2} + \dfrac{x^3}{3} + \dfrac{x^4}{4} + \dfrac{x^5}{20}$
7. $y = \dfrac{73}{12} - \dfrac{35}{6}x + \dfrac{7}{2}x^2 - \dfrac{5}{6}x^3 + \dfrac{x^4}{12}$; 2.914508

6.5 PICARD'S METHOD FOR SIMULTANEOUS FIRST ORDER DIFFERENTIAL EQUATIONS

Let $\dfrac{dy}{dx} = \phi(x, y, z)$ and $\dfrac{dz}{dx} = f(x, y, z)$ be the simultaneous differential equations with initial conditions $y(x_0) = y_0$; $z(x_0) = z_0$.

Picard's method gives

$$y^{(1)} = y_0 + \int_{x_0}^{x} \phi(x, y_0, z_0)\, dx \,;\, z^{(1)} = z_0 + \int_{x_0}^{x} f(x, y_0, z_0)\, dx$$

$$y^{(2)} = y_0 + \int_{x_0}^{x} \phi[x, y^{(1)}, z^{(1)}]\, dx \,;\, z^{(2)} = z_0 + \int_{x_0}^{x} f[x, y^{(1)}, z^{(1)}]\, dx$$

and so on.

ILLUSTRATIVE EXAMPLE

Example. Approximate y and z by using picard's method for the particular solution of $\dfrac{dy}{dx} = x + z$, $\dfrac{dz}{dx} = x - y^2$, given that $y = 2$, $z = 1$ when $x = 0$.

Solution. Let $\phi(x, y, z) = x + z$, $f(x, y, z) = x - y^2$
Here $x_0 = 0$, $y_0 = 2$, $z_0 = 1$.

$\therefore \quad \dfrac{dy}{dx} = \phi(x, y, z) \Rightarrow y = y_0 + \int_{x_0}^{x} f(x, y, z)\, dx \quad \ldots(1)$

Also $\quad \dfrac{dz}{dx} = f(x, y, z) \Rightarrow z = z_0 + \int_{x_0}^{x} f(x, y, z)\, dx \quad \ldots(2)$

First approximation,

$$y^{(1)} = y_0 + \int_{x_0}^{x} \phi(x, y_0, z_0)\, dx = 2 + \int_0^x (x + z_0)\, dx$$

$$y^{(1)} = 2 + \int_0^x (x + 1)\, dx = 2 + x + \dfrac{x^2}{2}$$

and $\quad z^{(1)} = z_0 + \int_{x_0}^{x} f(x, y_0, z_0)\, dx = 1 + \int_0^x (x - y_0)^2\, dx$

$$= 1 + \int_0^x (x - 4)\, dx = 1 - 4x + \dfrac{x^2}{2}$$

Second approximation

$$y^{(2)} = y_0 + \int_{x_0}^x \phi[x, y^{(1)}, z^{(1)}]\, dx$$

$$= 2 + \int_0^x [x + z^{(1)}]\, dx$$

$$= 2 + \int_0^x \left[x + 1 - 4x + \dfrac{x^2}{2}\right] dx$$

$$= 2 + x - \dfrac{3}{2}x^2 + \dfrac{x^3}{6}$$

and $\quad z^{(2)} = z_0 + \int_{x_0}^{x} f\left[x, y^{(1)}, z^{(1)}\right] dx$

$$= 1 + \int_0^x \left[x - \left(2 + x + \dfrac{x^2}{2}\right)^2\right] dx$$

$$= 1 - 4x - \dfrac{3}{2}x^2 - x^3 - \dfrac{x^4}{4} - \dfrac{x^5}{20}$$

EXERCISE 6.2

1. Approximate y and z by using picard's method for the solution of simultaneous differential equations $\dfrac{dy}{dx} = 2x + z$, $\dfrac{dz}{dx} = 3xy + x^2 z$ with $y = 2$, $z = 0$ at $x = 0$ up to third approximation.

$$y^{(3)} = 2 + x^2 + x^3 + \frac{3}{20}x^5 + \frac{1}{10}x^6;$$

$$z^{(3)} = 3x^2 + \frac{3}{4}x^3 + \frac{6}{5}x^5 + \frac{3}{20}x^7 + \frac{3}{40}x$$

6.6 TAYLOR'S SERIES METHOD

Aim: To solve the differential equation

$$\frac{dy}{dx} = f(x, y) \qquad \ldots(1)$$

with the initial condition $y(x_0) = y_0$ ...(2)

If $y(x)$ is the exact solution of (1), then the Taylor's series for $y(x)$ about the point $x = x_0$ is given by

$$y(x) = y_0 + \frac{(x-x_0)}{1!}y_0' + \frac{(x-x_0)^2}{2!}y_0'' + \frac{(x-x_0)^3}{3!}y_0''' + \cdots \quad \ldots(3)$$

Putting $x = x_1(= x_0 + h)$ in (3), we get

$$y_1 = y(x_1) = y_0 + \frac{h}{1!}y_0' + \frac{h^2}{2!}y_0'' + \frac{h^3}{3!}y_0''' + \frac{h^4}{4!}y_0^{iv} + \ldots \quad \ldots(4)$$

Here $y_0', y_0'', y_0''', \ldots$ can be found by using (1) and its derivatives at $x = x_0$. The series (4) can be truncated at any stage if h is small.

Now having got y_1, we can caculate $y_1', y_1'', y_1''', \ldots$ etc., by using $y' = f(x, y)$.

Now expanding $y(x)$ by Taylor' series about the point $x = x_1$, we get

$$y_2 = y_1 + \frac{h}{1!}y_1' + \frac{h^2}{2!}y_1'' + \frac{h^3}{3!}y_1''' + \frac{h^4}{4!}y_1^{iv} + \ldots$$

Proceeding like this, we get

$$y_{n+1} = y_n + \frac{h}{1!}y_n' + \frac{h^2}{2!}y_n'' + \frac{h^3}{3!}y_n''' + \ldots$$

6.7 TAYLOR'S METHOD FOR SIMULTANEOUS FIRST ORDER DIFFERENTIAL EQUATIONS

Aim: To solve the simultaneous differential equations of the type

$$\frac{dy}{dx} = f(x, y, z) \qquad \ldots(1)$$

and

$$\frac{dz}{dx} = \phi(x, y, z) \qquad \ldots(2)$$

with initial conditions $y(x_0) = y_0$ and $z(x_0) = z_0$
If h is the step size, then $y_1 = y(x_0 + h)$ and $z_1 = z(x_0 + h)$
Taylor's algorithm for (1) and (2) gives

$$y_1 = y_0 + \frac{h}{1!}y_0' + \frac{h^2}{2!}y_0'' + \frac{h^3}{3!}y_0''' + \ldots \qquad \ldots(3)$$

and
$$z_1 = z_0 + \frac{h}{1!}z_0' + \frac{h^2}{2!}z_0'' + \frac{h^3}{3!}z_0''' + \ldots \qquad \ldots(4)$$

Differentiating (1) and (2) successively, we get y'', y''', ... z'', z''' ... etc. so the values y_0'', y_0''', ... and z_0'', z_0''', ... can be obtained. Substituting these values in equations (3) and (4), we get y_1, z_1. Now having got y_1 and z_1 we obtain

$$y_2 = y_1 + \frac{h}{1!}y_1' + \frac{h^2}{2!}y_1'' + \frac{h^3}{3!}y_1''' + \ldots$$

and
$$z_2 = z_1 + \frac{h}{1!}z_1' + \frac{h^2}{2!}z_1'' + \frac{h^3}{3!}z_1''' + \ldots$$

Since y_1 and z_1 are known, we can calculate y_1', y_1'', y_1''', ... z_1', z_1'', z_1'''. Hence y_2 and z_2 can be obtained. Proceeding like this, we get other values of y and z.

ILLUSTRATIVE EXAMPLES

Example 1. Solve $\dfrac{dy}{dx} = x + y$, given $y(1) = 0$ and get $y(1.1)$, $y(1.2)$ by Taylor series method. Compare your result with the explicit solution.

Solution. Here $x_0 = 1$, $y_0 = 0$, $h = 0.1$

$y' = x + y$ $\qquad\qquad y_0' = x_0 + y_0 = 1 + 0 = 1$

$\Rightarrow \quad y'' = 1 + y' \qquad\qquad y_0'' = 1 + y_0' = 1 + 1 = 2$

$\Rightarrow \quad y''' = y'' \qquad\qquad\quad y_0''' = y_0'' = 2$

$\Rightarrow \quad y^{iv} = y''' \qquad\qquad\quad y_0^{iv} = y_0''' = 2$, etc.

By Taylor's series, we have

$$y_1 = y_0 + hy_0' + \frac{h^2}{2!}y_0'' + \frac{h^3}{3!}y_0''' + \frac{h^4}{4!}y_0^{iv} + \ldots$$

$\therefore \quad y_1 = y(1.1) = 0 + 0.1 \times 1 + \dfrac{(0.1)^2}{2} \times 2 + \dfrac{(0.1)^3}{6} \times 2 + \dfrac{(0.1)^4}{24} \times 2 + \ldots$

$y(1.1) = 0.1 + 0.01 + 0.00033 + 0.00000833 + \ldots$

$\Rightarrow \quad y(1.1) = 0.11033833$

Also, $\quad x_1 = x_0 + h = 1 + 0.1 = 1.1$
$\qquad y_1 = 0.11033833$

Again, $\quad y_1' = x_1 + y_1 = 1.1 + 0.11033833 = 1.21033833$

$\qquad y_1'' = 1 + y_1' = 1 + 1.21033833 = 2.21033833$

$\qquad y_1''' = y_1'' = 2.21033833$

$\qquad y_1^{iv} = y_1''' = y_1^{v} = \ldots = 2.21033833$

Now $\quad y_2 = y_1 + \dfrac{h}{1!}y_1' + \dfrac{h^2}{2!}y_1'' + \dfrac{h^3}{3!}y_1''' + \ldots$

$\Rightarrow \quad y_2 = y(1.2) = 0.11033833 + \dfrac{0.1}{1}(1.21033833) + \dfrac{(0.1)^2}{2}$

$\qquad (2.21033833) + \dfrac{(0.1)^3}{6}(2.21033833) + \dfrac{(0.1)^4}{24}(2.21033833) + \ldots$

$\qquad = 0.11033833 + 0.121033833 + 0.011051691 + .00036838$
$\qquad + .000009209743 + \ldots$

$\Rightarrow \quad y_2 = y(1.2) = 0.242801443$

$\therefore \quad y(1.2) = 0.242801443$

The exact solution of $\dfrac{dy}{dx} = x + y$ is

$\qquad y = -x - 1 + 2e^{x-1}$
$\qquad y(1.1) = -1.1 - 1 + 2e^{1.1-1} = -2.1 + 2e^{0.1} = -2.1 + 2(1.105170918)$
$\qquad y(1.1) = -2.1 + 2.210341836 = 0.110341836$
$\qquad y(1.2) = -1.2 - 1 + 2e^{1.2-1} = -2.2 + 2e^{0.2}$
$\qquad = -2.2 + 2(1.221402758)$
$\qquad = 0.242805516$

Example 2. Using Taylor's method, find the value of y (0.1) for the differential equation $\dfrac{dy}{dx} = x^2 + y^2$, $y(0) = 1$.

Solution. Here $x_0 = 0$, $y_0 = 1$, $h = 0.1$

$\qquad y' = x^2 + y^2$
$\qquad y'' = 2x + 2yy'$
$\qquad y''' = 2 + 2yy'' + 2(y')^2$
$\qquad y^{iv} = 2yy''' + 2y'y'' + 4y'y''$
$\Rightarrow \qquad y^{iv} = 2yy''' + 6y'y''$ etc.

Now
$$y_0' = x_0^2 + y_0^2 = 0 + (1)^2 = 1$$
$$y_0'' = 2x_0 + 2y_0 y_0' = 2$$
$$y_0''' = 2 + 2y_0 y_0'' + 2(y_0')^2 = 2 + 2(1)(2) + 2(1)^2$$
$$y_0''' = 8$$
$$y^{iv} = 2y_0 y_0''' + 6 y_0' y_0''$$
$$= 2(1)(8) + 6(1)(2) = 28$$

By Taylor series method, we have

$$y_1 = y_0 + hy_0' + \frac{h^2}{2!}y_0'' + \frac{h^3}{3!}y_0''' + \frac{h^4}{4!}y_0^{iv} + ...$$

$$y(0.1) = y_1 = 1 + 0.1(1) + \frac{(0.1)^2}{2}(2) + \frac{(0.1)^3}{6}(8) + \frac{(0.1)^4}{24}(28) + ...$$

$$y(0.1) = 1 + 0.1 + 0.01 + 0.001333333 + 0.000116666 + ...$$

$$y(0.1) = 1.111449999 \simeq 1.11145$$

Example 3. Using Taylor series method, find $y(1.1)$ and $y(1.2)$ correct to four decimal places, given that $\frac{dy}{dx} = xy^{1/3}$ and $y(1) = 1$.

Solution. Here $x_0 = 1$, $y_0 = 1$ and $h = 0.1$
$$y' = xy^{1/3}$$
$$y'' = \frac{1}{3}xy^{-2/3}y' + y^{1/3}$$
$$y'' = \frac{1}{3}x^2 y^{-1/3} + y^{1/3}$$
$$y''' = \frac{x^2}{3}\left(-\frac{1}{3}\right)y^{-4/3}y' + \frac{2x}{3}y^{-1/3} + \frac{1}{3}y^{-2/3}y'$$

Now $y_0' = 1 \, (1)^{1/3} = 1$
$$y_0'' = \frac{1}{3}x_0 y_0^{-2/3} y_0' + y_0^{1/3}$$
$$y_0'' = \frac{1}{3}(1)(1)(1) + 1 = 4/3$$
$$y_0''' = 8/9$$

By Taylor series method, we have

$$y_1 = y_0 + hy_0' + \frac{h^2}{2!}y_0'' + \frac{h^3}{3!}y_0''' + ...$$

$$y_1 = y(1.1) = 1 + (0.1) + \frac{(0.1)^2}{2}(4/3) + \frac{(0.1)^3}{6}(8/9) + \ldots$$
$$= 1 + 0.1 + 0.00666 + 0.000148 + \ldots$$
$$y(1.1) = 1.10681$$

Now $x_1 = 1.1, y_1 = 1.10681$

$$y_1' = x_1 y_1^{1/3} = (1.1)(1.10681)^{1/3} = 1.13785$$

$$y_1'' = \frac{1}{3} x_1 y_1^{-2/3} y_1' + y_1^{1/3}$$

$$= \frac{1}{3}(1.1)(1.10681)^{-2/3}(1.13785) + (1.10681)^{1/3}$$

$$= 0.38992 + 1.03441$$

$$y_1'' = 1.42433$$

$$y_1''' = 0.929787$$

$$\therefore \quad y_2 = y_1 + h y_1' + \frac{h^2}{2!} y_1'' + \frac{h^3}{3!} y_1''' + \ldots$$

$$y(1.2) = 1.10681 + (0.1)(1.13785) + \frac{0.01}{2}(1.42433) + \frac{0.001}{6}(0.929787) + \ldots$$

$$y(1.2) = 1.22772$$

Example 4. Using Taylor series method, find $y(0.2)$ and $y(0.4)$ correct to four decimal places given that $\dfrac{dy}{dx} + 2xy = 1$ and $y(0) = 0$.

Solution. Here $x_0 = 0, y_0 = 0, h = 0.2$

$$\frac{dy}{dx} + 2xy = 1$$

$$\Rightarrow \quad \frac{dy}{dx} = 1 - 2xy$$

i.e., $\quad y' = 1 - 2xy \qquad\qquad y_0' = 1 - 2x_0 y_0 = 1$

$\quad y'' = -2(xy' + y) \qquad\qquad y_0'' = -2(x_0 y_0' + y_0) = 0$

$\quad y''' = -2(xy'' + 2y') \qquad\qquad y_0''' = -4$

$\quad y^{iv} = -2(xy''' + 3y'') \qquad\qquad y_0^{iv} = 0$

By Taylor series method, we have

$$y_1 = y_0 + h y_0' + \frac{h^2}{2!} y_0'' + \frac{h^3}{3!} y_0''' + \frac{h^4}{4!} y_0^{iv} + \ldots$$

$$y_1 = y(0.2) = 0.2(1) + \frac{(0.2)^2}{2}(0) + \frac{(0.2)^3}{6}(-4) + \frac{(0.2)^4}{24}(0) + \ldots$$

Numerical Solution of Ordinary Differential Equations

$\Rightarrow y(0.2) = 0.2 - 0.005333333$

$y(0.2) = 0.19466667$

Now $x_1 = 0.2$, $y_1 = 0.19466667$, $h = 0.2$

$y'_1 = 1 - 2x_1 y_1 = 1 - 2(0.2)(0.19466667) = 1 - 0.077866668$

$y'_1 = 0.922133332$

$y''_1 = -2(x_1 y'_1 + y_1) = -2[(0.2)(0.922133332) + 0.19466667]$
$= 0.758186672$

$y'''_1 = 2[x_1 y''_1 + 2y'_1]$
$= -2[(0.2)(-0.758186672) + 2(0.922133332)]$

$y'''_1 = -2[-0.151637334 + 1.844266664]$

$y'''_1 = -3.38525866$

Again, By Taylor series method, we have

$$y_2 = y_1 + hy'_1 + \frac{h^2}{2!}y''_1 + \frac{h^3}{3!}y'''_1 + \frac{h^4}{4!}y_i^{iv} + \ldots$$

$$y_2 = y(0.4) = 0.19466667 + (0.2)(0.922133332) + \frac{(0.2)^2}{2}$$

$$(-0.758186672) + \frac{(0.2)^3}{6}(-3.38525866)$$

$y(0.4) = 0.359415924$

Example 5. Solve the simultaneous equations

$y' = 1 + xyz$, $y(0) = 0$
$z' = x + y + z$, $z(0) = 1$

Solution. $y' = 1 + xyz$

$y'' = yz + xy'z + xyz'$
$y''' = 2y'z + 2yz' + 2xy'z' + xy''z + xyz''$
$z' = x + y + z$
$z'' = 1 + y' + z'$
$z''' = y'' + z''$ etc.

Here $x = 0, y = 0, z = 1$

$\therefore$ $y' = 1, y'' = 0, y''' = 2$

and $z' = 1, z'' = 3, z''' = 3$

Hence by Taylor's series method, we get

$$y(x) = x + \frac{x^3}{3} \text{ and } z(x) = 1 + x + \frac{3}{2}x^2 + \frac{1}{2}x^3$$

Example 6. Solve $\dfrac{dy}{dx} = z - x, \dfrac{dz}{dx} = y + x$ with $y(0) = 1$, $z(0) = 1$, by taking $h = 0.1$ to get $y(0.1)$ and $z(0.1)$ by Taylor series method.

Solution. Here $x_0 = 0$, $y_0 = 1$, $h = 0.1$

$$y_1 = y(0.1) = ?$$

$$y' = z - x \Rightarrow y'_0 = z_0 - x_0 = 1$$

$$y'' = z' - 1 \Rightarrow y''_0 = z'_0 - 1 = 0$$

$$y''' = z'' \Rightarrow y'''_0 = z''_0 = 2$$

$$x_0 = 0, z_0 = 1, h = 0.1$$

$$z_1 = z(0.1) = ?$$

$$z' = x + y \Rightarrow z'_0 = x_0 + y_0 = 1$$

$$z'' = 1 + y' \Rightarrow z''_0 = 1 + y'_0 = 2$$

$$z''' = y'' \Rightarrow z'''_0 = y''_0 = 0$$

$$z^{iv} = y''' \Rightarrow z^{iv}_0 = y'''_0 = 2$$

By Taylor series method for y_1 and z_1, we have

$$y_1 = y_0 + h y'_0 + \dfrac{h^2}{2!} y''_0 + \dfrac{h^3}{3!} y'''_0 + \ldots \qquad \ldots(1)$$

and

$$z_1 = z_0 + h z'_0 + \dfrac{h^2}{2!} z''_0 + \dfrac{h^3}{3!} z'''_0 + \ldots \qquad \ldots(2)$$

$\therefore$

$$y_1 = y(0.1) = 1 + (0.1) 1 + \dfrac{(0.1)^2}{2} 0 + \dfrac{(0.1)^3}{6} (2) + \ldots$$

$$y_1 = y(0.1) = 1 + 0.1 + 0.000333 + \ldots$$

$$y(0.1) = 1.1003$$

and

$$z_1 = z(0.1) = 1 + (0.1)(1) + \dfrac{(0.1)^2}{2} (2) + \dfrac{(0.1)^3}{6} (0)$$

$$+ \dfrac{(0.1)^4}{24} (2) + \ldots$$

$$= 1 + 0.1 + .01 + 0.0000083 + \ldots$$

$$z(0.1) = 1.1100$$

Hence $y(0.1) = 1.1003$ and $z(0.1) = 1.1100$

EXERCISE 6.3

1. Find $y(0.1)$ given $\dfrac{dy}{dx} = x+y$, $y(0)=1$, by Taylor series method.

2. Find $y(0.1)$ given $\dfrac{dy}{dx} = x-y^2$ and $y(0) = 1$, by Taylor series method.

3. Compute y for $x = 0.1$ and 0.2 correct to four decimal places given $y' = y - x$, $y(0) = 2$, by Taylor series method.

4. Find $y(0.1)$ given $y' = x^2 y - 1$, $y(0) = 1$, by Taylor series method.

5. Find $y(0.1)$, $z(0.1)$ given

   $\dfrac{dy}{dx} = x + z$, $\dfrac{dz}{dx} = x - y^2$ and $y(0) = 2$, $z(0) = 1$.

6. Using Taylor series method, find the solution of the differential equation $xy' = x - y$, $y(2) = 2$ at $x = 2.1$ correct to five decimal places.

7. Find $y(1)$ for $\dfrac{dy}{dx} = 2y + 3e^x$, $y(0) = 0$

ANSWERS

1. 1.1103
2. 0.9138
3. 2.2052, 2.4214
4. 0.9003
5. 2.0845, 0.5867
6. 2.00238
7. 14.01

6.8 EULER'S METHOD

This method has limited application because of its low accuracy and give the solution in the form of a set of tabulated values.

Aim: To solve $\dfrac{dy}{dx} = f(x,y)$ with initial condition $y(x_0) = y_0$ ...(1)

Let $y = \phi(x)$ be the solution of (1). Let $x_0, x_1, x_2, \ldots$ be equidistant values of x. Since in a small interval, a curve is nearly a straight line. Thus at the point (x_0, y_0), we approximate the curve by the tangent at the point (x_0, y_0).

The equation of the tangent at $P_0 (x_0, y_0)$ is

$$y - y_0 = \left(\dfrac{dy}{dx}\right)_{P_0} (x - x_0)$$

$$y - y_0 = f(x_0, y_0)(x - x_0)$$

$\Rightarrow \qquad y = y_0 + (x - x_0) f(x_0, y_0)$ ...(2)

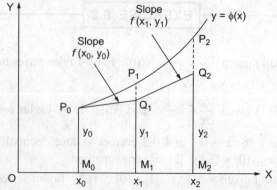

This gives the y-coordinate of any point on the tangent. Since the curve is approximated by the tangent in the interval (x_0, x_1). Therefore the value of y on the curve is approximately equal to the value of y on the tangent at (x_0, y_0) corresponding to $x = x_1$.

Putting $x = x_1 = (x_0 + h)$ in equation (2), we get
$$y_1 = y_0 + hf(x_0, y_0)$$
$$(M_1P_1 \approx M_1Q_1 = y_1)$$

Again, we approximate curve by the line through (x_1, y_1) and whose slope is $f(x_1, y_1)$, we get
$$y_2 = y_1 + hf(x_1, y_1)$$

In general, it can be shown that
$$y_{n+1} = y_n + hf(x_n, y_n) \; ; \; n = 0, 1, 2, 3, \ldots$$

This is called Euler's formula.

In this method, the actual curve is approximated by a sequence of short straight lines. As the intervals increases the straight line deviates much from the actual curve. Hence the accuracy cannot be obtained as the number of intervals in crease. Therefore the method is unsuitable for practical use and a modification of it, known as modified Euler's method which gives more accurate results.

$$Q_1P_1 = \text{error at } x = x_1$$
$$= \frac{(x_1 - x_0)^2}{2!} y''(x_1, y_1)$$
$$= \frac{h^2}{2} y''(x_1, y_1)$$

$\therefore$ It is of order h^2.

ILLUSTRATIVE EXAMPLES

Example 1. Using Euler's method to find $y(0.4)$ given $y' = xy$, $y(0) = 1$.

Solution. Here $x_0 = 0$, $y_0 = 1$ and take $h = 0.1$

By Euler's formula, we have

$$y_1 = y_0 + hf(x_0, y_0)$$
$$y_1 = 1 + 0.1 f(0, 1) = 1 + 0.1 (0 \times 1) = 1$$
$$y_2 = y_1 + hf(x_1, y_1)$$
$$= 1 + 0.1 f(0.1, 1) = 1 + 0.1 (0.1 \times 1)$$
$$y_2 = 1 + .01 = 1.01$$
$$y_3 = y_2 + hf(x_2, y_2) = 1.01 + 0.1 f(0.2, 1.01)$$
$$y_3 = 1.01 + 0.1 (0.2 \times 1.01) = 1.0302$$
$$y_4 = y_3 + hf(x_3, y_3) = 1.0302 + 0.1 f(0.3, 1.0302)$$
$$y_4 = 1.0302 + 0.1 (0.3 \times 1.0302) = 1.061106$$

Example 2. Solve the equation $\dfrac{dy}{dx} = 1 - y$ with $x = 0, y = 0$ by using Euler's method and tabulate the solutions at $x = 0.1, 0.2, 0.3$.

Solution. Here $x_0 = 0, y_0 = 0$ and take $h = 0.1$

By Euler's method, we have
$$y_1 = y_0 + hf(x_0, y_0)$$
$$y_1 = 0 + (0.1)(1 - 0) = 0.1$$
$\therefore \quad y(0.1) = 0.1$
$$y_2 = y_1 + hf(x_1, y_1) = 0.1 + 0.1(1 - 0.1) = 0.1 + 0.09 = 0.19$$
$\Rightarrow \quad y(0.2) = 0.19$
$$y_3 = y_2 + hf(x_2, y_2) = 0.19 + 0.1(1 - 0.19)$$
$$y_3 = 0.19 + (0.1)(0.81) = 0.271$$
$\therefore \quad y(0.3) = 0.271$

Example 3. Find y approximately for $x = 0.1$ by Euler's method: given $\dfrac{dy}{dx} = \dfrac{y-x}{y+x}$ with $x = 0, y = 1$.

Solution. Here $x_0 = 0, y_0 = 1$ and take $h = 0.02$

By Euler's method, we have
$$y_1 = y_0 + hf(x_0, y_0) = 1 + 0.02 \left(\frac{1-0}{1+0}\right) = 1.02$$
$$y_2 = y_1 + hf(x_1, y_1) = 1.02 + 0.02 \left(\frac{1.02 - 0.02}{1.02 + 0.02}\right) = 1.0392$$
$$y_3 = y_2 + hf(x_2, y_2) = 1.0392 + 0.02 \left(\frac{1.0392 - 0.04}{1.0392 + 0.04}\right)$$
$$y_3 = 1.0577$$
$$y_4 = y_3 + hf(x_3, y_3) = 1.0577 + 0.02 \left(\frac{1.0577 - 0.06}{1.0577 + 0.06}\right)$$
$$y_4 = 1.0756$$

$$y_5 = y_4 + hf(x_4, y_4) = 1.0756 + 0.02 \left(\frac{1.0756 - 0.08}{1.0756 + 0.08}\right)$$

$$y_5 = 1.0928$$

Hence $y = 1.0928$ when $x = 0.1$

Example 4. Using Euler's method, solve the equation $y' = x + y$, $y(0) = 1$, for $x = (0.0) \ (0.2) \ (1.0)$

Solution. Here $x_0 = 0$, $y_0 = 1$, $h = 0.2$

$$y' = x + y$$

By Euler's formula, we have

$$y_1 = y_0 + hf(x_0, y_0) = 1 + 0.2 \ (0 + 1) = 1.2$$
$$y_2 = y_1 + hf(x_1, y_1) = 1.2 + 0.2 \ (0.2 + 1.2) = 1.48$$
$$y_3 = y_2 + hf(x_2, y_2) = 1.48 + 0.2 \ (0.4 + 1.48) = 1.856$$
$$y_4 = y_3 + hf(x_3, y_3) = 1.856 + 0.2 \ (0.6 + 1.856)$$
$$y_4 = 2.3472$$
$$y_5 = y_4 + hf(x_4, y_4) = 2.3472 + 0.2 \ (0.8 + 2.3472)$$
$$y_5 = 2.97664$$

EXERCISE 6.4

1. Apply Euler's method to solve $\dfrac{dy}{dx} = x + y$, $y(0) = 0$ at $x = 0$ to $x = 1.0$ taking $h = 0.2$.

2. Using Euler's method, compute $y \ (0.04)$ for the differential equation $\dfrac{dy}{dx} = -y$; $y(0) = 1$, by taking $h = 0.01$.

3. Given $\dfrac{dy}{dx} = x^3 + y$; $y(0) = 1$. Compute y (0.02) by Euler's method by taking $h = 0.01$.

4. Compute $y \ (0.5)$ for the differential equation $\dfrac{dy}{dx} = y^2 - x^2$, with $y(0) = 1$ by using Euler's method.

ANSWERS

1. $y \ (0.2) = 0$, $y \ (0.4) = 0.04$, $y \ (0.6) = 0.128$
   $y \ (0.8) = 0.2736$, $y \ (1.0) = 0.48832$
2. 0.960596   3. $y(0.01) = 1.01$, $y(0.02) = 1.0201$
4. 1.76393

6.9 MODIFIED EULER'S METHOD

In this method, we will average the points.

Let $P(x_0, y_0)$ be the point on the solution curve.

Let PA be the tangent at (x_0, y_0) to the curve. Let this tangent meet the ordinate at $x = x_0 + \frac{1}{2}h$ at N.

y-Coordinate of $N = y_0 + \frac{1}{2}hf(x_0, y_0)$ ...(1)

Calculate the slope at N i.e

$$f\left(x_0 + \frac{1}{2}h, y_0 + \frac{1}{2}hf(x_0, y_0)\right)$$

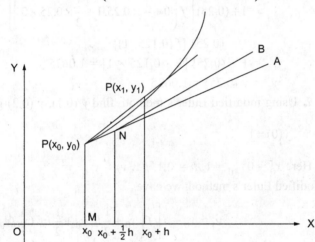

Now draw the line through $P(x_0, y_0)$ with this slope.

Let this line meet $x = x_1$ at $K(x_1, y_1^{(1)})$. This $y_1^{(1)}$ is taken as the approximate value of y at $x = x_1$.

$$\therefore \quad y_1^{(1)} = y_0 + h\left[f\left(x_0 + \frac{1}{2}h, y_0 + \frac{1}{2}hf(x_0, y_0)\right)\right]$$

In general,

$$y_{n+1} = y_n + h\left[f\left(x_n + \frac{1}{2}h, y_n + \frac{1}{2}hf(x_n, y_n)\right)\right], n = 0, 1, 2, ...$$

This is called **modified Euler's formula**.

Note: It can be observe that the error in Modified Euler's method is of order h^3.

ILLUSTRATIVE EXAMPLES

Example 1. Compute y at $x = 0.25$ by Modified Euler's method given that $y' = 2xy$, $y(0) = 1$.

Solution. Here $x_0 = 0$, $y_0 = 1$, $f(x, y) = y' = 2xy$, $h = 0.25$

By Modified Euler's method, we have

$$y_{n+1} = y_n + h\left[f\left(x_n + \frac{1}{2}h, y_n + \frac{1}{2}hf(x_n, y_n)\right)\right]$$

$$\therefore \quad y_1 = y_0 + h\left[f\left(x_0 + \frac{1}{2}h, y_0 + \frac{1}{2}hf(x_0, y_0)\right)\right]$$

$f(x_0, y_0) = 2x_0 y_0 = 2 \times 0 \times 1 = 0$

$$\therefore \quad y_1 = 1 + (0.25)\left[f\left(0 + \frac{1}{2} \times 0.25, 1 + \frac{1}{2} \times 0.25 \times 0\right)\right]$$

$\quad = 1 + (0.25)\ [f\ (0.125, 1)]$

$\quad = 1 + (0.25)\ [2 \times 0.125 \times 1] = 1.0625$

$\therefore \quad y_1 = y\ (0.25) = 1.0625$

Example 2. Using modified Euler's method, find $y\ (0.1)$, $y\ (0.2)$ given that $\frac{dy}{dx} = x^2 + y^2$, $y(0) = 1$.

Solution. Here $x_0 = 0$, $y_0 = 1$, $h = 0.1$, $f(x, y) = x^2 + y^2$

By Modified Euler's method, we have

$$y_1 = y_0 + h\left[f\left(x_0 + \frac{1}{2}h, y_0 + \frac{1}{2}hf(x_0, y_0)\right)\right] \quad ...(1)$$

Now $y_0 + \frac{1}{2}hf(x_0, y_0) = y_0 + \frac{1}{2}h(x_0^2 + y_0^2)$

$$= 1 + \frac{0.1}{2}(0 + 1) = 1.05$$

Put in equation (1), we get

$y_1 = 1 + (0.1)\ [f\ (0.05, 1.05)]$

$y_1 = 1 + (0.1)\ [(0.05)^2 + (1.05)^2] = 1.1105$

$\therefore \quad y_1 = y(0.1) = 1.1105$

Again $\quad y_2 = y_1 + h\left[f\left(x_1 + \frac{1}{2}h, y_1 + \frac{1}{2}hf(x_1, y_1)\right)\right] \quad ...(2)$

$f\ (x_1, y_1) = f\ (0.1, 1.1105) = (0.1)^2 + (1.1105)^2 = 1.24321$

$y_1 + \frac{1}{2}hf(x_1, y_1) = 1.1105 + (0.05)\ (1.24321) = 1.172660$

∴ From equation (2), we get
$$y_2 = 1.1105 + (0.1) [f(0.15, 1.172660)]$$
$$y_2 = 1.1105 + (0.1) [(0.15)^2 + (1.172660)^2]$$
$$y_2 = 1.25026$$
∴ $y_2 = y(0.2) = 1.25026$

Example 3. Using modified Euler's method, obtain a solution of the equation,

$$\frac{dy}{dx} = x + |\sqrt{y}| = f(x, y)$$ with initial condition $y = 1$ at $x = 0$ for the range $0 \leq x \leq 0.4$ in steps of 0.2.

Solution. Here, $x_0 = 0$, $y_0 = 1$, $h = 0.2$, $x_1 = 0.2$, $x_2 = 0.4$

$$f(x, y) = x + |\sqrt{y}|$$

∴ $f(x_0, y_0) = x_0 + |\sqrt{y_0}| = 0 + 1 = 1$

By modified Euler's method, we have

$$y_1 = y_0 + h \left[f \left(x_0 + \frac{1}{2}h, y_0 + \frac{1}{2}hf(x_0, y_0) \right) \right] \quad ...(1)$$

$$y_0 + \frac{1}{2}hf(x_0, y_0) = 1 + \frac{0.2}{2}(1) = 1.1$$

Putting in (1), we get
$$y_1 = 1 + (0.2) [f(0.1, 1.1)]$$
$$y_1 = 1 + (0.2) [0.1 + |\sqrt{1.1}|] = 1.2297618$$

∴ $y_1 = y(0.2) = 1.2297618$

Again,
$$y_2 = y_1 + h \left[f \left(x_1 + \frac{1}{2}h, y_1 + \frac{1}{2}hf(x_1, y_1) \right) \right] \quad ...(2)$$

$$y_1 + \frac{1}{2}hf(x_1, y_1) = 1.2297618 + \frac{0.2}{2} f(0.2, 1.2297618)$$

$$= 1.2297618 + 0.1 \left[(0.2) + |\sqrt{1.2297618}| \right]$$

$$= 1.3606564$$

∴ From equation (2), we get
$$y_2 = 1.2297618 + 0.2 [f(0.3, 1.3606564)]$$
$$= 1.2297618 + (0.2) \left[(0.3) + |\sqrt{1.3606564}| \right]$$
$$y_2 = 1.5230562$$

∴ $y_2 = y(0.4) = 1.5230562$

Example 4. Solve $\dfrac{dy}{dx} = 1 - y$, $y(0) = 0$, in the range $0 \leq x \leq 0.3$ using Modified Euler's method by taking $h = 0.1$

Solution. Here $x_0 = 0$, $y_0 = 0$, $h = 0.1$, $x_1 = 0.1$, $x_2 = 0.2$, $x_3 = 0.3$

$$\dfrac{dy}{dx} = f(x, y) = 1 - y$$

$\therefore \quad f(x_0, y_0) = 1 - y_0 = 1 - 0 = 1$

$\therefore$ By Modified Euler's method, we have

$$y_1 = y_0 + h\left[f\left(x_0 + \dfrac{1}{2}h, y_0 + \dfrac{1}{2}hf(x_0, y_0) \right) \right] \quad \ldots(1)$$

Now

$$y_0 + \dfrac{1}{2}hf(x_0, y_0) = 0 + \dfrac{0.1}{2} \times 1 = 0.05, \text{ Put in (1)}$$

$$y_1 = 0 + 0.1 \ [f(0.05, 0.05)]$$
$$y_1 = 0.1 \ [1 - 0.05] = 0.095$$
$\therefore \quad y_1 = y(0.1) = 0.095$

Again, $\quad x_1 = 0.1, \ y_1 = 0.095$

$$y_2 = y_1 + h\left[f\left(x_1 + \dfrac{1}{2}h, y_1 + \dfrac{1}{2}hf(x_1, y_1) \right) \right] \quad \ldots(2)$$

Now

$$y_1 + \dfrac{1}{2}hf(x_1, y_1) = 0.095 + \dfrac{0.1}{2} f(0.1, 0.095)$$

$$= 0.095 + 0.05 \ (1 - 0.095)$$
$$= 0.14025, \text{ Put in equation (2), we get}$$
$$y_2 = 0.095 + 0.1 \ [f(0.15, 0.14025)]$$
$$= 0.095 + 0.1 \ (1 - 0.14025) = 0.095 + 0.085975$$
$$y_2 = 0.180975$$

$\therefore \quad y_2 = y(0.2) = 0.180975$

Again, $\quad x_2 = 0.2, \ y_2 = 0.180975$

$$y_3 = y_2 + h\left[f\left(x_2 + \dfrac{1}{2}h, y_2 + \dfrac{1}{2}hf(x_2, y_2) \right) \right] \quad \ldots(3)$$

Now

$$y_2 + \dfrac{1}{2}hf(x_2, y_2) = 0.180975 + \dfrac{0.1}{2}f \ (0.2, 0.180975)$$

$$= 0.180975 + 0.05 \ (1 - 0.180975)$$
$$= 0.180975 + 0.0409512$$
$$= 0.2219262, \text{ put in equation (3), we get}$$

$$y_3 = 0.180975 + 0.1\ [f\ (0.25,\ 0.2219262)]$$
$$= 0.180975 + 0.1\ (1 - 0.2219262)$$
$$= 0.180975 + 0.0778073$$
$$y_3 = 0.2587823$$
$$\therefore \quad y_3 = (0.3) = 0.2587823$$

EXERCISE 6.5
---

1. Using Modified Euler's method to obtain $y\ (0.2)$, for the differential equation $\dfrac{dy}{dx} = \log(x+y)$ with $y(0) = 1$, $h = 0.2$

2. Using Modified Euler's method to obtain $y\ (0.2)$, $y\ (0.4)$, $y\ (0.6)$, for the differential equation $\dfrac{dy}{dx} = y - x^2, y(0) = 1$.

3. Solve the equation $\dfrac{dy}{dx} = 1 - y$, given $y(0) = 0$, using Modified Euler's method and tabulate the solutions at $x = 0.1, 0.2$ and 0.3. compare your results with the exact solutions.

4. Taking $h = 0.05$, determine the value of y at $x = 0.1$ by modified Euler's method, given that $\dfrac{dy}{dx} = x^2 + y; y(0) = 1$.

5. Using Modified Euler's method to compute y for $x = 0.05$ and 0.10. Given that $\dfrac{dy}{dx} = x + y$ with $x_0 = 0$, $y_0 = 1$.

6. Solve $\dfrac{dy}{dx} = y - \dfrac{2x}{y}, y(0) = 1$ in the range $0 \le x \le 0.2$ by using Modified Euler's method with $h = 0.1$

ANSWERS
---

1. $y\ (0.2) = 1.0095$ 2. $1.218, 1.467, 1.737$
3. $y\ (0.1) = 0.095$, $y\ (0.2) = 0.18098$, $y\ (0.3) = 0.258787$
   Exact solution
   $y\ (0.1) = 0.09516258$, $y\ (0.2) = 0.181269247$
   $y\ (0.3) = 0.259181779$
4. $y\ (0.05) = 1.0513$, $y\ (0.10) = 1.1055$
5. $1.0526, 1.1104$
6. $y\ (0.1) = 1.0954762$
   $y\ (0.2) = 1.1832984$

6.10 RUNGE-KUTTA METHOD

From the previous discussion it is clear that the Euler's method suffers from in accuracy whereas Taylor's method is a time consuming and complicated procedure for most of the problems. But, in Runge-Kutta methods, the derivatives of higher order are not required and give greater accuracy. Runge-kutta methods are known by their order.

6.11 FIRST ORDER RUNGE-KUTTA METHOD

Consider the differential equation

$\frac{dy}{dx} = f(x, y)$ with initial condition $y(x_0) = y_0$.

By Euler's method, we have

$$y_1 = y_0 + hf(x_0, y_0) = y_0 + hy'_0 \qquad ...(1)$$

$[\because y' = f(x, y)]$

and from Taylor's series method

$$y_1(x) = y_1(x_0 + h) = y_0 + hy'_0 + \frac{h^2}{2}y''_0 + ... \qquad ...(2)$$

comparing (1) and (2) we observe that Euler's method and Taylor's series method agrees upto the term h.

Hence, Runge-kutta method of order 1 is nothing but Euler's method.

6.12 SECOND ORDER RUNGE-KUTTA METHOD

Consider the differential equation $\frac{dy}{dx} = f(x, y)$ with initial condition

$y(x_0) = y_0$

Runge-kutta method of second order is given by

$$k_1 = hf(x_0, y_0)$$
$$k_2 = hf(x_0 + h, y_0 + k_1)$$
$$\Delta y = \frac{1}{2}(k_1 + k_2)$$

and $\qquad y_1 = y_0 + \Delta y = y_0 + \frac{1}{2}(k_1 + k_2)$

Here, we can observe that Runge-kutta method of second order is nothing but Euler's modified method. The inherent error in the second order Runge-kutta method is of order h^3.

6.13 THIRD ORDER RUNGE-KUTTA METHOD

Consider the differential equation $\frac{dy}{dx} = f(x, y)$ with initial condition $y(x_0) = y_0$

Runge-kutta method of third order is given by

$$y_1 = y_0 + \Delta y \qquad ...(1)$$

Numerical Solution of Ordinary Differential Equations 273

where
$$\Delta y = \frac{1}{6}(k_1 + 4k_2 + k_3)$$
$$k_1 = hf(x_0, y_0)$$
$$k_2 = hf\left(x_0 + \frac{h}{2}, y_0 + \frac{k_1}{2}\right)$$
$$k_3 = hf(x_0 + h, y_0 + 2k_2 - k_1)$$

Formula (1) can be generalized for successive approximations. Equation (1) agrees with Taylor's series expansion for y_1 upto and including terms in h^3. This method is also known as Runge's method. The inherent error in the third order Runge-kutta method is of order h^4.

6.14 FOURTH ORDER RUNGE-KUTTA METHOD

Consider the differential equation $\frac{dy}{dx} = f(x, y)$ with initial condition $y(x_0) = y_0$.

The fourth order Runge-kutta method is given by
$$k_1 = hf(x_0, y_0)$$
$$k_2 = hf\left(x_0 + \frac{h}{2}, y_0 + \frac{k_1}{2}\right)$$
$$k_3 = hf\left(x_0 + \frac{h}{2}, y_0 + \frac{k_2}{2}\right)$$
$$k_4 = hf(x_0 + h, y_0 + k_3)$$
$$\Delta y = \frac{1}{6}(k_1 + 2k_2 + 2k_3 + k_4)$$

and
$$y_1 = y_0 + \Delta y, \; x_1 = x_0 + h$$

similarly for the next intervals.

The inherent error in the fourth order Runge-kutta method is of order h^5.

ILLUSTRATIVE EXAMPLES

Example 1. Given $y' = x^2 - y$, $y(0) = 1$, find $y(0.1)$, $y(0.2)$ by using Runge-kutta methods of (i) second order (ii) third order, and (iii) fourth order.

Solution. Given $y' = f(x, y) = x^2 - y$, $x_0 = 0$, $y_0 = 1$

$\therefore \quad f(x_0, y_0) = x_0^2 - y_0 = 0 - 1 = -1$

let $h = 0.1$

(i) Runge-kutta method of 2nd order

Here, $k_1 = hf(x_0, y_0) = 0.1(-1) = -0.1$
$k_2 = hf(x_0 + h, y_0 + k_1)$
$= hf(0.1, 0.9)$

$$k_2 = (0.1) [(0.1)^2 - 0.9] = -0.089$$

$$\therefore \quad \Delta y = \frac{1}{2}(k_1 + k_2) = \frac{1}{2}[-0.1 - 0.089] = -0.0945$$

$$\therefore \quad y_1 = y(0.1) = y_0 + \Delta y = 1 - 0.0945 = 0.9055$$

Again, taking $x_1 = 0.1, y_1 = 0.9055$ in place of (x_0, y_0) repeating the process, we get

$$k_1 = hf(x_1, y_1) = h(x_1^2 - y_1)$$
$$k_1 = (0.1) [(0.1)^2 - 0.9055] = -0.08955$$
$$k_2 = hf[x_1 + h, y_1 + k_1] = hf[0.2, 0.81595]$$
$$k_2 = 0.1 [(0.2)^2 - 0.81595] = -0.077595$$

$$\therefore \quad \Delta y = \frac{1}{2}(k_1 + k_2) = \frac{1}{2}[-0.08955 - 0.077595]$$

$$\Delta y = -0.0835725$$

$$\therefore \quad y_2 = y(0.2) = y_1 + \Delta y = 0.9055 - 0.0835725$$
$$y(0.2) = 0.8219275$$

(ii) Runge-kutta method of 3rd order

Here $x_0 = 0, y_0 = 1, h = 0.1$

$$k_1 = hf(x_0, y_0) = (0.1) f(0, 1) = (0.1) [0 - 1] = -0.1$$

$$k_2 = hf\left(x_0 + \frac{h}{2}, y_0 + \frac{k_1}{2}\right) = (0.1) f\left(\frac{0.1}{2}, 1 - \frac{0.1}{2}\right)$$

$$= (0.1) f(0.05, 0.95)$$
$$= (0.1) [(0.05)^2 - 0.95] = (0.1) [0.0025 - 0.95]$$
$$k_2 = -0.09475$$
$$k_3 = hf(0.1, 1 + 2(-0.09475) + 0.1)$$
$$k_3 = (0.1) [(0.1)^2 - 0.9105] = -0.09005$$

$$\Delta y = \frac{1}{6} [k_1 + 4k_2 + k_3]$$

$$= \frac{1}{6} [-0.1 + 4(-0.09475) - 0.09005]$$

$$\Delta y = -0.0917083$$

$$\therefore \quad y_1 = y_0 + \Delta y = 1 - 0.0917083$$
$$y_1 = y(0.1) = 0.9082916$$

Again, taking $x_1 = 0.1, y_1 = 0.9082916$ in place of (x_0, y_0) and repeating the process, we get

$$k_1 = hf(x_1, y_1) = (0.1) f(0.1, 0.9082916)$$
$$k_1 = (0.1) [(0.1)^2 - 0.9082916] = -0.0898291$$

$$k_2 = hf\left(x_1 + \frac{h}{2}, y_1 + \frac{k_1}{2}\right) = hf(0.15, 0.863377)$$
$$k_2 = (0.1)\,[(0.15)^2 - 0.863377] = -0.0840877$$
$$k_3 = hf(x_1 + h, y_1 + 2k_2 - k_1)$$
$$= hf(0.2, 0.8299453) = (0.1)\,[(0.2)^2 - 0.8299453]$$
$$k_3 = -0.0789945$$

$\therefore \quad \Delta y = \dfrac{1}{6}(k_1 + 4k_2 + k_3)$

$$= \frac{1}{6}\,(-0.0898291 + 4\,(-0.0840877)$$
$$+ (-0.0789945)]$$
$$\Delta y = -0.0841957$$

$\therefore \quad y_2 = y_1 + \Delta y = 0.9082916 - 0.0841957$
$\quad y_2 = y(0.2) = 0.8240958$

(iii) Runge-kutta method of 4th order

Here $\quad k_1 = hf(x_0, y_0) = 0.1\,f(0, 1) = 0.1\,[0 - 1] = -0.1$

$$k_2 = hf\left(x_0 + \frac{h}{2}, y_0 + \frac{k_1}{2}\right) = (0.1)\,f(0.05, 0.95)$$
$$k_2 = (0.1)\,[(0.05)^2 - 0.95] = -0.09475$$

$$k_3 = hf\left(x_0 + \frac{h}{2}, y_0 + \frac{k_2}{2}\right)$$
$$= (0.1)\,f(0.05, 0.952625)$$
$$= (0.1)\,[(0.05)^2 - 0.952625]$$
$$k_3 = -0.0950125$$
$$k_4 = hf(x_0 + h, y_0 + k_3)$$
$$= (0.1)\,f[0.1, 0.9049875]$$
$$= (0.1)\,[(0.1)^2 - 0.9049875]$$
$$k_4 = -0.0894987$$

$\therefore \quad \Delta y = \dfrac{1}{6}(k_1 + 2k_2 + 2k_3 + k_4)$

$$\Delta y = \frac{1}{6}\,(-0.1 + 2\,(-0.09475) + 2$$
$$(-0.0950125 - 0.0894987)$$
$$\Delta y = -0.0948372$$

$\therefore \quad y_1 = y(0.1) = y_0 + \Delta y$
$\quad = 1 - 0.0948372$

$\therefore \quad y(0.1) = 0.9051628$

Taking $x_1 = 0.1$, $y_1 = 0.9051628$ in place of (x_0, y_0) and repeating the process, we get

$$k_1 = hf(x_1, y_1) = hf(0.1, 0.9051628)$$
$$= 0.1 f(0.1, 0.9051628)$$
$$k_1 = (0.1)[(0.1)^2 - 0.9051628] = -0.0895162$$
$$k_1 = -0.0895162$$

$$k_2 = hf\left(x_1 + \frac{h}{2}, y_1 + \frac{k_1}{2}\right) = hf(0.15, 0.8604046)$$
$$k_2 = (0.1)[(0.15)^2 - 0.8604046] = -0.0837904$$

$$k_3 = hf\left(x_1 + \frac{h}{2}, y_1 + \frac{k_2}{2}\right) = hf(0.15, 0.8632674)$$
$$k_3 = (0.1)[(0.15)^2 - 0.8632674] = -0.0840767$$

$$k_4 = hf(x_1 + h, y_1 + k_3) = hf(0.2, 0.8210859)$$
$$= (0.1)[(0.2)^2 - 0.8210859] = -0.0781085$$

$\therefore \quad \Delta y = \frac{1}{6}(k_1 + 2k_2 + 2k_3 + k_4)$

$$= \frac{1}{6}[-0.0895162 + 2(-0.0837904) + 2(-0.0840767) - 0.0781085]$$

$\Delta y = -0.0838931$

$\therefore \quad y_2 = y(0.2) = y_1 + \Delta y = 0.9051628 - 0.0838931$

$y(0.2) = 0.8212697$

Example 2. Obtain the values of y at $x = 0.1$, $x = 0.2$ by using R.K. Method of (*i*) second order (*ii*) third order and (*iii*) fourth order for the differential equation $y' = -y$ with $y(0) = 1$.

Solution. Here $x_0 = 0$, $y_0 = 1$, $y' = -y$ and $h = 0.1$

(i) Runge-kutta method of 2nd order

$$k_1 = hf(x_0, y_0) = 0.1 f(0, 1)$$
$$k_1 = 0.1[-1] = -0.1$$
$$k_2 = hf(x_0 + h, y_0 + k_1)$$
$$= (0.1) f(0.1, 1 - 0.1) = (0.1) f[0.1, 0.90]$$
$$k_2 = (0.1)[-0.90] = -0.09$$

$\therefore \quad \Delta y = \frac{1}{2}(k_1 + k_2)$

$$= \frac{1}{2}(-0.1 - 0.09)$$

$\Delta y = -0.095$

$\therefore$ $\qquad y_1 = y(0.1) = y_0 + \Delta y$
$\qquad y(0.1) = 1 - 0.095 = 0.905$
Again $\qquad x_1 = 0.1, y_1 = 0.905$ and $h = 0.1$
$\qquad k_1 = hf(x_1, y_1) = (0.1) f(0.1, 0.905)$
$\qquad k_1 = (0.1)[-0.905] = -0.0905$
$\qquad k_2 = hf(x_1 + h, y_1 + k_1)$
$\qquad\quad = (0.1) f(0.1 + 0.1, 0.905 - 0.0905)$
$\qquad\quad = (0.1) f(0.2, 0.8145]$
$\qquad k_2 = (0.1)[-0.8145] = -0.08145$

$$\Delta y = \frac{1}{2}(k_1 + k_2) = \frac{1}{2}(-0.0905 - 0.08145)$$

$\qquad \Delta y = -0.085975$
$\therefore \qquad y_2 = y(0.2) = y_1 + \Delta y$
$\Rightarrow \qquad y(0.2) = 0.905 - 0.085975$
$\qquad y(0.2) = 0.819025$

(ii) Runge-kutta method of 3rd order

Here $\qquad k_1 = hf(x_0, y_0) = -0.1$

$$k_2 = hf\left(x_0 + \frac{h}{2}, y_0 + \frac{k_1}{2}\right) = -0.095$$

$\qquad k_3 = hf(x_0 + h, y_0 + 2k_2 - k_1)$
$\qquad k_3 = 0.1 f(0.1, 0.9) = (0.1)(-0.9) = -0.09$

$$\Delta y = \frac{1}{6}(k_1 + 4k_2 + k_3)$$

$\qquad y(0.1) = y_1 = y_0 + \Delta y = 1 - 0.09 = 0.91$
Again, $\qquad k_1 = hf(x_1, y_1) = (0.1)(-0.91) = -0.091$

$$k_2 = hf\left(x_1 + \frac{h}{2}, y_1 + \frac{k_1}{2}\right)$$

$\qquad k_2 = (0.1) f(0.15, 0.865) = (0.1)(-0.865) = -0.0865$
$\qquad k_3 = hf(x_1 + h, y_1 + 2k_2 - k_1)$
$\qquad k_3 = (0.1) f(0.2, 0.828) = (0.1)(-0.828) = -0.0828$

$$\Delta y = \frac{1}{6}(k_1 + 4k_2 + k_3) = \frac{1}{6}(-0.091 - 0.3460 - 0.0828)$$

$\qquad \Delta y = -0.086633333$
$\therefore \qquad y_2 = y_1 + \Delta y = 0.91 - 0.086633333$
$\therefore \qquad y_2 = y(0.2) = 0.823366666$

(iii) Runge-kutta method of 4th order

Here $\quad k_1 = hf(x_0, y_0) = (0.1) f(0, 1) = (0.1)(-1) = -0.1$

$$k_2 = hf\left(x_0 + \frac{h}{2}, y_0 + \frac{k_1}{2}\right)$$
$$= (0.1) f(0.05, 0.95) = (0.1)(-0.95)$$
$$k_2 = -0.095$$

$$k_3 = hf\left(x_0 + \frac{h}{2}, y_0 + \frac{k_2}{2}\right) = (0.1) f(0.05, 0.9525)$$
$$k_3 = -0.09525$$
$$k_4 = hf(x_0 + h, y_0 + k_3) = (0.1) f(0.1, 0.90475)$$
$$= (0.1)(-0.90475) = -0.090475$$

$$\Delta y = \frac{1}{6}(k_1 + 2k_2 + 2k_3 + k_4)$$

$$\Delta y = \frac{1}{6}(-0.1 + 2(-0.095) + 2(-0.09525) - 0.090475)$$

$$\Delta y = \frac{1}{6}(-0.1 - 0.19 - 0.1905 - 0.090475)$$

$$\Delta y = -0.0951625$$

$\therefore \quad y_1 = y(0.1) = y_0 + \Delta y = 1 - 0.0951625$

$\Rightarrow \quad y(0.1) = 0.9048375$

Again $\quad x_1 = 0.1, y_1 = 0.9048375$

$k_1 = hf(x_1, y_1) = (0.1) f(0.1, 0.9048375)$
$k_1 = (0.1)(-0.9048375) = -0.09048375$

$$k_2 = hf\left(x_1 + \frac{h}{2}, y_1 + \frac{k_1}{2}\right) = (0.1) f(0.15, 0.8595956)$$
$$k_2 = (0.1)(-0.8595956) = -0.08595956$$

$$k_3 = hf\left(x_1 + \frac{h}{2}, y_1 + \frac{k_2}{2}\right)$$
$$= (0.1) f(0.15, 0.8618577) = (0.1)(-0.8618577)$$
$$k_3 = -0.08618577$$
$$k_4 = hf(x_1 + h, y_1 + k_3)$$
$$= (0.1) f(0.2, 0.8186517) = (0.1)(-0.8186517)$$
$$k_4 = -0.08186517$$

$$\Delta y = \frac{1}{6}(k_1 + 2k_2 + 2k_3 + k_4)$$

$$\Delta y = \frac{1}{6} \{-0.09048375 + 2(-0.08595956)$$
$$+ 2(-0.08618577) - 0.08186517)\}$$
$$\Delta y = -0.0861066067$$
∴ $y_2 = y(0.2) = y_1 + \Delta y = 0.9048375 - 0.0861066067$
⇒ $y(0.2) = 0.81873089$

Example 3. Apply the Fourth order Runge-kutta method to find y(0.2) for the differential equation $y' = x + y$ with $y(0) = 1$.

Solution. Here $x_0 = 0$, $y_0 = 1$ and take $h = 0.1$

By Runge kutta method of 4th order, we have

$$k_1 = hf(x_0, y_0) = (0.1) f(0, 1) = (0.1)(0 + 1) = 0.1$$

$$k_2 = hf\left(x_0 + \frac{h}{2}, y_0 + \frac{k_1}{2}\right) = (0.1) f(0.05, 1.05)$$

$$k_2 = (0.1)(0.05 + 1.05) = 0.11$$

$$k_3 = hf\left(x_0 + \frac{h}{2}, y_0 + \frac{k_2}{2}\right)$$

$$k_3 = (0.1) f(0.05, 1.055) = (0.1)(0.05 + 1.055) = 0.1105$$

$$k_4 = hf(x_0 + h, y_0 + k_3) = (0.1) f(0.1, 1.1105)$$
$$= (0.1)(0.1 + 1.1105) = 0.12105$$

∴ $$\Delta y = \frac{1}{6}(k_1 + 2k_2 + 2k_3 + k_4)$$

$$= \frac{1}{6}(0.1 + 0.22 + 0.2210 + 0.12105) = 0.110341667$$

∴ $y(0.1) = y_1 = y_0 + \Delta y = 1 + 0.110341667 = 1.110341667$
∴ $y(0.1) \approx 1.110342$

Again, $x_1 = 0.1$, $y_1 = 1.110342$
$$k_1 = hf(x_1, y_1) = (0.1)(x_1 + y_1) = (0.1)(0.1 + 1.110342)$$
$$k_1 = 0.1210342$$

$$k_2 = hf\left(x_1 + \frac{h}{2}, y_1 + \frac{k_1}{2}\right) = (0.1) f(0.15, 1.170859)$$
$$k_2 = (0.1)(0.15 + 1.170859) = 0.1320859$$

$$k_3 = hf\left(x_1 + \frac{h}{2}, y_1 + \frac{k_2}{2}\right) = (0.1) f(0.15, 1.1763848)$$

$$k_3 = (0.1)(0.15 + 1.1763848) = 0.13263848$$
$$k_4 = hf(x_1 + h, y_1 + k_3) = (0.1) f(0.2, 1.24298048)$$
$$k_4 = 0.144298048$$

$\therefore \quad \Delta y = \dfrac{1}{6}(k_1 + 2k_2 + 2k_3 + k_4) = \dfrac{1}{6}(0.1210342 + 2(0.1320859)$

$\qquad\qquad + 2\,(0.13263848) + 0.144298048)$

$\qquad \Delta y = \dfrac{1}{6}(0.794781008) = 0.13244663501$

$\therefore \quad y_2 = y\,(0.2) = y_1 + \Delta y = 1.10342 + 0.132463501$

$\qquad\qquad = 1.242805501$

$\Rightarrow \quad y\,(0.2) \approx 1.2428055$

Example 4. Using R.K. Method of fourth order, find $y\,(0.2)$ correct to 4 decimal places if $y' = y - x,\ y(0) = 2$ taking $h = 0.1$.

Solution. Here $x_0 = 0,\ y_0 = 2,\ h = 0.1$

$\qquad k_1 = hf\,(x_0, y_0) = 0.1\,f\,(0.2) = (0.1)\,(2 - 0) = 0.2$

$\qquad k_2 = hf\left(x_0 + \dfrac{h}{2},\ y_0 + \dfrac{k_1}{2}\right) = (0.1)\,f\,(0.05, 2.1)$

$\qquad k_2 = (0.1)\,(2.1 - 0.05) = 0.205$

$\qquad k_3 = hf\left(x_0 + \dfrac{h}{2},\ y_0 + \dfrac{k_2}{2}\right) = (0.1)\,f\,(0.05, 2.1025)$

$\qquad k_3 = (0.1)\,(2.1025 - 0.05) = 0.20525$

$\qquad k_4 = hf\,(x_0 + h,\ y_0 + k_3) = (0.1)\,f\,(0.1, 2.20525)$

$\qquad k_4 = (0.1)\,(2.20525 - 0.1) = 0.210525$

$\qquad \Delta y = \dfrac{1}{6}(k_1 + 2k_2 + 2k_3 + k_4)$

$\qquad\qquad = \dfrac{1}{6}\{0.2 + 2\,(0.205) + 2\,(0.20525) + 0.210525\}$

$\qquad \Delta y = \dfrac{1}{6}(1.23102525) = 0.205170833$

$\therefore \quad y_1 = y\,(0.1) = y_0 + \Delta y = 2 + 0.205170833 = 2.205170833$

$\therefore \quad y_1 \approx 2.2052$

Again, $\quad x_1 = 0.1,\ y_1 = 2.205170833$

$\qquad k_1 = hf\,(x_1, y_1) = (0.1)\,f\,(0.1, 2.205170833)$

$\qquad k_1 = (0.1)\,(2.205170833 - 0.1) = 0.2105170833$

$\qquad k_2 = hf\left(x_1 + \dfrac{h}{2},\ y_1 + \dfrac{k_1}{2}\right)(0.1)f\,(0.15, 2.310429375)$

$\qquad k_2 = (0.1)\,(2.310429375 - 0.15) = 0.216042937$

$\qquad k_3 = hf\left(x_1 + \dfrac{h}{2},\ y_1 + \dfrac{k_2}{2}\right) = (0.1)\,f\,(0.15, 2.313192302)$

$$k_3 = (0.1)(2.313192307 - 0.15) = 0.21631923$$
$$k_4 = hf(x_1 + h, y_1 + k_3) = (0.1)f(0.2, 2.421490063)$$
$$k_4 = (0.1)(2.421490063 - 0.2) = 0.222149006$$

$$\therefore \quad \Delta y = \frac{1}{6}(k_1 + 2k_2 + 2k_3 + k_4)$$

$$= \frac{1}{6}\{0.210517083 + 2(0.216042937) + 2(0.216319232) + 0.222149006\}$$

$$\Delta y = \frac{1}{6}(1.297390423) = 0.276231737$$

$$\therefore \quad y_2 = y(0.2) = y_1 + \Delta y = 2.205170833 + 0.216231737$$
$$\Rightarrow \quad y(0.2) = 2.42140257$$
$$\Rightarrow \quad y(0.2) \approx 2.4214$$

Example 5. Using R.K. method of fourth order, find $y(0.8)$ correct to 4 decimal places if $y' = y - x^2$, $y(0.6) = 1.7379$.

Solution. Here $x_0 = 0.6$, $y_0 = 1.7379$, $h = 0.1$, $x_1 = 0.7$, $x_2 = 0.8$

$$f(x, y) = y^1 = y - x^2$$
$$k_1 = hf(x_0, y_0) = (0.1)(0.6, 1.7379)$$
$$k_1 = (0.1)[1.7379 - (0.6)^2] = 0.1378$$

$$k_2 = hf\left(x_0 + \frac{h}{2}, y_0 + \frac{k_1}{2}\right) = (0.1)f(0.65, 1.8068)$$
$$= (0.1)[1.8068 - (0.65)^2] = 0.1384$$

$$k_3 = hf\left(x_0 + \frac{h}{2}, y_0 + \frac{k_2}{2}\right)$$
$$= (0.1)f(0.65, 1.8071)$$
$$k_3 = (0.1)[1.8071 - (0.65)^2] = 0.1385$$
$$k_4 = hf(x_0 + h, y_0 + k_3) = (0.1)f(0.7, 1.8764)$$
$$k_4 = (0.1)[1.8764 - (0.7)^2] = 0.1386$$

$$\Delta y = \frac{1}{6}[k_1 + 2k_2 + 2k_3 + k_4]$$

$$= \frac{1}{6}\{0.1378 + 2(0.1384) + 2(0.1385) + 0.1386\}$$

$$\Delta y = 0.1383666$$
$$y_1 = y(0.7) = y_0 + \Delta y = 1.7379 + 0.1383666$$
$$\Rightarrow \quad y(0.7) = 1.8762667$$
$$y(0.7) \approx 1.8763$$

Again, $\quad x_1 = 0.7, y_1 = 1.8763$
$$k_1 = hf(x_1, y_1) = (0.1)[1.8763 - (0.7)^2] = 0.1386$$

$$k_2 = hf\left(x_1 + \frac{h}{2}, y_1 + \frac{k_1}{2}\right) = (0.1)\,f\,(0.75, 1.9456)$$

$$k_2 = (0.1)\,[1.9456 - (0.75)^2] = 0.1383$$

$$k_3 = hf\left(x_1 + \frac{h}{2}, y_1 + \frac{k_2}{2}\right)$$

$$= (0.1)\,f\,(0.75, 1.9455)$$

$$k_3 = (0.1)\,[1.9455 - (0.75)^2] = 0.1383$$

$$k_4 = hf\,(x_1 + h, y_1 + k_3)$$

$$= (0.1)\,f\,(0.8, 2.0146)$$

$$k_4 = (0.1)\,[2.0146 - (0.8)^2] = 0.1375$$

$\therefore \quad \Delta y = \dfrac{1}{6}[k_1 + 2k_2 + 2k_3 + k_4]$

$$= \frac{1}{6}\,[0.1386 + 2\,(0.1383) + 2\,(0.1383) + 0.1375]$$

$$\Delta y = 0.1382166$$

$\therefore \quad y_2 = y\,(0.8) = y_1 + \Delta y$

$y\,(0.8) = 1.8763 + 0.1382166$

$y\,(0.8) = 2.0145167$

$y\,(0.8) \approx 2.0145$

Example 6. Given $\dfrac{dy}{dx} = \dfrac{y-x}{y+x}$, $y(0) = 1$, $h = 0.2$. Find $y\,(0.2)$ by using Runge-kutta method of order 4.

Solution. Here $x_0 = 0$, $y_0 = 1$, $h = 0.2$

$$f(x, y) = \frac{y-x}{y+x} \Rightarrow f(x_0, y_0) = \frac{y_0 - x_0}{y_0 + x_0} = \frac{1-0}{1+0} = 1$$

$$k_1 = hf\,(x_0, y_0) = 0.2 \times 1 = 0.2$$

$$k_2 = hf\left(x_0 + \frac{h}{2}, y_0 + \frac{k_1}{2}\right) = (0.2)\,f\,(0.1, 1.1)$$

$$k_2 = (0.2)\left[\frac{1.1 - 0.1}{1.1 + 0.1}\right] = 0.2 \times 0.8333 = 0.1667$$

$$k_3 = hf\left(x_0 + \frac{h}{2}, y_0 + \frac{k_2}{2}\right) = (0.2)\,f\,(0.1, 1.0834)$$

$$= (0.2)\left[\frac{1.0834 - 0.1}{1.0834 + 0.1}\right] = 0.1661906$$

$$k_3 = 0.1662$$

$$k_4 = hf(x_0 + h, y_0 + k_3)$$
$$= (0.2) f(0.2, 1.1662) = (0.2) \left[\frac{1.1662 - 0.2}{1.1662 + 0.2}\right]$$
$$k_4 = (0.2) \left[\frac{0.9662}{1.3662}\right] = 0.1414434$$
$$\Delta y = \frac{1}{6}[k_1 + 2k_2 + 2k_3 + k_4]$$
$$= \frac{1}{6}[0.2 + 2(0.1667) + 2(0.1662) + 0.1414434]$$
$$\Delta y = 0.16787339$$
$$y_1 = y(0.2) = y_0 + \Delta y = 1 + 0.1678739$$
$$y(0.2) = 1.1678739$$

Example 7. Given $\frac{dy}{dx} = xy^{1/3}$, $y(1) = 1$. Find $y(1.1)$ by Runge-kutta method.

Solution. Here $x_0 = 1$, $y_0 = 1$, $h = 0.1$
$$f(x, y) = xy^{1/3}$$
$$k_1 = hf(x_0, y_0) = (0.1)(1)(1)^{1/3} = 0.1$$
$$k_2 = hf\left(x_0 + \frac{h}{2}, y_0 + \frac{k_1}{2}\right) = (0.1) f(1.05, 1.05)$$
$$= (0.1)(1.05)(1.05)^{1/3} = 0.10672$$
$$k_3 = hf\left(x_0 + \frac{h}{2}, y_0 + \frac{k_2}{2}\right)$$
$$= (0.1) f(1.05, 1.05336)$$
$$= (0.1)(1.05)(1.05336)^{1/3}$$
$$k_3 = 0.10684$$
$$k_4 = hf(x_0 + h, y_0 + k_3)$$
$$= (0.1) f(1.1, 1.10684)$$
$$= (0.1)(1.1)(1.10684)^{1/3}$$
$$k_4 = 0.11379$$
$$\Delta y = \frac{1}{6}(k_1 + 2k_2 + 2k_3 + k_4)$$
$$= \frac{1}{6}\{0.1 + 2(0.10672) + 2(0.10684) + 0.11379\}$$
$$\Delta y = 0.106882$$
$$\therefore y_1 = y(1.1) = y_0 + \Delta y$$
$$= 1 + 0.10682$$
$$y(1.1) = 1.10682$$

6.15 RUNGE-KUTTA METHOD FOR SIMULTANEOUS FIRST ORDER DIFFERENTIAL EQUATIONS

Consider the simultaneous equations

$$\frac{dy}{dx} = f_1(x, y, z) \quad \ldots(1)$$

$$\frac{dz}{dx} = f_2(x, y, z) \quad \ldots(2)$$

with the initial conditions $y(x_0) = y_0$ and $z(x_0) = z_0$.

Now, starting from (x_0, y_0, z_0), the increments Δy and Δz in y and z are given by the following formulae:

$$k_1 = hf_1(x_0, y_0, z_0)$$

$$k_2 = hf_1\left(x_0 + \frac{h}{2}, h + \frac{k_1}{2}, z_0 + \frac{l_1}{2}\right)$$

$$k_3 = hf_1\left(x_0 + \frac{h}{2}, y_0 + \frac{k_2}{2}, z_0 + \frac{l_2}{2}\right)$$

$$k_4 = hf_1(x_0 + h, y_0 + k_3, z_0 + l_3)$$

$$\Delta y = \frac{1}{6}[k_1 + 2k_2 + 2k_3 + k_4]$$

$$l_1 = hf_2(x_0, y_0, z_0)$$

$$l_2 = hf_2\left(x_0 + \frac{h}{2}, y_0 + \frac{k_1}{2}, z_0 + \frac{l_1}{2}\right)$$

$$l_3 = hf_2\left(x_0 + \frac{h}{2}, y_0 + \frac{k_2}{2}, z_0 + \frac{l_2}{2}\right)$$

$$l_4 = hf_2(x_0 + h, y_0 + k_3, z_0 + l_3)$$

$$\Delta z = \frac{1}{6}(l_1 + 2l_2 + 2l_3 + l_4)$$

Hence $y_1 = y_0 + \Delta y$,
$z_1 = z_0 + \Delta z$

To compute y_2, z_2 we simply replace x_0, y_0, z_0 by x_1, y_1, z_1 in the aboue formulae.

ILLUSTRATIVE EXAMPLES

Example 1. Find $y(0.1)$, $z(0.1)$ from the system of equations $\frac{dy}{dx} = yz + x$

$\frac{dz}{dx} = xz + y$; given that $y(0) = 1$, $z(0) = -1$ by using Runge-kutta method of 4th order.

Solution. Here $x_0 = 0, y_0 = 1, z_0 = -1, h = 0.1$

$f_1(x, y, z) = yz + x$
$f_2(x, y, z) = xz + y$

$k_1 = hf_1(x_0, y_0, z_0) = h(y_0 z_0 + x_0) = -0.1$
$l_1 = hf_2(x_0, y_0, z_0) = h(x_0 z_0 + y_0) = 0.1$

$k_2 = hf_1\left(x_0 + \dfrac{h}{2}, y_0 + \dfrac{k_2}{2}, z_0 + \dfrac{l_2}{2}\right)$

$k_2 = hf_1(0.05, 0.95, -0.95) = -0.08525$

$l_2 = hf_2\left(x_0 + \dfrac{h}{2}, y_0 + \dfrac{k_1}{2}, z_0 + \dfrac{l_1}{2}\right)$

$l_2 = hf_2(0.05, 0.95, -0.95)$
$= 0.09025$

$k_3 = hf_1\left(x_0 + \dfrac{h}{2}, y_0 + \dfrac{k_2}{2}, z_0 + \dfrac{l_2}{2}\right)$

$k_3 = hf_1(0.05, 0.957375, -0.954875) = -0.0864173$

$l_3 = hf_2\left(x_0 + \dfrac{h}{2}, y_0 + \dfrac{k_2}{2}, z_0 + \dfrac{l_2}{2}\right)$

$l_3 = hf_2(0.05, 0.957375, -0.954875) = -0.0864173$
$k_4 = hf_1(x_0 + h, y_0 + k_3, z_0 + l_3) = -0.073048$
$l_4 = hf_2(x_0 + h, y_0 + k_3, z_0 + l_3) = 0.0822679$

$\therefore \quad \Delta y = \dfrac{1}{6}(k_1 + 2k_2 + 2k_3 + k_4) = -0.0860637$

$\Delta z = \dfrac{1}{6}(l_1 + 2l_2 + 2l_3 + l_4) = 0.0907823$

$\therefore \quad y_1 = y(0.1) = y_0 + \Delta y = 1 - 0.086637 = 0.9139363$
$z_1 = z(0.1) = z_0 + \Delta z = -1 + 0.097823 = -0.9092176$

Example 2. Find $y(0.1), z(0.1)$ from the system of equations, $\dfrac{dy}{dx} = x + z$

$\dfrac{dz}{dx} = x - y^2$, given that $y(0) = 2, z(0) = 1$ by using Runge-kutta method of 4th order.

Solution. Here $x_0 = 0, y_0 = 2, z_0 = 1, h = 0.1$

$f_1(x, y, z) = x + z$
$f_2(x, y, z) = x - y^2$

$k_1 = hf_1(x_0, y_0, z_0) = (0.1) f_1(0, 2, 1) = (0.1)(0 + 1) = 0.1$
$l_1 = hf_2(x_0, y_0, z_0) = (0.1) f_2(0, 2, 1) = (0.1)(0 - 2^2) = -0.4$

$$k_2 = hf_1\left(x_0+\frac{h}{2}, y_0+\frac{k_1}{2}, z_0+\frac{l_1}{2}\right) = (0.1)\, f_1\,(0.05, 2.05, 0.8)$$

$$k_2 = (0.1)\,(0.05 + 0.8) = 0.085$$

$$l_2 = hf_2\left(x_0+\frac{h}{2}, y_0+\frac{k_1}{2}, z_0+\frac{l_1}{2}\right) = (0.1)\, f_2\,(0.05, 2.05, 0.8)$$

$$l_2 = (0.1)\,[0.05 - (2.05)^2] = -0.41525$$

$$k_3 = hf_1\left(x_0+\frac{h}{2}, y_0+\frac{k_2}{2}, z_0+\frac{l_2}{2}\right)$$

$$= (0.1)\, f_1\,(0.05, 2.0425, 0.79238)$$

$$k_3 = (0.1)\,[0.05 + 0.792238] = 0.084238$$

$$l_3 = hf_2\left(x_0+\frac{h}{2}, y_0+\frac{k_2}{2}, z_0+\frac{l_2}{2}\right)$$

$$l_3 = (0.1)\, f_2\,(0.05, 2.0425, 0.79238)$$

$$l_3 = (0.1)\,[0.05 - (2.0425)^2] = -0.4122$$

$$k_4 = hf_1\,(x_0+h, y_0+k_3, z_0+l_3)$$

$$= (0.1)\, f_1\,(0.1, 2.084238, 0.5878)$$

$$k_2 = (0.1)\,[(0.1) - (2.084238)^2] = -0.073048$$

$$l_4 = hf_2\,(x_0\,h, y_0+k_3, z_0+l_3) = 0.0822679$$

$$\Delta y = \frac{1}{6}(k_1 + 2k_2 + 2k_3 + k_4) = -0.0860637$$

$$\Delta z = \frac{1}{6}(l_1 + 2l_2 + 2l_3 + l_4) = 0.0907823$$

$$\therefore \quad y_1 = y\,(0.1) = y_0 + \Delta y = 1 - 0.0860637 = 0.9139363$$

$$z_1 = z\,(0.1) = z_0 + \Delta z = -1 + 0.0907823 = -0.9092176$$

EXERCISE 6.6

1. Use Runge-Kutta 4th order formula to find $y\,(1.4)$, if $y(1) = 2$ and $\dfrac{dy}{dx} = xy$ by taking $h = 0.2$

2. Evaluate $y\,(1.4)$ given that $\dfrac{dy}{dx} = x + y$, $y\,(1.2) = 2$ by using Runge-kutta method of order 4.

3. Given $\dfrac{dy}{dx} = x^2 + y^2$, $y(1) = 1.5$, $h = 0.1$. Find $y\,(1.2)$ by using Runge-kutta method of order 4.

4. Given that $\frac{dy}{dx} = y - x$, $y(0) = 2$, $h = 0.1$ Find $y(0.2)$ by using Runge-kutta method of order 4.

5. Using Runge-kutta method of fourth order, solve $\frac{dy}{dx} = \frac{y^2 - x^2}{y^2 + x^2}$, given $y(0) = 1$ at $x = 0.2$, $x = 0.4$

6. Compute $y(0.3)$ given $\frac{dy}{dx} + y + xy^2 = 0$, $y(0) = 1$ by taking $h = 0.1$ by using R.K. method of fourth order.

7. Solve $\frac{dy}{dx} = x + z, \frac{dz}{dx} = x - y$, given $y(0) = 0$, $z(0) = 1$, for $x = 0.0$ to 0.2 taking $h = 0.1$ by R.K. method of order 4.

8. Solve the system of equations; $\frac{dy}{dx} = xz + 1$, $\frac{dz}{dx} = -xy$ for $x = 0.3\ (0.3)(0.9)$ taking $x = 0$, $y = 0$, $z = 1$, by using R.K. method of 4th order.

9. Solve $\frac{dy}{dx} = x + z, \frac{dz}{dx} = x - y^2$ for $y(0.1)$, $z(0.1)$ given that $y(0) = 2$, $z(0) = 1$ by R.K. Method of 4th order.

ANSWERS

1. 2.994858
2. 2.7299
3. $y(1.1) = 1.8954$, $y(1.2) = 2.5041$
4. 2.4214
5. 1.19598, 1.3751
6. 0.9006, 0.8046, 0.7144
7. $y(0.1) = 0.1050$, $z(0.1) = 0.9998$, $y(0.2) = 0.2199$, $z(0.2) = 0.9986$
8. $y(0.3) = 0.3448$, $z(0.3) = 0.99$, $y(0.6) = 0.7738$, $z(0.6) = 0.9121$
 $y(0.9) = 1.255$, $z(0.9) = 0.6806$
9. $y(0.1) = 2.0845$, $z(0.1) = 0.5868$

Chapter – 7

Curve Fitting

7.1 INTRODUCTION

Let (x_i, y_i) $i = 1, 2,..., n$ be a given set of n pairs of values, where x be an independent variable and y be a dependent variable. These pairs of values of x and y give us n points on the known curve whose equation is $y = f(x)$.

Curve fitting means an exact relationship between two variables by algebraic equation, this relationship is the equation of the curve. Therefore, curve fitting means to form an equation of the curve from the given data. Curve fitting has very much importance in theoretical as well as practical statistics.

Theoretically, it is useful in study of correlation and regression and particularly it enables us to represent the relationship between two variables by simple algebraic expressions, e.g., polynomials, exponential or logarithmic functions.

7.2 METHOD OF LEAST SQUARES

The method of least squares is the most systematic procedure to fit a unique curve through given data points and is widely used in practical computations. Let (x_i, y_i) $i = 1, 2, ..., n$ be a given set of n pairs of values and suppose we want to fit a curve $y = f(x)$ to the given n pairs of values. Let Y_i be the value of y corresponding to the value of x_i of x as determined by $y = f(x)$. The value Y_i is called the estimated value of the given value y_i corresponding to x_i. If e_i is the error of approximation at $x = x_i$, then

$$e_i = y_i - Y_i = y_i - f(x_i)$$

If we minimize the sum of the squares of the errors then it is called least squares method. Let S be the sum of squares of errors then,

$$S = \sum_{i=1}^{n} e_i^2 = \sum_{i=1}^{n} [y_i - f(x_i)]^2$$

Now, we have to minimize S.

7.3 FITTING OF A STRAIGHT LINE BY METHOD OF LEAST SQUARES

Let $y = a + bx$ be the straight line to be fitted to the given data (x_i, y_i); $i = 1, 2, ..., n$. The problem is to determine 'a' and 'b' so that the line is the line of best fit.

Let $P_i(x_i, y_i)$ be any general point in the scatter diagram. Draw P_iR perpendicular to x-axis meeting the line at Q_i. Abscissa of Q_i is x_i and since Q_i lies on the line, its ordinate is $a + bx_i$. Hence the coordinate of Q_i are $(x_i, a + bx_i)$.

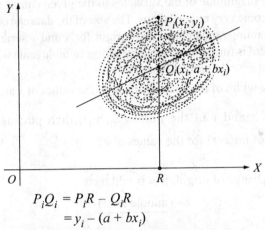

$$P_iQ_i = P_iR - Q_iR$$
$$= y_i - (a + bx_i)$$

is called the error of estimate or the residual for y_i. According to the principle of least squares, we have to determine 'a' and 'b' so that

$$S = \sum_{i=1}^{n}(P_iQ_i)^2 = \sum_{i=1}^{n}(y_i - a - bx_i)^2$$

is minimum.

From the principle of maxima and minima, the partial derivatives of S, with respect to a and b should vanish separately, i.e.,

$$\frac{\partial S}{\partial a} = 0 \Rightarrow -2\sum_{i=1}^{n}(y_i - a - bx_i) = 0$$

$$\frac{\partial S}{\partial b} = 0 \Rightarrow -2\sum_{i=1}^{n}(y_i - a - bx_i)(x_i) = 0$$

We have, $\sum_{i=1}^{n} y_i = na + b\sum_{i=1}^{n} x_i$

and $\sum_{i=1}^{n} x_i y_i = a\sum_{i=1}^{n} x_i + b\sum_{i=1}^{n} x_i^2$

These equations are known as the normal equations for estimating 'a' and 'b'.

All the quantities $\sum_{i=1}^{n} x_i$, $\sum_{i=1}^{n} x_i^2$, $\sum_{i=1}^{n} y_i$ and $\sum_{i=1}^{n} x_i y_i$ can be obtained from the given set of points (x_i, y_i) $i = 1, 2, ..., n$ and the normal equations can be solved for 'a' and 'b'. With the values of 'a' and 'b' so obtained equation $y = a + bx$ is the line of best fit to the given set of points (x_i, y_i).

7.4 CHANGE OF ORIGIN AND SCALE

Some times the magnitude of the variables in the given data is so big that the calculations become very much tedious. The size of the data can be considerably reduced by assuming some convenient origin for x and y series in the given data. The problem is further simplified by taking suitable scale when the values of x are given at equally spaced intervals.

Let h be the width of the interval at which the values of x are given and let the origins of x and y at the point (x_0, y_0), then putting $u = \dfrac{x - x_0}{h}$ (h is the width of interval for the values of x)

and $\quad v = y - y_0$.

In case of change of origin, if n is odd then

$$u = \frac{x - (\text{middle term})}{h}$$

and if n is even then

$$u = \frac{x - (\text{mean of two middle terms})}{h/2}$$

Similar transformations can be applied to polynomials of higher degree.

7.5 NORMAL EQUATIONS FOR DIFFERENT FORMS OF CURVE

1. **Fitting of a Straight Line:**

   Let the straight line of best fit be
   $$y = a + bx$$
   Then the normal equations are

   $$\sum_{i=1}^{n} y_i = na + b \sum_{i=1}^{n} x_i$$

   and $\quad \sum_{i=1}^{n} x_i y_i = a \sum_{i=1}^{n} x_i + b \sum_{i=1}^{n} x_i^2$

   On solving these equations, we can get a and b.

2. **Fitting of a Second Degree Parabola:**

   Let the second degree parabola to be fitted be
   $$y = a + bx + cx^2$$

Then the normal equations are

$$\sum_{i=1}^{n} y_i = na + b\sum_{i=1}^{n} x_i + c\sum_{i=1}^{n} x_i^2$$

$$\sum_{i=1}^{n} x_i y_i = a\sum_{i=1}^{n} x_i + b\sum_{i=1}^{n} x_i^2 + c\sum_{i=1}^{n} x_i^3$$

and $$\sum_{i=1}^{n} x_i^2 y_i = a\sum_{i=1}^{n} x_i^2 + b\sum_{i=1}^{n} x_i^3 + c\sum_{i=1}^{n} x_i^4$$

On solving these equations, we can get a, b and c.

3. **Fitting of the Curve $y = ab^x$:**

Taking logarithm on both the sides, we have
$$\log_{10} y = \log_{10} a + x \log_{10} b$$
$\Rightarrow \qquad Y = A + Bx$

where $Y = \log_{10} y$, $A = \log_{10} a$, $B = \log_{10} b$

Then the normal equations are

$$\sum_{i=1}^{n} Y_i = nA + B\sum_{i=1}^{n} x_i$$

and $$\sum_{i=1}^{n} x_i Y_i = A\sum_{i=1}^{n} x_i + B\sum_{i=1}^{n} x_i^2$$

On solving these equations, we can get A and B. Now, $a =$ antilog (A) and $b =$ anitlog (B).

4. **Fitting of the Curve $y = ax^b$:**

Taking logarithm on both the sides, we have
$$\log_{10} y = \log_{10} a + b \log_{10} x$$
$\Rightarrow \qquad Y = A + bX$

where $Y = \log_{10} y$, $A = \log_{10} a$, $X = \log_{10} x$

Then the normal equations are

$$\sum_{i=1}^{n} Y_i = nA + b\sum_{i=1}^{n} X_i$$

and $$\sum_{i=1}^{n} X_i Y_i = A\sum_{i=1}^{n} X_i + b\sum_{i=1}^{n} X_i^2$$

On solving these equations, we can get A and b. Now, $a =$ antilog (A).

5. **Fitting of the Curve $y = ae^{bx}$:**

Taking logarithm on both the sides, we have
$$\log_{10} y = \log_{10} a + bx \log_{10} e$$
$\Rightarrow \qquad Y = A + Bx$

where $Y = \log_{10} y$, $A = \log_{10} a$, $B = b \log_{10} e$

Then the normal equations are

$$\sum_{i=1}^{n} Y_i = nA + B\sum_{i=1}^{n} x_i$$

and $\quad \sum_{i=1}^{n} x_i Y_i = A\sum_{i=1}^{n} x_i + B\sum_{i=1}^{n} x_i^2$

On solving these equations, we can get A and B. Now, $a =$ antilog (A)

and $b = \dfrac{B}{\log_{10} e}$.

6. **Fitting of the Curve $y = a + bx^2$:**

Let the curve to be fitted be

$$y = a + bx^2$$

Then the normal equations are

$$\sum_{i=1}^{n} y_i = na + b\sum_{i=1}^{n} x_i^2$$

and $\quad \sum_{i=1}^{n} x_i^2 y_i = a\sum_{i=1}^{n} x_i^2 + b\sum_{i=1}^{n} x_i^4$

On solving these equations, we can get a and b.

7. **Fitting of the Curve $PV^r = k$:**

Here, $\quad PV^r = k \Rightarrow V = k^{1/r} P^{-1/r}$

Taking logarithm on both the sides, we have

$$\log_{10} V = \dfrac{1}{r}\log_{10} k - \dfrac{1}{r}\log_{10} P$$

$\Rightarrow \quad Y = A + BX$

where $\quad Y = \log_{10} V, A = \dfrac{1}{r}\log_{10} k, B = -\dfrac{1}{r}, X = \log_{10} P$

Then the normal equations are

$$\sum_{i=1}^{n} Y_i = nA + B\sum_{i=1}^{n} X_i$$

and $\quad \sum_{i=1}^{n} X_i Y_i = A\sum_{i=1}^{n} X_i + B\sum_{i=1}^{n} X_i^2$

On solving these equations, we can get A and B. Now using the values of A and B we can find the values of r and k.

8. **Fitting of the Curve $y = \dfrac{a}{x} + b\sqrt{x}$:**

Let the curve to be fitted be

$$y = \frac{a}{x} + b\sqrt{x}$$

Then the normal equations are

$$\sum_{i=1}^{n} \frac{y_i}{x_i} = a \sum_{i=1}^{n} \frac{1}{x_i^2} + b \sum_{i=1}^{n} \frac{1}{\sqrt{x_i}}$$

and

$$\sum_{i=1}^{n} \sqrt{x_i} y_i = a \sum_{i=1}^{n} \frac{1}{\sqrt{x_i}} + b \sum_{i=1}^{n} x_i$$

On solving these equations, we can get a and b.

ILLUSTRATIVE EXAMPLES

Example 1. By the method of least squares, find the straight line that best fits the following data:

x	1	2	3	4	5
y	14	13	9	5	2

Solution: Let the straight line of best fit be

$$y = a + bx \qquad \ldots(1)$$

The normal equations are

$$\sum_{i=1}^{n} y_i = na + b \sum_{i=1}^{n} x_i \qquad \ldots(2)$$

and

$$\sum_{i=1}^{n} x_i y_i = a \sum_{i=1}^{n} x_i + b \sum_{i=1}^{n} x_i^2 \qquad \ldots(3)$$

where $n = 5$ and the values of $\sum x, \sum y, \sum x^2$ and $\sum xy$ are calculated in the following table:

x	y	x^2	xy
1	14	1	14
2	13	4	26
3	9	9	27
4	5	16	20
5	2	25	10
$\sum x = 15$	$\sum y = 43$	$\sum x^2 = 55$	$\sum xy = 97$

Now the equations (2) and (3) becomes

$$43 = 5a + 15b$$
$$97 = 15a + 55b$$

On solving these equations, we have $a = 18.2$ and $b = -3.2$
Putting these values in (1), we get the line of best fit as
$$y = 18.2 - 3.2x$$

Example 2. Fit a straight line by the method of least squares to the following data:

x	0	1	2	3	4
y	1	1.8	3.3	4.5	6.3

Solution: Let the straight line of best fit be
$$y = a + bx \qquad \ldots(1)$$
The normal equations are
$$\sum_{i=1}^{n} y_i = na + b\sum_{i=1}^{n} x_i \qquad \ldots(2)$$
and
$$\sum_{i=1}^{n} x_i y_i = a\sum_{i=1}^{n} x_i + b\sum_{i=1}^{n} x_i^2 \qquad \ldots(3)$$

where $n = 5$ and the values of $\sum x, \sum y, \sum x^2$ and $\sum xy$ are calculated in the following table:

x	y	x^2	xy
0	1	0	0
1	1.8	1	1.8
2	3.3	4	6.6
3	4.5	9	13.5
4	6.3	16	25.2
$\sum x = 10$	$\sum y = 16.9$	$\sum x^2 = 30$	$\sum xy = 47.1$

Now the equations (2) and (3) becomes
$$16.9 = 5a + 10b$$
$$47.1 = 10a + 30b$$

On solving these equations, we have $a = 0.72$ and $b = 1.33$
Putting these values in (1), we get the straight line as
$$y = 0.72 + 1.33x$$

Example 3. The weight of a calf taken at weekly intervals is given below. Fit a straight line by using the least square method.

Age (x)	1	2	3	4	5	6	7	8	9	10
Weight (y)	52.5	58.7	65	70.2	75.4	81.1	87.2	95.5	102.2	108.4

Solution. Here, $n = 10$ i.e., even and $h = 1$. Thus, we take

$$u = \frac{x - \left(\frac{5+6}{2}\right)}{\frac{1}{2}} = 2x - 11$$

Let the equation of straight line be
$$y = a + bu \qquad \ldots(1)$$

The normal equations are

$$\sum_{i=1}^{n} y_i = na + b\sum_{i=1}^{n} u_i \qquad \ldots(2)$$

and

$$\sum_{i=1}^{n} u_i y_i = a\sum_{i=1}^{n} u_i + b\sum_{i=1}^{n} u_i^2 \qquad \ldots(3)$$

Now, the values of $\sum u, \sum y, \sum u^2$ and $\sum uy$ are calculated in the following table:

x	y	u	u^2	uy
1	52.5	–9	81	–472.5
2	58.7	–7	49	–410.9
3	65.0	–5	25	–325.0
4	70.2	–3	9	–210.6
5	75.4	–1	1	–75.4
6	81.1	1	1	81.1
7	87.2	3	9	261.6
8	95.5	5	25	477.5
9	102.2	7	49	715.4
10	108.4	9	81	975.6
$\sum x = 55$	$\sum y = 796.2$	$\sum u = 0$	$\sum u^2 = 330$	$\sum uy = 1016.8$

Now the equations (2) and (3) becomes
$$796.2 = 10a + 0 \times b$$
$$1016.8 = 0 \times a + 330b$$

On solving these equations, we have $a = 79.62$ and $b = 3.0812$
Putting these values in (1), we get the straight line as
$$y = 79.62 + 3.0812u$$

Now, the required line to the given data be
$$y = 79.62 + 3.0812(2x - 11)$$
$$y = 45.7268 + 6.1624x$$

Example 4. Fit a straight line by the method of least squares to the following data:

x	1	2	3	4	5
y	15	70	140	250	380

Solution. Here, $n = 5$ i.e., odd and $h = 1$.

Thus, we take $u = \dfrac{x-3}{1} = x - 3$

Let the equation of straight line be

$$y = a + bu \qquad \ldots(1)$$

The normal equations are

$$\sum_{i=1}^{n} y_i = na + b \sum_{i=1}^{n} u_i \qquad \ldots(2)$$

and

$$\sum_{i=1}^{n} u_i y_i = a \sum_{i=1}^{n} u_i + b \sum_{i=1}^{n} u_i^2 \qquad \ldots(3)$$

Now, the values of $\sum u, \sum y, \sum u^2$ and $\sum uy$ are calculated in the following table:

x	y	u	u^2	uy
1	15	−2	4	−30
2	70	−1	1	−70
3	140	0	0	0
4	250	1	1	250
5	380	2	4	760
$\sum x = 15$	$\sum y = 855$	$\sum u = 0$	$\sum u^2 = 10$	$\sum uy = 910$

Now the equations (2) and (3) becomes

$$855 = 5a + 0 \times b$$
$$910 = 0 \times a + 10b$$

On solving these equations, we have $a = 171$ and $b = 91$

Putting these values in (1), we get the straight line as

$$y = 171 + 91u$$

Now, the required line to the given data be

$$y = 171 + 91(x - 3)$$
$$y = -102 + 91x$$

Example 5. Fit a parabola of the form $y = a + bx + cx^2$ to the following data:

x	1	2	3	4	5
y	2	6	7	8	10

Solution. We have to fit a parabola of the form
$$y = a + bx + cx^2 \qquad \ldots(1)$$
The normal equations are
$$\sum_{i=1}^{n} y_i = na + b\sum_{i=1}^{n} x_i + c\sum_{i=1}^{n} x_i^2 \qquad \ldots(2)$$
and
$$\sum_{i=1}^{n} x_i y_i = a\sum_{i=1}^{n} x_i + b\sum_{i=1}^{n} x_i^2 + c\sum_{i=1}^{n} x_i^3 \qquad \ldots(3)$$
$$\sum_{i=1}^{n} x_i^2 y_i = a\sum_{i=1}^{n} x_i^2 + b\sum_{i=1}^{n} x_i^3 + c\sum_{i=1}^{n} x_i^4 \qquad \ldots(4)$$

x	y	xy	x^2	$x^2 y$	x^3	x^4
1	2	2	1	2	1	1
2	6	12	4	24	8	16
3	7	21	9	63	27	81
4	8	32	16	128	64	256
5	10	50	25	250	125	625
$\sum x = 15$	$\sum y = 33$	$\sum xy = 117$	$\sum x^2 = 55$	$\sum x^2 y = 467$	$\sum x^3 = 225$	$\sum x^4 = 979$

Now the equations (2), (3) and (4) becomes
$$33 = 5a + 15b + 55c$$
$$117 = 15a + 55b + 225c$$
$$467 = 55a + 225b + 979c$$
On solving these equations, we have $a = -0.8$, $b = 3.514$ and $c = -0.286$
Putting these values in (1), we get the required parabola as
$$y = -0.8 + 3.514x - 0.286x^2$$

Example 6. Fit a curve of the form $y = ab^x$ to the following data:

x	2	3	4	5	6
y	144	172.8	207.4	248.8	298.5

Solution. We have to fit the curve of the form
$$y = ab^x \qquad \ldots(1)$$
The normal equations are

$$\sum_{i=1}^{n} Y_i = nA + B\sum_{i=1}^{n} x_i \qquad \ldots(2)$$

and
$$\sum_{i=1}^{n} x_i Y_i = A\sum_{i=1}^{n} x_i + B\sum_{i=1}^{n} x_i^2 \qquad \ldots(3)$$

where $Y = \log_{10} y$, $A = \log_{10} a$, $B = \log_{10} b$

x	y	x^2	$Y = \log_{10} y$	xY
2	144.0	4	2.1584	4.3168
3	172.8	9	2.2375	6.7125
4	207.4	16	2.3168	9.2672
5	248.8	25	2.3959	11.9795
6	298.5	36	2.4749	14.8490
$\sum x = 20$		$\sum x^2 = 90$	$\sum Y = 11.5835$	$\sum xY = 47.1254$

Now, the equations (2) and (3) becomes

$11.5835 = 5A + 20B$

$47.1254 = 20A + 90B$

On solving these equations, we have $A = 2$ and $B = 0.0792$

Now, $a =$ antilog $(2) = 100$ and $b =$ anitlog $(0.0792) = 1.2$

Putting these values in (1), we have

$$y = 100(1.2)^x$$

Example 7. Fit a curve of the form $y = ax^b$ to the following data:

x	2	3	4	5
y	27.8	62.1	110	161

Solution. We have to fit the curve of the form
$$y = ax^b \qquad \ldots(1)$$
The normal equations are

$$\sum_{i=1}^{n} Y_i = nA + b\sum_{i=1}^{n} X_i \qquad \ldots(2)$$

and
$$\sum_{i=1}^{n} X_i Y_i = A\sum_{i=1}^{n} X_i + b\sum_{i=1}^{n} X_i^2 \qquad \ldots(3)$$

where $Y = \log_{10} y$, $A = \log_{10} a$, $X = \log_{10} x$

x	y	$X = \log_{10} x$	$Y = \log_{10} Y$	X^2	XY
2	27.8	0.3010	1.4440	0.0906	0.4346
3	62.1	0.4771	1.7931	0.2276	0.8555
4	110.0	0.6021	2.0414	0.3625	1.2291
5	161.0	0.6990	2.2068	0.4886	1.5426
		$\sum X = 2.0792$	$\sum Y = 7.4853$	$\sum X^2 = 1.1693$	$\sum XY = 4.0618$

Now, the equations (2) and (3) becomes
$$7.4853 = 4A + 2.0792b$$
$$4.0618 = 2.0792A + 1.1693b$$
On solving these equations, we have $A = 0.8678$ and $b = 1.9311$
Now, $\quad a = $ antilog $(0.8678) = 7.376$
Putting these values in (1), we have
$$y = 7.376 x^{1.9311}$$

Example 8. Fit a curve of the form $y = ae^{bx}$ to the following data:

x	1	2	3	4
y	1.65	2.70	4.50	7.35

Solution. We have to fit the curve of the form
$$y = ae^{bx} \qquad \ldots(1)$$
The normal equations are

$$\sum_{i=1}^{n} Y_i = nA + B \sum_{i=1}^{n} x_i \qquad \ldots(2)$$

$$\sum_{i=1}^{n} x_i Y_i = A \sum_{i=1}^{n} x_i + B \sum_{i=1}^{n} x_i^2 \qquad \ldots(3)$$

where $Y = \log_{10} y$, $A = \log_{10} a$, $B = b \log_{10} e$

x	y	$Y = \log_{10} y$	x^2	xY
1	1.65	0.2175	1	0.2175
2	2.70	0.4314	4	0.8628
3	4.50	0.6532	9	1.9596
4	7.35	0.8663	16	3.4652
$\sum x = 10$		$\sum Y = 2.1684$	$\sum x^2 = 30$	$\sum xY = 6.5051$

Now, the equations (2) and (3) becomes

$$2.1684 = 4A + 10B$$
$$6.5051 = 10A + 30B$$
On solving these equations, we have $A = 0$ and $B = 0.2168$
Now, $a =$ antilog $(0) = 1.0$ and $b \log_{10} e = 0.2168$ or $b = 0.4992$
Putting these values in (1), we have
$$y = 1.0 e^{0.4992x}$$
$$y = e^{0.4992x}$$

Example 9. Fit a curve of the form $y = ae^{bx}$ to the following data:

x	1	5	7	9	12
y	10	15	12	15	21

Solution. We have to fit the curve of the form
$$y = ae^{bx} \qquad ...(1)$$
The normal equations are
$$\sum_{i=1}^{n} Y_i = nA + B \sum_{i=1}^{n} x_i \qquad ...(2)$$
and
$$\sum_{i=1}^{n} x_i Y_i = A \sum_{i=1}^{n} x_i + B \sum_{i=1}^{n} x_i^2 \qquad ...(3)$$
where $Y = \log_{10} y$, $A = \log_{10} a$, $B = b \log_{10} e$

x	y	$Y = \log_{10} y$	x^2	xY
1	10	1.0000	1	1.0000
5	15	1.1761	25	5.8805
7	12	1.0792	49	7.5544
9	15	1.1761	81	10.5849
12	21	1.3222	144	15.8664
$\sum x = 34$		$\sum Y = 5.7536$	$\sum x^2 = 300$	$\sum xY = 40.8862$

Now, the equations (2) and (3) becomes
$$5.7536 = 5A + 34B$$
$$40.8862 = 34A + 300B$$
On solving these equations, we have $A = 0.9766$ and $B = 0.02561$
Now, $a =$ antilog $(0.9766) = 9.4754$ and $b \log_{10} e = 0.02561$ or $b = 0.059$
Putting these values in (1), we have
$$y = 9.4754 e^{0.059x}$$

EXERCISE 7.1

1. By the method of least squares, find the straight line that best fits the following data:

x	0	1	2	3	4
y	1.0	2.9	4.8	6.7	8.6

2. By the method of least squares, find the straight line that best fits the following data:

x	1	2	3	4	5
y	14	27	40	55	68

3. Fit a curve of the form $y = ab^x$ to the following data:

x	2	3	4	5	6
y	8.3	15.4	33.1	65.2	126.4

4. Fit a curve of the form $y = ab^x$ in least square sense to the following data:

x	1	2	3	4	5	6	7	8
y	1	1.2	1.8	2.5	3.6	4.7	6.6	9.1

5. Fit a curve of the form $y = ax^b$ to the following data:

x	1	2	3	4	5
y	7.1	27.8	62.1	110	161

6. Fit a curve of the form $y = ax^b$ to the following data:

x	1	2	3	4	5	6
y	2.98	4.26	5.21	6.1	6.8	7.5

7. Fit a curve of the form $y = ae^{bx}$ to the following data:

x	1	2	3	4	5	6
y	1.6	4.5	13.8	40.2	125	300

8. Fit a curve of the form $y = ae^{bx}$ to the following data:

x	0	2	4
y	5.012	10	31.62

9. Fit a parabola of the form $y = a + bx + cx^2$ to the following data:

x	0	1	2	3	4
y	-4	-1	4	11	20

10. Fit a parabola of the form $y = a + bx + cx^2$ to the following data:

x	-2	-1	0	1	2
y	15	1	1	3	19

11. Fit a curve of the form $y = a + bx^2$ to the following data:

x	-1	0	1	2
y	2	5	3	0

12. Fit a curve of the form $y = a + bx^2$ to the following data:

x	0	0.1	0.2	0.3	0.4	0.5
y	1	1.01	0.99	0.85	0.81	0.75

13. Fit a curve of the form $PV^r = k$ to the following data:

$P(kg/cm^2)$	0.5	1.0	1.5	2.0	2.5	3.0
V(Litres)	1620	1000	750	620	520	460

14. Fit a curve of the form $y = \dfrac{a}{x} + b\sqrt{x}$ to the following data:

x	0.1	0.2	0.4	0.5	1	2
y	21	11	7	6	5	6

ANSWERS

1. $y = 1.0 + 1.9x$
2. $y = 13.6x$
3. $y = 2.04(1.995)^x$
4. $y = 0.6823(1.384)^x$
5. $y = 7.173x^{1.952}$
6. $y = 2.978x^{0.5143}$
7. $y = 0.558e^{1.0631x}$
8. $y = 4.642e^{0.46x}$
9. $y = -4 + 2x + x^2$
10. $y = -1.057 + x + 4.43x^2$
11. $y = 4.167 - 1.111x^2$
12. $y = 1.003 - 1.108x^2$
13. $PV^{1.41} = 16980$
14. $y = \dfrac{1.9733}{x} + 3.2878\sqrt{x}$

Chapter – 8

Regression Analysis

8.1 REGRESSION

Regression analysis attempts to establish the nature of relationship between the variables. It also helps to determine the functional relationship between the variables so that one can predict or estimate the value of one variable for the given value of the other variable. Regression measures the nature and extent of correlation.

8.2 LINEAR REGRESSION

If the variable in a bivariate distribution are correlated, then points in scatter diagram will be more or less concentrated round a curve. This curve is called the curve of regression. If the curve is a straight line, it is called a line of regression and the regression is said to be linear. Since the line of regression gives the best estimate to the value of dependent variable for any given value of the independent variable, therefore, it is called the line of best fit which is obtained by the method of least squares. Since any one of the two variables x and y can be taken as the independent variable and the other as a dependent variable. Therefore, there are two regression lines, one as the line of regression of y on x and the other as the line of regression of x on y. The linear regression does not test whether the data are linear. It assumes that the data are linear, and finds the slope and intercept that make a straight line best fit to our data.

8.3 LINES OF REGRESSION

Let the equation of line of regression of y on x be

$$y = a + bx \qquad \ldots(1)$$

then

$$\bar{y} = a + b\bar{x} \qquad \ldots(2)$$

Now subtracting (2) from (1), we have

$$y - \bar{y} = b(x - \bar{x}) \qquad \ldots(3)$$

The normal equations for the equation (1) are

$$\sum y = na + b\sum x$$
$$\sum xy = a\sum x + b\sum x^2 \qquad \ldots(4)$$

Shifting the origin to $(\bar{x}, \bar{y})$, (4) becomes

$$\sum(x-\bar{x})(y-\bar{y}) = a\sum(x-\bar{x}) + b\sum(x-\bar{x})^2 \qquad \ldots(5)$$

We know that

$$r = \frac{Cov.(x,y)}{\sigma_x \sigma_y} = \frac{\frac{1}{n}\sum(x-\bar{x})(y-\bar{y})}{\sigma_x \sigma_y},$$

$$\sum(x-\bar{x}) = 0 \text{ and } \sigma_x^2 = \frac{1}{n}\sum(x-\bar{x})^2$$

From (5), we have

$$nr\sigma_x \sigma_y = a\cdot 0 + b\cdot n\sigma_x^2 \text{ or } b = r\frac{\sigma_y}{\sigma_x}$$

Hence, from (3), the line of regression of y on x is given by

$$y - \bar{y} = r\frac{\sigma_y}{\sigma_x}(x - \bar{x})$$

Similarly, the line of regression of x on y is given by

$$x - \bar{x} = r\frac{\sigma_x}{\sigma_y}(y - \bar{y})$$

$r\dfrac{\sigma_y}{\sigma_x}$ is called the regression coefficient of y on x and is denoted by b_{yx}

$$b_{yx} = r\frac{\sigma_y}{\sigma_x} = \frac{Cov(x,y)}{\sigma_x^2} = \frac{n\sum xy - \sum x \sum y}{n\sum x^2 - (\sum x)^2}$$

$r\dfrac{\sigma_x}{\sigma_y}$ is called the regression coefficient of x on y and is denoted by b_{xy}

$$b_{xy} = r\frac{\sigma_x}{\sigma_y} = \frac{Cov(x,y)}{\sigma_y^2} = \frac{n\sum xy - \sum x \sum y}{n\sum y^2 - (\sum y)^2}$$

Note: The line of regression of y on x is used to estimate the value of y for given value of x. The line of regression of x on y is used to estimate the value of x for given value of y.

8.4 PROPERTIES OF REGRESSION COEFFICIENTS

1. The geometric mean of the two regression coefficients is the coefficient of correlation i.e.,

$$r = \sqrt{b_{yx} \times b_{xy}}$$

Proof: We know that the two regression coefficients are

$$b_{yx} = r\frac{\sigma_y}{\sigma_x} \quad \text{and} \quad b_{xy} = r\frac{\sigma_x}{\sigma_y}$$

Now, Geometric mean $= \sqrt{b_{yx} \times b_{xy}}$

$$= \sqrt{r\frac{\sigma_y}{\sigma_x} \times r\frac{\sigma_x}{\sigma_y}} = \sqrt{r^2} = \pm r$$

Note: r will be positive (+) if b_{yx} and b_{yx} are positive, r will be negative (–) if b_{yx} and b_{xy} are negative.

2. The arithmetic mean of the regression coefficients is greater than or equal to the coefficient of correlation i.e.,

$$\frac{b_{yx} + b_{xy}}{2} \geq r$$

Proof: We know that the two regression coefficients are

$$b_{yx} = r\frac{\sigma_y}{\sigma_x} \quad \text{and} \quad b_{xy} = r\frac{\sigma_x}{\sigma_y}$$

Now, Arithmetic mean $\dfrac{b_{yx} + b_{xy}}{2} \geq r$

or $\dfrac{r\dfrac{\sigma_y}{\sigma_x} + r\dfrac{\sigma_x}{\sigma_y}}{2} \geq r$ or $\dfrac{\sigma_y}{\sigma_x} + \dfrac{\sigma_x}{\sigma_y} - 2 \geq 0$

or $\dfrac{1}{\sigma_x \sigma_y}\left[\sigma_x^2 + \sigma_y^2 - 2\sigma_x \sigma_y\right] \geq 0$

or $\dfrac{1}{\sigma_x \sigma_y}[\sigma_x - \sigma_y]^2 \geq 0$ which is true

∴ $\sigma_x, \sigma_y \geq 0$

∴ $\dfrac{b_{yx} + b_{xy}}{2} \geq r$

3. If one of the regression coefficient is greater than unity, the other must be less than unity i.e., If $b_{yx} > 1$ then $b_{xy} < 1$.

Proof: The two regression coefficients are b_{yx} and b_{xy}.

Let $b_{yx} > 1$ then $\dfrac{1}{b_{yx}} < 1$

We know that $b_{yx} \times b_{xy} = r^2 \leq 1$ or $b_{yx} \times b_{xy} \leq 1$ or $b_{xy} \leq \dfrac{1}{b_{yx}} < 1$

Similarly, if $b_{xy} > 1$ then $b_{yx} < 1$.

4. Regression coefficients are independent of the origin but not of scale.

Proof: Let $u = \dfrac{x-a}{h}$ and $v = \dfrac{y-b}{k}$ where a, b, h and k are constants.

Now, $b_{yx} = r\dfrac{\sigma_y}{\sigma_x} = r\dfrac{k\sigma_v}{h\sigma_u} = \dfrac{k}{h}\left(r\dfrac{\sigma_v}{\sigma_u}\right) = \dfrac{k}{h}b_{vu}$

$(\because \sigma_x^2 = h^2\sigma_u^2,\ \sigma_y^2 = k^2\sigma_v^2)$

Similarly, $b_{xy} = \dfrac{h}{k}b_{uv}$

8.5 ANGLE BETWEEN TWO LINES OF REGRESSION

If θ be the acute angle between the two lines of regression in case of two variables x and y, then

$$\tan\theta = \dfrac{1-r^2}{r} \cdot \dfrac{\sigma_x \sigma_y}{\sigma_x^2 + \sigma_y^2}$$

where r, σ_x and σ_y have their usual meanings.

Explain the significance when $r = 0$ and $r = \pm 1$.

Proof: Equation of the line of regression of y on x is

$$y - \bar{y} = r\dfrac{\sigma_y}{\sigma_x}(x - \bar{x})$$

and the equation of the line of regression of x on y is

$$x - \bar{x} = r\dfrac{\sigma_x}{\sigma_y}(y - \bar{y})$$

Their slopes are $m_1 = r\dfrac{\sigma_y}{\sigma_x}$ and $m_2 = \dfrac{\sigma_y}{r\sigma_x}$

$$\tan\theta = \pm\dfrac{m_2 - m_1}{1 + m_1 m_2}$$

$$= \pm\dfrac{\dfrac{\sigma_y}{r\sigma_x} - r\dfrac{\sigma_y}{\sigma_x}}{1 + \dfrac{\sigma_y^2}{\sigma_x^2}} = \pm\dfrac{1-r^2}{r} \cdot \dfrac{\sigma_y}{\sigma_x} \cdot \dfrac{\sigma_x^2}{\sigma_x^2 + \sigma_y^2}$$

$$= \pm\dfrac{1-r^2}{r} \cdot \dfrac{\sigma_x \sigma_y}{\sigma_x^2 + \sigma_y^2}$$

Since $r^2 \leq 1$ and σ_x, σ_y are positive.

∴ Positive sign gives the acute angle between the lines.

Hence, $\tan \theta = \dfrac{1-r^2}{r} \cdot \dfrac{\sigma_x \sigma_y}{\sigma_x^2 + \sigma_y^2}$

When $r = 0$, $\tan \theta = \infty$ so that $\theta = \dfrac{\pi}{2}$

So the two lines of regression are perpendicular to each other.

When $r = \pm 1$, $\tan \theta = 0$ so that $\theta = 0$ or π

So the two lines of regression coincide and there is perfect correlation between the two variables x and y.

8.6 NONLINEAR REGRESSION

Nonlinear regression is a form of regression analysis in which observational data are modeled by a function which is a nonlinear combination of the model parameters and depends on one or more independent variables.

8.7 LINEARIZATION

Some nonlinear regression problems can be transformed as a linear form by a suitable transformation of the problem.

8.8 MULTIPLE REGRESSION

Multiple regression is a method used to examine the relationship between one dependent variable and two or more independent variables x_i. The regression coefficients a_i in the regression equation

$$y = a_0 + a_1 x_1 + a_2 x_2 + \ldots + a_n x_n$$

are estimated using the method of least squares. In this method, the sum of squared residuals between the regression plane and the observed values of the dependent variable are minimized. The regression equation represents a (hyper) plane in a $(n+1)$ dimensional space in which n is the number of independent variables $x_1, x_2, \ldots, x_n$ plus one dimension for the dependent variable y. The independent variables are some times referred to as explanatory variables, because of their use in explaining the variation in y or as the predictor variables, because of their use in predicting y.

Let us consider a two independent variable regression equation as

$$y = a_0 + a_1 x_1 + a_2 x_2$$

To estimate the regression coefficients a_0, a_1, and a_2, we use the least square method as follows:

According to the principle of least squares, we have to determine a_0, a_1, and a_2 so that

308 *Numerical and Statistical Techniques*

$$S = \sum_{i=1}^{n} [y_i - (a_0 + a_1 x_{i1} + a_2 x_{i2})]^2$$

is minimum.

From the principle of maxima and minima, the partial derivatives of S, with respect to a_0, a_1, and a_2 should vanish separately, i.e.,

$$\frac{\partial S}{\partial a_0} = 0 \implies -2\sum_{i=1}^{n} [y_i - (a_0 + a_1 x_{i1} + a_2 x_{i2})] = 0$$

$$\frac{\partial S}{\partial a_1} = 0 \implies \sum_{i=1}^{n} [y_i - (a_0 + a_1 x_{i1} + a_2 x_{i2})] x_{i1} = 0$$

$$\frac{\partial S}{\partial a_2} = 0 \implies -2\sum_{i=1}^{n} [y_i - (a_0 + a_1 x_{i1} + a_2 x_{i2})] x_{i2} = 0$$

We have, $\quad \sum_{i=1}^{n} y_i = na_0 + a_1 \sum_{i=1}^{n} x_{i1} + a_2 \sum_{i=1}^{n} x_{i2}$

and $\quad \sum_{i=1}^{n} x_{i1} y_i = a_0 \sum_{i=1}^{n} x_{i1} + a_1 \sum_{i=1}^{n} x_{i1}^2 + a_2 \sum_{i=1}^{n} x_{i1} x_{i2}$

$$\sum_{i=1}^{n} x_{i2} y_i = a_0 \sum_{i=1}^{n} x_{i2} + a_1 \sum_{i=1}^{n} x_{i1} x_{i2} + a_2 \sum_{i=1}^{n} x_{i2}^2$$

These normal equations can be solved for estimating the values of a_0, a_1 and a_2.

All the quantities $\sum_{i=1}^{n} x_{i1}$, $\sum_{i=1}^{n} x_{i2}$, $\sum_{i=1}^{n} x_{i1}^2$, $\sum_{i=1}^{n} x_{i2}^2$, $\sum_{i=1}^{n} x_{i1} x_{i2}$, $\sum_{i=1}^{n} y_i$, $\sum_{i=1}^{n} x_{i1} y_i$,

$\sum_{i=1}^{n} x_{i2} y_i$ can be obtained from the given set of points (x_{i1}, x_{i2}, y_i) $i = 1, 2, ..., n$ and the values of a_0, a_1, and a_2 are obtained from the above equations.

With the values of a_0, a_1 and a_2 we get the required two independent variable regression equation as

$$y = a_0 + a_1 x_1 + a_2 x_2.$$

Note: Similarly, we can find more than two independent variable regression equation.

---------------| **ILLUSTRATIVE EXAMPLES** |---------------

Example 1. Find the equation of two lines of regression for the following data:

x	1	2	3	4	5
y	7	6	5	4	3

Find an estimate of y for x = 3.5 from the appropriate line of regression.

Solution.

x	y	x^2	y^2	xy
1	7	1	49	7
2	6	4	36	12
3	5	9	25	15
4	4	16	16	16
5	3	25	9	15
$\sum x = 15$	$\sum y = 25$	$\sum x^2 = 55$	$\sum y^2 = 135$	$\sum xy = 65$

Here, n = 5

$$\bar{x} = \frac{1}{n}\sum x = \frac{15}{5} = 3, \quad \bar{y} = \frac{1}{n}\sum y = \frac{25}{5} = 5$$

Now, $b_{yx} = \dfrac{n\sum xy - \sum x \sum y}{n\sum x^2 - (\sum x)^2} = \dfrac{5 \times 65 - 15 \times 25}{5 \times 55 - (15)^2} = \dfrac{13-15}{11-9} = -1$

$b_{xy} = \dfrac{n\sum xy - \sum x \sum y}{n\sum y^2 - (\sum y)^2} = \dfrac{5 \times 65 - 15 \times 25}{5 \times 135 - (25)^2} = \dfrac{13-15}{27-25} = -1$

So, the line of regression of y on x is

$$y - \bar{y} = b_{yx}(x - \bar{x})$$
$$y - 5 = -1(x - 3)$$
$$y = -x + 8$$

and the line of regression of x on y is

$$x - \bar{x} = b_{xy}(y - \bar{y})$$
$$x - 3 = -1(y - 5)$$
$$x = -y + 8$$

To estimate the value of y when x is given, we use the line of regression of y on x, i.e.

$$y = -x + 8$$

Now, substitute x = 3.5, we have

$$y = -3.5 + 8 = 4.5$$

Example 2. The following table gives age (x) in years of cars and annual maintenance cost (y) in hundred rupees:

x	1	3	5	7	9
y	15	18	21	23	22

Estimate the maintenance cost for a 4 year old car after finding the appropriate line of regression.

Solution.

x	y	x^2	xy
1	15	1	15
3	18	9	54
5	21	25	105
7	23	49	161
9	22	81	198
$\sum x = 25$	$\sum y = 99$	$\sum x^2 = 165$	$\sum xy = 533$

Here, $n = 5$

$$\bar{x} = \frac{1}{n}\sum x = \frac{25}{5} = 5, \quad \bar{y} = \frac{1}{n}\sum y = \frac{99}{5} = 19.8$$

Now,
$$b_{yx} = \frac{n\sum xy - \sum x \sum y}{n\sum x^2 - (\sum x)^2}$$

$$= \frac{5 \times 533 - 25 \times 99}{5 \times 165 - (25)^2} = \frac{2665 - 2475}{825 - 625} = \frac{190}{200} = 0.95$$

The line of regression of y on x is given by

$$y - \bar{y} = b_{yx}(x - \bar{x})$$
$$y - 19.8 = 0.95(x - 5)$$
$$y = 0.95x + 15.05$$

When $x = 4$ years, we have
$$y = 0.95 \times 4 + 15.05 = 18.85 \text{ hundred rupees} = \text{Rs. } 1885$$

Example 3. From the following information on values of two variables x and y, find the two regression lines and the correlation coefficient between x and y.
$n = 10$, $\Sigma x = 20$, $\Sigma y = 40$, $\Sigma x^2 = 240$, $\Sigma y^2 = 410$, $\Sigma xy = 200$.

Solution. We know that

$$b_{yx} = \frac{n\sum xy - \sum x \sum y}{n\sum x^2 - (\sum x)^2} = \frac{10 \times 200 - 20 \times 40}{10 \times 240 - (20)^2} = \frac{20 - 8}{24 - 4} = \frac{3}{5}$$

and
$$b_{xy} = \frac{n\sum xy - \sum x \sum y}{n\sum y^2 - (\sum y)^2} = \frac{10 \times 200 - 20 \times 40}{10 \times 410 - (40)^2} = \frac{20 - 8}{41 - 16} = \frac{12}{25}$$

$$\bar{x} = \frac{1}{n}\sum x = \frac{20}{10} = 2, \quad \bar{y} = \frac{1}{n}\sum y = \frac{40}{10} = 4$$

The two regression lines are

$$y - \bar{y} = b_{yx}(x - \bar{x})$$

$$y - 4 = \frac{3}{5}(x - 2)$$

and
$$y = 0.6x + 2.8$$
$$x - \bar{x} = b_{xy}(y - \bar{y})$$
$$x - 2 = \frac{12}{25}(y - 4)$$
$$x = 0.48y + 0.08$$

We know that $r = \pm\sqrt{b_{yx} \times b_{xy}}$

$$r = \sqrt{\frac{3}{5} \times \frac{12}{25}} = \sqrt{\frac{36}{125}} = 0.536$$

Example 4. For 100 students of a class, the regression equation of marks in Statistics (x) on the marks in Mathematics (y) is $3y - 5x + 180 = 0$. The mean marks in Mathematics is 50 and variance of marks in Statistics is $\frac{4}{9}$ th of the variance of marks in Mathematics. Find the mean marks in Statistics and the coefficient of correlation between marks in the two subjects.

Solution. Since the given line of regression is x on y so we have

$$x = \frac{3}{5}y + \frac{180}{5} = \frac{3}{5}y + 36$$

We have $b_{xy} = \frac{3}{5} = r \cdot \frac{\sigma_x}{\sigma_y}$

Given variance of $x = \frac{4}{9}$ variance of y

$$\frac{\text{variance of } x \, (\sigma_x^2)}{\text{variance of } y \, (\sigma_y^2)} = \frac{4}{9} \Rightarrow \frac{\sigma_x}{\sigma_y} = \frac{2}{3}$$

So, $\frac{3}{5} = r \times \frac{2}{3} \Rightarrow r = \frac{9}{10} = 0.9$

($\because b_{xy}$ is positive, r is positive)

Since the mean of x and mean of y lie on the regression lines, we have

$$\bar{x} = \frac{3}{5}\bar{y} + 36 \Rightarrow \bar{x} = \frac{3}{5} \times 50 + 36 = 66 \quad (\because \bar{y} = 50)$$

Example 5. The lines of regression of y on x and x on y are $y = x + 5$ and $16x - 9y = 94$ respectively. Find the variance of x if the variance of y is 16. Also find the covariance of x and y.

Solution. Regression equation of y on x is $y = x + 5 \Rightarrow b_{yx} = 1$

(coefficient of x)

Regression equation of x on y is $16x - 9y = 94$ i.e. $x = \frac{9}{16}y + \frac{94}{16}$

$$\Rightarrow \qquad b_{xy} = \frac{9}{16} \qquad \text{(coefficient of } y\text{)}$$

We know that $r = \pm\sqrt{b_{yx} \times b_{xy}}$

$$r = \sqrt{1 \times \frac{9}{16}} = \frac{3}{4} = 0.75$$

(r is positive, since b_{yx} and b_{xy} are positive)

$$b_{xy} = r\frac{\sigma_x}{\sigma_y} \Rightarrow \sigma_x = \frac{b_{xy} \times \sigma_y}{r} = \frac{\left(\frac{9}{16}\right) \times 4}{\left(\frac{3}{4}\right)} = 3$$

$$(\because \sigma_y^2 = 16)$$

$$r = \frac{\text{Cov.}(x, y)}{\sigma_x \sigma_y}$$

$$\Rightarrow \qquad \text{Cov}(x, y) = r\sigma_x\sigma_y = \frac{3}{4} \times 3 \times 4 = 9$$

Example 6. The equations of two lines of regression are $4x + 3y + 7 = 0$ and $3x + 4y + 8 = 0$.

Find

(i) the mean values of x and y

(ii) the regression coefficients b_{yx} and b_{xy}

(iii) the correlation coefficient between x and y

(iv) the standard deviation of y, if the variance of x is 4

(v) the value of y for $x = 5$.

Solution.

(i) Since the mean of x and mean of y lie on the regression lines, we have

$$4\bar{x} + 3\bar{y} + 7 = 0 \quad \text{or} \quad 4\bar{x} + 3\bar{y} = -7$$

and $3\bar{x} + 4\bar{y} + 8 = 0 \quad$ or $\quad 3\bar{x} + 4\bar{y} = -8$

Now, on solving the above equations for $\bar{x}$ and $\bar{y}$ we have

$$\bar{x} = -\frac{4}{7} \quad \text{and} \quad \bar{y} = -\frac{11}{7}$$

Mean of $x = -\frac{4}{7}$ and mean of $y = -\frac{11}{7}$

(ii) Let the regression line of y on x be

$$3x + 4y + 8 = 0 \quad \text{or} \quad y = -\frac{3}{4}x - 2$$

$$\therefore \qquad b_{yx} = -\frac{3}{4} \qquad \text{(coefficient of } x\text{)}$$

and the regression line of x on y be

$$4x + 3y + 7 = 0 \quad \text{or} \quad x = -\frac{3}{4}y - \frac{7}{4}$$

$$\therefore \quad b_{xy} = -\frac{3}{4} \qquad \text{(coefficient of } y)$$

Since $b_{yx} \times b_{xy} = -\frac{3}{4} \times -\frac{3}{4} = \frac{9}{16} < 1$

Hence, the choice of regression lines is correct.

So $b_{yx} = -\frac{3}{4}$ and $b_{xy} = -\frac{3}{4}$

(iii) We know that $r = \pm\sqrt{b_{yx} \times b_{xy}}$

$$\therefore \quad r = \pm\sqrt{\left(\frac{-3}{4}\right) \times \left(\frac{-3}{4}\right)} = \pm\frac{3}{4} = -0.75$$

(r is negative, since b_{yx} and b_{xy} are negative)

(iv) We have $\sigma_x^2 = 4 \implies \sigma_x = 2$

Now, $b_{yx} = -\frac{3}{4}$ or $r\frac{\sigma_y}{\sigma_x} = -\frac{3}{4}$

$$\left(-\frac{3}{4}\right) \times \frac{\sigma_y}{2} = -\frac{3}{4} \implies \sigma_y = 2$$

(v) Since we have to find y when x is given, we use line of regression of y on x

$$y = -\frac{3}{4}x - 2$$

Putting $x = 5$, we have

$$y = -\frac{3}{4} \times 5 - 2 = -\frac{23}{4} = -5.75$$

Example 7. The two regression lines of the variables x and y are:
$$x = 19.13 - 0.87y \quad \text{and} \quad y = 11.64 - 0.50x$$

Find
(i) mean of x and mean of y
(ii) correlation coefficient between x and y (IU, 2007-08)

Solution: (i) Since the mean of x and mean of y lie on the regression lines, we have

$$\bar{x} = 19.13 - 0.87\bar{y} \quad \text{or} \quad \bar{x} + 0.87\bar{y} = 19.13$$

and $\quad \bar{y} = 11.64 - 0.50\bar{x} \quad \text{or} \quad 0.50\bar{x} + \bar{y} = 11.64$

Now, on solving the above equations for $\bar{x}$ and $\bar{y}$, we have

$$\bar{x} = 15.935 \quad \text{and} \quad \bar{y} = 3.67$$

Mean of $x = 15.935$ and mean of $y = 3.67$

(ii) Let the regression line of y on x be
$$y = 11.64 - 0.50x$$
$$\therefore b_{yx} = -0.50 \quad \text{(coefficient of } x\text{)}$$
and the regression line of x on y be
$$x = 19.13 - 0.87y$$
$$\therefore b_{xy} = -0.87 \quad \text{(coefficient of } y\text{)}$$
We know that $r = \pm\sqrt{b_{yx} \times b_{xy}}$
$$\therefore r = \pm\sqrt{-0.50 \times -0.87} = -0.66$$
(r is negative, since b_{yx} and b_{xy} are negative)

Example 8. For two random variables, x and y with the same mean, the two regression equations are $y = ax + b$ and $x = cy + d$. Show that

(i) $\bar{x} = \dfrac{bc+d}{1-ac}$ and $\bar{y} = \dfrac{ad+b}{1-ac}$ (ii) $\dfrac{\sigma_x}{\sigma_y} = \sqrt{\dfrac{c}{a}}$ (iii) $r(x, y) = \sqrt{ac}$

Solution: (i) Since the mean of x and mean of y lie on the regression lines, we have
$$\bar{y} = a\bar{x} + b \quad \text{and} \quad \bar{x} = c\bar{y} + d$$
Now, on solving the above equations for $\bar{x}$ and $\bar{y}$, we have
$$\bar{x} = \dfrac{bc+d}{1-ac} \quad \text{and} \quad \bar{y} = \dfrac{ad+b}{1-ac}$$

(ii) Here $b_{yx} = \dfrac{r\sigma_y}{\sigma_x} = a$ and $b_{xy} = \dfrac{r\sigma_x}{\sigma_y} = c$

Now on dividing b_{xy} by b_{yx} we have
$$\dfrac{r\sigma_x}{\sigma_y} \times \dfrac{\sigma_x}{r\sigma_y} = \dfrac{c}{a} \quad \text{or} \quad \dfrac{\sigma_x^2}{\sigma_y^2} = \dfrac{c}{a} \quad \text{or} \quad \dfrac{\sigma_x}{\sigma_y} = \sqrt{\dfrac{c}{a}}$$

(iii) $r = \sqrt{b_{yx} \times b_{xy}} = \sqrt{ac}$

Example 9. Fit a parabola of the form $y = a + bx + cx^2$ to the following data:

x	1	2	3	4	5
y	2	6	7	8	10

Estimate y for $x = 7$.

Solution: We have to fit a parabola of the form
$$y = a + bx + cx^2 \quad \text{...(1)}$$
The normal equations are

$$\sum_{i=1}^{n} y_i = na + b\sum_{i=1}^{n} x_i + c\sum_{i=1}^{n} x_i^2 \qquad ...(2)$$

and
$$\sum_{i=1}^{n} x_i y_i = a\sum_{i=1}^{n} x_i + b\sum_{i=1}^{n} x_i^2 + c\sum_{i=1}^{n} x_i^3 \qquad ...(3)$$

$$\sum_{i=1}^{n} x_i^2 y_i = a\sum_{i=1}^{n} x_i^2 + b\sum_{i=1}^{n} x_i^3 + c\sum_{i=1}^{n} x_i^4 \qquad ...(4)$$

x	y	xy	x^2	x^2y	x^3	x^4
1	2	2	1	2	1	1
2	6	12	4	24	8	16
3	7	21	9	63	27	81
4	8	32	16	128	64	256
5	10	50	25	250	125	625
$\sum x = 15$	$\sum y = 33$	$\sum xy = 117$	$\sum x^2 = 55$	$\sum x^2 y = 467$	$\sum x^3 = 225$	$\sum x^4 = 979$

Now the equations (2), (3) and (4) becomes
$$33 = 5a + 15b + 55c$$
$$117 = 15a + 55b + 225c$$
$$467 = 55a + 225b + 979c$$

On solving these equations, we have $a = -0.8$, $b = 3.514$ and $c = -0.286$

Putting these values in (1), we get the required parabola as
$$y = -0.8 + 3.514x - 0.286x^2$$

When $x = 7$
$$y = -0.8 + 3.514 \times 7 - 0.286 \times 49$$
$$y = 9.784$$

Example 10. Fit a curve of the form $y = ae^{bx}$ to the following data:

x	1	2	3	4
y	1.65	2.70	4.50	7.35

Estimate y for $x = 6$.

Solution. We have to fit the curve of the form
$$y = ae^{bx} \qquad ...(1)$$

The normal equations are
$$\sum_{i=1}^{n} Y_i = nA + B\sum_{i=1}^{n} x_i \qquad ...(2)$$

and
$$\sum_{i=1}^{n} x_i Y_i = A\sum_{i=1}^{n} x_i + B\sum_{i=1}^{n} x_i^2 \qquad ...(3)$$

where $Y = \log_{10} y$, $A = \log_{10} a$, $B = b \log_{10} e$

x	y	$Y = \log_{10} y$	x^2	xY
1	1.65	0.2175	1	0.2175
2	2.70	0.4314	4	0.8628
3	4.50	0.6532	9	1.9596
4	7.35	0.8663	16	3.4652
$\sum x = 10$		$\sum Y = 2.1684$	$\sum x^2 = 30$	$\sum xY = 6.5051$

Now, the equations (2) and (3) becomes
$$2.1684 = 4A + 10B$$
$$6.5051 = 10A + 30B$$
On solving these equations, we have $A = 0$ and $B = 0.2168$
Now, $a = $ antilog $(0) = 1.0$ and $b \log_{10} e = 0.2168$ or $b = 0.4992$
Putting these values in (1), we have
$$y = 1.0 e^{0.4992x}$$
$$y = e^{0.4992x}$$
When $x = 6$
$$y = e^{0.4992 \times 6}$$
$$y = 19.99$$

Example 11. Fit an equation of the form $y = a_0 + a_1 x_1 + a_2 x_2$ from the following data:

x_1	1	3	4	5
x_2	2	3	5	6
y	4	5	6	7

Estimate the value of y when $x_1 = 6$ and $x_2 = 7$.

Solution: Let the regression equation be
$$y = a_0 + a_1 x_1 + a_2 x_2 \qquad \ldots(1)$$
The normal equations for equation (1) are

$$\sum_{i=1}^{n} y_i = n a_0 + a_1 \sum_{i=1}^{n} x_{i1} + a_2 \sum_{i=1}^{n} x_{i2} \qquad \ldots(2)$$

$$\sum_{i=1}^{n} x_{i1} y_i = a_0 \sum_{i=1}^{n} x_{i1} + a_1 \sum_{i=1}^{n} x_{i1}^2 + a_2 \sum_{i=1}^{n} x_{i1} x_{i2} \qquad \ldots(3)$$

and $\sum_{i=1}^{n} x_{i2} y_i = a_0 \sum_{i=1}^{n} x_{i2} + a_1 \sum_{i=1}^{n} x_{i1} x_{i2} + a_2 \sum_{i=1}^{n} x_{i2}^2 \qquad \ldots(4)$

x_1	x_2	y	x_1^2	x_2^2	$x_1 x_2$	$x_1 y$	$x_2 y$
1	2	4	1	4	2	4	8
3	3	5	9	9	9	15	15
4	5	6	16	25	20	24	30
5	6	7	25	36	30	35	42

Here, $\sum x_1 = 13$, $\sum x_2 = 16$, $\sum y = 22$, $\sum x_1^2 = 51$, $\sum x_2^2 = 74$, $\sum x_1 x_2 = 61$, $\sum x_1 y = 78$ and $\sum x_2 y = 95$

Now the equations (2), (3) and (4) becomes

$$22 = 4a_0 + 13a_1 + 16a_2$$
$$78 = 13a_0 + 51a_1 + 61a_2$$
$$95 = 16a_0 + 61a_1 + 74a_2$$

On solving these equations, we have $a_0 = 2.702$, $a_1 = 0.005141$ and $a_2 = 0.6953$

Putting these values in equation (1) we have

$$y = 2.702 + 0.005141 x_1 + 0.6953 x_2$$

When $x_1 = 6$ and $x_2 = 7$

$$y = 2.702 + 0.005141 \times 6 + 0.6953 \times 7$$
$$y = 7.60$$

Example 12. Obtain a regression plane by using multiple linear regression to fit the data given below:

x_1	1	2	3	4
x_2	0	1	2	3
y	12	18	24	30

Solution: Let the equation of regression plane be

$$y = a_0 + a_1 x_1 + a_2 x_2 \qquad \ldots(1)$$

The normal equations for equation (1) are

$$\sum_{i=1}^{n} y_i = na_0 + a_1 \sum_{i=1}^{n} x_{i1} + a_2 \sum_{i=1}^{n} x_{i2} \qquad \ldots(2)$$

$$\sum_{i=1}^{n} x_{i1} y_i = a_0 \sum_{i=1}^{n} x_{i1} + a_1 \sum_{i=1}^{n} x_{i1}^2 + a_2 \sum_{i=1}^{n} x_{i1} x_{i2} \qquad \ldots(3)$$

and $\sum_{i=1}^{n} x_{i2} y_i = a_0 \sum_{i=1}^{n} x_{i2} + a_1 \sum_{i=1}^{n} x_{i1} x_{i2} + a_2 \sum_{i=1}^{n} x_{i2}^2 \qquad \ldots(4)$

x_1	x_2	y	x_1^2	x_2^2	$x_1 x_2$	$x_1 y$	$x_2 y$
1	0	12	1	0	0	12	0
2	1	18	4	1	2	36	18
3	2	24	9	4	6	72	48
4	3	30	16	9	12	120	90

Here, $\sum x_1 = 10$, $\sum x_2 = 6$, $\sum y = 84$, $\sum x_1^2 = 30$, $\sum x_2^2 = 14$, $\sum x_1 x_2 = 20$, $\sum x_1 y = 240$ and $\sum x_2 y = 156$

Now the equations (2), (3) and (4) becomes
$$84 = 4a_0 + 10a_1 + 6a_2$$
$$240 = 10a_0 + 30a_1 + 20a_2$$
$$156 = 6a_0 + 20a_1 + 14a_2$$

On solving these equations, we have $a_0 = 10$, $a_1 = 2$ and $a_2 = 4$
Putting these values in equation (1) we have
$$y = 10 + 2x_1 + 4x_2$$

EXERCISE 8.1

1. Find the regression coefficient b_{xy} for the following data:
   $n = 6$, $\Sigma x = 30$, $\Sigma y = 42$, $\Sigma x^2 = 184$, $\Sigma y^2 = 318$, $\Sigma xy = 199$.

2. Find the regression coefficient b_{yx} and b_{xy} between x and y for the following data:
   $n = 7$, $\Sigma x = 24$, $\Sigma y = 12$, $\Sigma x^2 = 374$, $\Sigma y^2 = 97$, $\Sigma xy = 157$.
   Also, find the coefficient of correlation between x and y.

3. Find the regression line of x on y and estimate the value of x, when $y = 5$ from the following data:
   $n = 25$, $\Sigma x = 125$, $\Sigma y = 100$, $\Sigma x^2 = 1650$, $\Sigma y^2 = 1500$, $\Sigma xy = 50$.

4. Find the line of regression of y on x for the following data:

x	10	9	8	7	6	4	3
y	8	12	7	10	8	9	6

5. Find the line of regression of y on x for the following data:

x	1	3	4	6	8	9	11	14
y	1	2	4	4	5	7	8	9

Estimate the value of y, when $x = 10$.

6. Find the regression lines for the following data:

x	6	2	10	4	8
y	9	11	5	8	7

7. Find the regression lines for the following data:

x	1	2	3	4	5
y	2	5	3	8	7

Estimate the value of y, when $x = 10$.

8. The equations of two lines of regression are $8x - 10y + 66 = 0$ and $40x - 18y = 214$.
Find
   (i) the mean values of x and y
   (ii) the correlation coefficient between x and y
   (iii) the standard deviation of y, if the variance of x is 9.

9. Two random variables have the regression lines $3x + 2y - 26 = 0$ and $6x + y - 31 = 0$. Find the mean values of x and y and the coefficient of correlation. If the variance of x is 25, find standard deviation of y from the given data.

10. Find the multiple linear regression equation of y on x_1 and x_2 from the following data:

x_1	15	12	8	6	4	3
x_2	30	24	20	14	10	4
y	4	6	7	9	13	15

Estimate the value of y when $x_1 = 16$ and $x_2 = 32$.

11. Find the multiple linear regression equation of y on x_1 and x_2 from the following data:

x_1	4	7	9	12
x_2	1	2	5	8
y	7	12	17	20

Estimate the value of y when $x_1 = 13$ and $x_2 = 9$.

ANSWERS

1. -0.46

2. $b_{yx} = 0.397$; $b_{xy} = 1.516$; $r = 0.776$

3. $x = -\dfrac{9}{22}y + \dfrac{146}{22}$; 4.5911

4. $y = \dfrac{1}{3}x + \dfrac{133}{21}$

5. $y = \dfrac{7}{11}x + \dfrac{6}{11}$; 6.91

6. $y = 11.9 - 0.65x$; $x = 16.4 - 1.3y$

7. $y = 1.1 + 1.3x$; $x = 0.5 + 0.5y$; $y = 14.1$

8. (i) $\bar{x} = 13$, $\bar{y} = 17$, (ii) 0.6, (iii) 4

9. $\bar{x} = 4$, $\bar{y} = 7$, $\sigma_y = 15$, $r = -0.5$

10. $y = 16.479 + 0.389x_1 - 0.623x_2$; $y = 2.767$

11. $y = 0.6444 + 1.661x_1 + 0.0169x_2$; $y = 22.3895$

Chapter – 9

Time Series and Forecasting

9.1 INTRODUCTION

An ordered sequence of values of a variable at equally spaced time intervals is called a time series. *e.g.*, the yearly demand for a product, yearly production of a certain commodity, the annual enrollment in a department of the university, weekly prices of a certain commodity, number of phone calls per hour etc. Data collected irregularly or only once are not time series.

Some of the important definitions given by different statisticians and economists are as follows:

- "A time series is a set of statistical observations arranged in chronological order" by **Moris Hamburg**
- "A time series is a set of observations taken at specified times, usually at equal intervals. Mathematically, a time series is defined by the values y_1, y_2, ... of a variable y (temperature, closing price of a share, etc.) at time t_1, t_2, ... Thus, y is function of t symbolized by $y = f(t)$" by **Spiegel**
- "A time series may be defined as a sequence of repeated meas]urements of a variable made periodically through time" by **Cecil H. Meyers**
- "When quantitative data are arranged in the order of their occurrence, the resulting statistical series is called a time series" by **Wessel and Wellet**

9.2 ANALYSIS OF TIME SERIES

Time series analysis provides tools for selecting a model that describes the time series and using the model to forecast future events. The impact of time series analysis on scientific applications is inevitable, with many of the most intensive and sophisticated applications of time series methods especially to problems in the physical and environmental sciences.

A time series analysis consists of two steps:

1. building a model that represents a time series
2. using the model to predict (forecast) future values.

9.3 APPLICATIONS OF TIME SERIES

The time series analysis is used in many areas such as:
- Economic forecasting, sales forecasting, budgetary analysis, stock market analysis, yield projections, process and quality control, inventory studies, census analysis, data mining etc.

Time series analysis is useful for the following reasons:
1. It gives a general description of the past behavior of the series.
2. It helps in evaluation of current performances.
3. It is useful in planning the future operations.
4. It facilitates the comparison between current performances and expected performances on the basis of the past performances.
5. It is helpful in forecasting.

9.4 COMPONENTS OF TIME SERIES

There are four components of a time series:
1. Secular trend or long term
2. Seasonal variations
3. Cyclical variations
4. Random or Irregular variations

1. **Secular Trend (T):** Secular trend is that characteristic of a time series which extends consistently throughout the entire period of time under consideration. It gives a long term tendency of an activity to grow or decline. It can also be defined as the long term movement in a time series without irregular effects. The factors which remain more or less constant over a long period also produce a trend. There is no specific period which can be called as a long period. It depends upon the different situations. The trend may be upwards and downwards. It is the result of influences such as population growth, price inflation and general economic changes.

2. **Seasonal Variations (S):** In weekly or monthly data, the seasonal component, often referred to as seasonality, is the component of variation in a time series which is dependent on the time of year. It describes any regular fluctuations with a period of less than one year. It may be daily, weekly, monthly, quarterly or half yearly. These variations occur due to weather conditions, traditions, seasons etc., e.g., the costs of various types of fruits and vegetables, the costs of agricultural commodities during the harvest period, the demand of woolen clothes in winter, all show marked seasonal variation. Seasonal variations are called the short term variations. Seasonality in a time series can be identified by regularly spaced peaks and troughs which have a consistent direction and approximately the same magnitude every year, relative to the trend.

3. **Cyclical Variations (C):** Cyclical variations are oscillatory or recurrent merits in a time series, which normally last more than a year. Cyclical variations do not describe any regular pattern. The cycles in a time series shows the ups and downs, prosperity and recession of a business. Cyclical variations are very useful in forecasting the turning point in business activity and help in framing the suitable policies for establishing the business activity. Cycles related to business are called business cycles or trade cycles.

4. **Random or Irregular Variations (I):** Random variations are those variations which do not repeat in a definite pattern. Here, every series have occasional influences which occur just once or more but without regularity. We can not predict the time of their occurrence. The irregular variations also called as the residual. It results from short term fluctuations in the series which are neither systematic nor predictable. These variations are caused by unexpected factors as wars, floods, earthquakes etc.

9.5 FORECASTING

Forecasting is the estimation of the value of a variable (or set of variables) at some future point in time. A forecasting exercise is usually carried out in order to provide an aid to decision-making and in planning the future. If we can predict what the future will be like we can modify our behavior now to be in a better position, than we otherwise would have been, when the future arrives.

9.6 FORECASTING MODELS

There are many mathematical models for forecasting trends and cycles. Choosing an appropriate model for a particular forecasting application depends on the historical data. The study of the historical data is called exploratory data analysis. Its purpose is to identify the trends and cycles in the data so that appropriate model can be chosen.

The common feature of these mathematical models is that historical data is the only criteria for producing a forecast. Mathematical models involve smoothing constants, coefficients and other parameters that must decided by the forecaster. To a large degree, the choice of these parameters determines the forecast.

The most common mathematical models are follows:
1. **Additive Model:** In some time series, the amplitude of both the seasonal and irregular variations do not change as the level of the trend rises or falls. In such cases, an additive model is appropriate. This model is used when it is assumed that the four components are independent of one another, *i.e.*, when the pattern of occurrence and the magnitude of movement in any particular component are not affected by other components under this assumption, the magnitude

of time series y_t at any time t is the sum of the separate influences of its four components, i.e.,
$$y_t = T_t + S_t + C_t + I_t$$
Each of the four components has the same units as the original series. When the time series data are recorded against years, the seasonal components vanish and in that case
$$y_t = T_t + C_t + I_t$$

2. **Multiplicative Model:** In many time series, the amplitude of both the seasonal and irregular variations increase as the level of the trend rises. In this situation, a multiplicative model is usually appropriate. In this model it is assumed that the forces giving rise to four types of variations of time series are independent. Here y_t at any time t is the multiplication of its four variations, i.e.,
$$y_t = T_t \times S_t \times C_t \times I_t$$
On taking logarithm on both the sides we have,
$$\log y_t = \log T_t + \log S_t + \log C_t + \log I_t$$
This model is suitable for projections.

The units of measurements are the same for all the four components. When the time series data are recorded against years, the seasonal components vanish and in that case
$$y_t = T_t \times C_t \times I_t$$
On taking logarithm on both the sides we have,
$$\log y_t = \log T_t + \log C_t + \log I_t$$

Note. To choose an appropriate model, we have to examine a graph of the original series and try a range of models, selecting the one which provides the most stable seasonal component. If the magnitude of the seasonal component is relatively constant regardless of changes in the trend, an additive model is suitable. If it varies with changes in the trend, a multiplicative model is preferable.

9.7 FORECASTING METHODS

There are basically three types of forecasting methods:
1. Judgmental Methods – subjective, qualitative.
2. Time Series Methods (extrapolation methods) – quantitative, use a series of past data points to predict the future.
3. Causal Methods – quantitative, forecast the quantity of interest by relating it directly to one or more other quantities.

Here, we will discuss the quantitative methods only, but recognize no method can accurately forecast the future every time.

Time Series and Forecasting

9.8 MEASUREMENT OF TREND

The trend can be determined in the following four ways:
1. Free Hand Method or Graphical Method
2. Semi Average Method
3. Method of Moving Averages
4. Method of Least Squares

1. **Free Hand Method or Graphical Method:** It is the simplest method for trend. In this method actual points are plotted on the graph paper, *i.e.*, time on x-axis and the quantity of measurement on y-axis. Join these points to get the actual data line. Now, to get the trend line use free hand in such a way that the line passes roughly through the mid way of each peak and closely represent the time series. Now, at any point of time we can find the trend value by reading the value of measurement on y-axis.

Example 1. Fit a trend line to the following data by free hand method:

Year	2000	2001	2002	2003	2004	2005	2006	2007
Sales	70	120	90	140	130	124	160	155

Solution. Plot the actual points on the graph paper, *i.e.*, time (year) on x-axis and the sales on y-axis. Join these points to get the actual data line. Now, to get the trend line use free hand in such a way that the line passes roughly through the mid way of each peak and closely represent the time series.

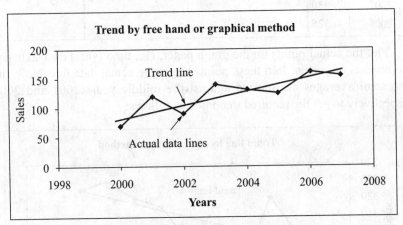

2. **Semi Average Method:** In this method actual points are plotted on the graph paper, *i.e.*, time on x-axis and the quantity of measurement on y-axis. Join these points to get the actual data line. Now, the time series is divided into equal halves. Find the total of first half and the total of second half. These totals are called the semi totals. Now, find the average of the first and second half series. These averages are

called the semi averages. These average values are plotted on the graph paper against the mid points of the corresponding half series. The line joining to these two points gives the trend line.

Note: If the number of time periods is odd then find the average of two half series by ignoring the middle period.

Example 2. Fit a trend line to the following data by semi average method:

Year	1999	2000	2001	2002	2003	2004	2005	2006	2007	2008
Imports	290	300	320	280	290	295	315	320	330	325

Solution. Here, the number of time periods is even, so find the total of imports for the first five years and then find the total of imports for the last five years. These totals are called the semi totals. Now, find the semi average as follows:

Year	Imports	Semi total of imports	Semi averages
1999	290		
2000	300		
2001	320	$290 + 300 + 320 + 280 + 290 = 1480$	$\dfrac{1480}{5} = 296$
2002	280		
2003	290		
2004	295		
2005	315		
2006	320	$295 + 315 + 320 + 330 + 325 = 1585$	$\dfrac{1585}{5} = 317$
2007	330		
2008	325		

Plot the actual points on the graph paper, *i.e.*, time (year) on *x*-axis and the imports on *y*-axis. Join these points to get the actual data line. Now, the two semi averages are plotted against the middle years 2001 and 2006 respectively to get the required trend line as follows:

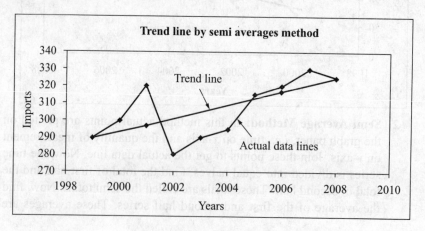

Time Series and Forecasting

Example 3. Fit a trend line to the following data by semi average method:

Year	2000	2001	2002	2003	2004	2005	2006	2007	2008
Sales	100	95	110	115	120	95	120	115	130

Solution. Here, the number of time periods is odd, so find the total of sales for the first four years, leave the sales of the year 2004 and then find the total of sales for the last four years. These totals are called the semi totals. Now, find the semi average as follows:

Year	Imports	Semi total of imports	Semi averages
2000	100		
2001	95		
2002	110	$100 + 95 + 110 + 115 = 420$	$\dfrac{420}{4} = 105$
2003	115		
2004	120		
2005	95		
2006	120		
2007	115	$95 + 120 + 115 + 130 = 460$	$\dfrac{460}{4} = 115$
2008	130		

Plot the actual points on the graph paper, *i.e.*, time (year) on *x*-axis and the sales on *y*-axis. Join these points to get the actual data line. Plot the first semi average against the middle of years 2001 and 2002, plot the second semi average against the middle of years 2006 and 2007 to get the required trend line as follows:

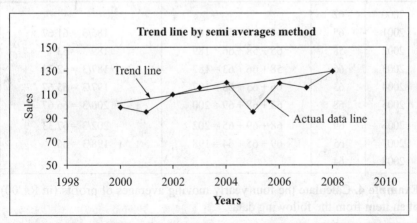

3. **Method of Moving Averages:** This is the simplest and most commonly used forecasting method. It works best when the time series fluctuates around a constant base level.

In this method we find the average of the n most recent observations to forecast the next period. From one period to the next, the average 'moves' by replacing the oldest observation in the average with the most recent observation. These averages are placed against the middle of the time period of each group. In the process, short-term irregularities in the data series are smoothed out.

If the number of periods in a group is odd, we can locate the middle period easily. If the number of periods in a group is even, no single period is middle. In this case, the average of the averages in pairs is calculated and placed against the mid period of the two. This is called the centering of moving averages. Plot these moving averages on the graph paper to get the trend line.

Example 3. Calculate the three yearly moving averages of sales (in Rs.'000) of an item from the following data:

Year	2000	2001	2002	2003	2004	2005	2006	2007	2008
Sales	62	65	58	66	63	68	69	65	64

Solution. Find the total of sales for the first three years 2000, 2001, 2002 and write it against the mid year 2001 in the column for three yearly moving totals. Now, leave the first value of the sale and add the next three values and write it against the mid year 2002 in the column for three yearly moving totals. Continue this process up to last three values by dropping second, third value of the sales and so on. Now, write the three yearly moving averages against the three yearly moving totals by dividing each total by three. We have the following table:

Year	Sales	Three yearly moving total	Three yearly moving average
2000	62		
2001	65	62 + 65 + 58 = 185	185/3 = 61.67
2002	58	65 + 58 + 66 = 189	189/3 = 63
2003	66	58 + 66 + 63 = 187	187/3 = 62.33
2004	63	66 + 63 + 68 = 197	197/3 = 65.67
2005	68	63 + 68 + 69 = 200	200/3 = 66.67
2006	69	68 + 69 + 65 = 202	202/3 = 67.33
2007	65	69 + 65 + 64 = 198	198/3 = 66
2008	64		

Example 4. Calculate the four yearly moving averages of profits (in Rs.'00) of an item from the following data:

Year	2000	2001	2002	2003	2004	2005	2006	2007
Profit	67	70	63	71	68	73	74	70

Solution. Find the total of profit for the first four years 2000, 2001, 2002, 2003 and write it against the mid of the years 2001 and 2002 in the column for four yearly moving totals. Now, leave the first value of the profit and add the next four values and write it against the mid of the years 2002 and 2003 in the column for four yearly moving totals. Continue this process up to last four values by dropping second, third value of the profit and so on.

Now, write the four yearly moving averages against the four yearly moving totals by dividing each total by four. We have the following table:

Year	Profit	Four yearly moving totals	Four yearly moving averages	Four yearly moving averages centred
2000	67			
2001	70			
2002	63	67 + 70 + 63 + 71 = 271	271/4 = 67.75	(67.75 + 68)/2 = 67.875
2003	71	70 + 63 + 71 + 68 = 272	272/4 = 68	(68 + 68.75)/2 = 68.375
2004	68	63 + 71 + 68 + 73 = 275	275/4 = 68.75	(68.75 + 71.5)/2 = 70.125
2005	73	71 + 68 + 73 + 74 = 286	286/4 = 71.5	(71.5 + 71.25)/2 = 71.375
2006	74	68 + 73 + 74 + 70 = 285	285/4 = 71.25	
2007	70			

4. **Method of Least Squares:** This method is a version of the linear regression technique. It attempts to draw a straight line through the time series data points in such a way that it comes as close to the points as possible. *i.e.*, It attempts to reduce the vertical deviations of the points from the trend line, and does this by minimizing the squared values of the deviations of the points from the line.

Let us suppose that we want to fit the trend line $y = a + bx$ to the given time series (x_i, y_i); $i = 1, 2,..., n$. Where x_i represents the values on the horizontal axis (time), and y_i represents the values of the observed data on the vertical axis *e.g.*, demand, sales, profit etc.

Now, we want to compute a slope 'b' for the trend line and the point 'a' where the line crosses the y-axis.

According to the principle of least squares, we have to determine 'a' and 'b' so that

$$S = \sum_{i=1}^{n} (P_i Q_i)^2 = \sum_{i=1}^{n} (y_i - a - bx_i)^2$$

is minimum.

From the principle of maxima and minima, the partial derivatives of S, with respect to a and b should vanish separately, *i.e.*,

$$\frac{\partial S}{\partial a} = 0 \implies -2\sum_{i=1}^{n}(y_i - a - bx_i) = 0$$

$$\frac{\partial S}{\partial b} = 0 \implies -2\sum_{i=1}^{n}(y_i - a - bx_i)(x_i) = 0$$

We have, $\quad \sum_{i=1}^{n} y_i = na + b\sum_{i=1}^{n} x_i$

and $\quad \sum_{i=1}^{n} x_i y_i = a\sum_{i=1}^{n} x_i + b\sum_{i=1}^{n} x_i^2$

These equations are known as the normal equations for estimating 'a' and 'b'.

All the quantities $\sum_{i=1}^{n} x_i$, $\sum_{i=1}^{n} x_i^2$, $\sum_{i=1}^{n} y_i$ and $\sum_{i=1}^{n} x_i y_i$ can be obtained from the given set of points $(x_i, y_i) i = 1, 2,..., n$ and the normal equations can be solved for 'a' and 'b'. With the values of 'a' and 'b' so obtained equation $y = a + bx$ is the line of best fit to the given set of points (x_i, y_i).

Example 5. Fit a trend line by the method of least square to the following data:

Year	2001	2002	2003	2004	2005	2006	2007
Sales	160	180	184	166	188	198	184

Solution. Let the equation of trend line be

$$y = a + bu \quad \text{where } u = x - 2004 \quad \ldots (1)$$

Now, the normal equations for equation (1) are as:

$$\Sigma y_i = na + b\Sigma u_i \quad \ldots (2)$$
$$\Sigma u_i y_i = a\Sigma u_i + b\Sigma u_i^2 \quad \ldots (3)$$

Year (x_i)	Sales (y_i)	$u_i = x_i - 2004$	u_i^2	$u_i y_i$	$y_i = a + bu_i$
2001	160	-3	9	-480	$180 + 4(-3) = 168$
2002	180	-2	4	-360	$180 + 4(-2) = 172$
2003	184	-1	1	-184	$180 + 4(-1) = 176$
2004	166	0	0	0	$180 + 4(0) = 180$
2005	188	1	1	188	$180 + 4(1) = 184$
2006	198	2	4	396	$180 + 4(2) = 188$
2007	184	3	9	552	$180 + 4(3) = 192$
Total	1260	0	28	112	

Time Series and Forecasting

Now, the equations (2) and (3) become
$$1260 = 7a + b \times 0$$
$$112 = a \times 0 + 28b$$

On solving these equations, we have $a = 180$ and $b = 4$.

Putting these values in (1), we get the equation of trend line as
$$y = 180 + 4u$$
$$y = 180 + 4(x - 2004)$$

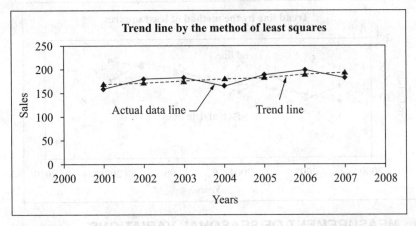

Example 6. Fit a trend line by the method of least square to the following data:

Year	2001	2002	2003	2004	2005	2006	2007	2008
Production	12	10	14	11	13	15	16	13

Solution. Let the equation of trend line be
$$y = a + bu \text{ where } u = 2(x - 2004.5) \quad \ldots (1)$$

Now, the normal equations for equation (1) are as:
$$\Sigma y_i = na + b\Sigma u_i \quad \ldots (2)$$
$$\Sigma u_i y_i = a\Sigma u_i + b\Sigma u_i^2 \quad \ldots (3)$$

Year (x_i)	Production (y_i)	$u_i = 2(x_i - 2004.5)$	u_i^2	$u_i y_i$	$y_i = a + bu_i$
2001	12	-7	49	-84	$13 + 0.25\,(2)(-7) = 09.50$
2002	10	-5	25	-50	$13 + 0.25\,(2)(-5) = 10.50$
2003	14	-3	9	-42	$13 + 0.25\,(2)(-3) = 11.50$
2004	11	-1	1	-11	$13 + 0.25\,(2)(-1) = 12.50$
2005	13	1	1	13	$13 + 0.25\,(2)(1) = 13.50$
2006	15	3	9	45	$13 + 0.25\,(2)(3) = 14.50$
2007	16	5	25	80	$13 + 0.25\,(2)(5) = 15.50$
2008	13	7	49	91	$13 + 0.25\,(2)(7) = 16.50$
Total	104	0	168	42	

Now, the equations (2) and (3) become
$$104 = 8a + b \times 0$$
$$42 = a \times 0 + 168b$$
On solving these equations, we have $a = 13$ and $b = 0.25$.
Putting these values in (1), we get the equation of trend line as
$$y = 13 + 0.25u$$
$$y = 13 + 0.25 \times 2(x - 2004.5)$$

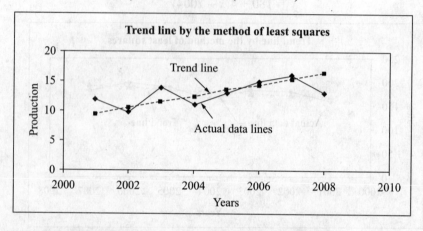

9.9 MEASUREMENT OF SEASONAL VARIATIONS

The seasonal variations can be determined in the following four ways:
1. Simple Average Method
2. Ratio to Trend Method
3. Ratio to Moving Average Method
4. Link Relative Method

1. **Simple Average Method:** This is the simplest method for obtaining seasonal index. We know that a season is a part of the year, the season may be half yearly, quarterly, monthly, weekly etc. To calculate seasonal index, first find the average of each period (season) and then the average of the averages. The seasonal index for any period is obtained by

$$\text{Seasonal index for a period} = \frac{\text{Average of that period}}{\text{Average of all the period averages}} \times 100$$

If $\bar{x}_1, \bar{x}_2, ..., \bar{x}_k$ be the seasonal averages and $\bar{x}$ be the average of the k seasonal averages then the

$$\text{Seasonal index for first period} = \frac{\text{Average of first period}}{\text{Average of all the } k \text{ periods averages}} \times 100$$

$$\text{Seasonal index for second period} = \frac{\text{Average of second period}}{\text{Average of all the } k \text{ periods averages}} \times 100$$

and so on.

Time Series and Forecasting

Example 7. Calculate the seasonal indices for the following data related to quarterly sales using simple average method:

Year	Quarterly sales			
	Q1	Q2	Q3	Q4
2004	200	205	210	220
2005	230	215	205	230
2006	220	220	215	240
2007	230	230	220	230

Solution:

Year	Quarterly sales			
	Q_1	Q_2	Q_3	Q_4
2004	200	205	210	220
2005	230	215	205	230
2006	220	220	215	240
2007	230	230	220	230
Quarterly Totals	880	870	850	920

Quarterly averages for:

$$Q_1 = \frac{\text{Total for } Q_1}{4} = \frac{880}{4} = 220$$

$$Q_2 = \frac{\text{Total for } Q_2}{4} = \frac{870}{4} = 217.5$$

$$Q_3 = \frac{\text{Total for } Q_3}{4} = \frac{850}{4} = 212.5$$

$$Q_4 = \frac{\text{Total for } Q_4}{4} = \frac{920}{4} = 230$$

$$\text{Average of the quarterly averages} = \frac{\text{Total of averages}}{\text{Number of averages}}$$

$$= \frac{220 + 217.5 + 212.5 + 230}{4}$$

$$= \frac{880}{4} = 220$$

$$\text{Seasonal index for } Q_1 = \frac{\text{Average of } Q_1}{\text{Average of the quarterly averages}} \times 100$$

$$= \frac{220}{220} \times 100 = 100$$

Seasonal index for $Q_2 = \dfrac{217.5}{220} \times 100 = 98.86$

Seasonal index for $Q_3 = \dfrac{212.5}{220} \times 100 = 96.59$

Seasonal index for $Q_4 = \dfrac{230}{220} \times 100 = 104.54$

2. Ratio to Trend Method: In this method the trend values for each month (quarter) are obtained by least square method. Divide actual data of each month (quarter) by the corresponding trend value and write them in percentages. Now, find the average of the different values for each month (quarter). Find the median or mean to remove the cyclic or irregular variations. These median or mean values are the seasonal indices. If the seasons are months the sum of seasonal indices should be 1200 and in case of quarters the sum of seasonal indices should be 400. If the sum is not comes out to be as required, then we have to make an adjustment in the seasonal indices multiplying each of them by (1/sum of seasonal indices) × 1200 or (1/sum of seasonal indices) × 400 as the case may be. These adjusted seasonal indices are the required seasonal indices.

Example 8. Calculate the seasonal indices for the following data using ratio to trend method:

Year	Quarterly sales			
	Q_1	Q_2	Q_3	Q_4
2004	30	40	36	34
2005	34	52	50	44
2006	40	58	54	48
2007	54	76	68	62
2008	80	92	86	82

Solution. First we will find annual trend values by using least square method as follows:

Year (x_i)	Annual totals	Average of quarterly values (y_i)	$u_i = x_i - 2006$	u_i^2	$u_i y_i$	Annual trend values $y_i = a + b u_i$
2004	140	35	−2	4	−70	56 + 12 (−2) = 32
2005	180	45	−1	1	−45	56 + 12 (−1) = 44
2006	200	50	0	0	0	56 + 12 (0) = 56
2007	260	65	1	1	65	56 + 12 (1) = 68
2008	340	85	2	4	170	56 + 12 (2) = 80
Total		280	0	10	120	

Time Series and Forecasting

Let the equation of trend line be
$$y = a + bu \quad \text{where } u = x - 2006 \quad \ldots (1)$$
The normal equations for equation (1) are:
$$\Sigma y_i = na + b\Sigma u_i \quad \ldots (2)$$
$$\Sigma u_i y_i = a\Sigma u_i + b\Sigma u_i^2 \quad \ldots (3)$$
Now, the equations (2) and (3) become
$$280 = 5a + b \times 0$$
$$120 = a \times 0 + 10b$$
On solving these equations, we have $a = 56$ and $b = 12$.

Putting these values in (1), we get the equation of trend line as
$$y = 56 + 12u$$
$$y = 56 + 12(x - 2006)$$
The annual increment is $b = 12$, the quarterly increment = $12/4 = 3$

Now, find the quarterly trend values as follows:

The trend value for the year 2004 is 32 it will be placed in the middle of Q_2 and Q_3. The quarterly increment is 3, so the trend value for Q_3 will be $32 + 3/2 = 33.50$ and for Q_2 trend value will be $32 - 3/2 = 30.50$. The trend values for Q_1 and Q_4 will be $30.50 - 3 = 27.50$ and $33.50 + 3 = 36.50$ respectively. Similarly, we can find trend values for the remaining years. Then we have

Year	Quarterly sales			
	Q_1	Q_2	Q_3	Q_4
2004	27.50	30.50	33.50	36.50
2005	39.50	42.50	45.50	48.50
2006	51.50	54.50	57.50	60.50
2007	63.50	66.50	69.50	72.50
2008	75.50	78.50	81.50	84.50

Now, we will find the ratios of actual values to trend values and express them in percentages as:

Year	Ratios to trend for quarters			
	Q_1	Q_2	Q_3	Q_4
2004	109.09	131.15	107.46	93.15
2005	86.08	122.35	109.89	90.72
2006	77.67	106.42	93.91	79.34
2007	85.04	114.29	97.84	85.52
2008	105.96	117.20	105.52	97.04
Totals	463.84	591.41	514.62	445.77

Averages	92.77	118.28	102.92	89.15
Adjusted seasonal indices	$\dfrac{92.77}{403.12} \times 400$ $= 92.05$	$\dfrac{118.28}{403.12} \times 400$ $= 117.37$	$\dfrac{102.92}{403.12} \times 400$ $= 102.12$	$\dfrac{89.15}{403.12} \times 400$ $= 88.46$

Note: In case of quarters the sum of seasonal indices should be 400; here the sum of seasonal indices (averages) is 403.12. So we have to make an adjustment in the seasonal indices multiplying each of them by (1/sum of seasonal indices) × 400.

3. **Ratio to Moving Average Method:** In this method, the trend of the data is calculated by the moving averages. If data is month wise, 12 months centered moving averages are calculated, and if the data is quarterly, 4 quarters centered moving averages are calculated. Now, we remove the trend and cyclic variations from the data, by dividing actual data by moving averages entered against it and expressed these ratios as percentages. Find the median or mean to remove the cyclic or irregular variations. These median or mean values are the seasonal indices. If the seasons are months the sum of seasonal indices should be 1200 and in case of quarters the sum of seasonal indices should be 400. If the sum is not comes out to be as required, then we have to make an adjustment in the seasonal indices multiplying each of them by (1/sum of seasonal indices) × 1200 or (1/sum of seasonal indices) × 400 as the case may be. These adjusted seasonal indices are the required seasonal indices.

Example 9. Calculate the seasonal indices for the following data using ratio to moving averages method:

Year	Quarterly			
	Q_1	Q_2	Q_3	Q_4
2006	40	58	54	48
2007	54	76	68	62
2008	80	92	86	82

Solution. First we will find four quarterly moving averages and ratios to moving averages as follows:

Time Series and Forecasting

Year and quarter		Actual values (y_i)	Four quarterly moving totals	Four quarterly moving averages	Four quaterly moving averages centered	Ratio to moving averages (percentages)
2006	Q_1	40				
	Q_2	58				
			200	50		
	Q_3	54			51.75	104.35
			214	53.50		
	Q_4	48			55.75	86.10
2007	Q_1	54			59.75	90.38
			246	61.50		
	Q_2	76				
			260	65		
	Q_3	68			68.25	99.63
			286	71.50		
	Q_4	62			73.50	84.35
			302	75.50		
2008	Q_1	80			77.75	102.89
			320	80		
	Q_2	92			82.50	111.52
			340	85		
	Q_3	86				
	Q_4	82				

Now, the percentages of original data to moving averages are arranged according to years and quarters to find the seasonal indices as:

Year	Ratios to trend for quarters			
	Q_1	Q_2	Q_3	Q_4
2006	–	–	104.35	86.10
2007	90.38	120.16	99.63	84.36
2008	102.89	111.52	–	–
Totals	193.27	231.68	203.98	170.45
Averages	96.635	115.84	101.99	85.225
Adjusted seasonal indices	$\dfrac{96.635}{399.69} \times 400$ = 96.71	$\dfrac{115.84}{399.69} \times 400$ = 115.93	$\dfrac{101.99}{399.69} \times 400$ = 102.07	$\dfrac{85.225}{399.69} \times 400$ = 85.29

Note: In case of quarters the sum of seasonal indices should be 400; here the sum of seasonal indices (averages) is 399.69. So we have to make an adjustment in the seasonal indices multiplying each of them by (1/sum of seasonal indices) × 400.

4. Link Relative Method: This method was given by Karl Pearson. This method is least used due to complicated calculations. In this method, first we find the link relatives of seasonal figures as

$$\text{Link relative } (LR) = \frac{\text{Current season's figure}}{\text{Previous season's figure}} \times 100$$

Calculate the average of link relatives for each season. Convert link relatives (LR) into chain relatives (CR) on the basis of first season. Assume the chain relative of first season as 100.

$$\text{Chain relative } (CR) = \frac{\text{Average LR of current season} \times \text{CR of previous season}}{100}$$

Find the new chain relative (NCR) for first season on the basis of previous chain relative which would be

$$\text{New chain relative } (NCR) = \frac{\text{LR of first season} \times \text{CR of previous season}}{100}$$

Find the correction factor for each season as:

In case of months,

$$\text{Correction factor } (CF) = \frac{\text{NCR for first month - Old CR i.e., 100}}{12}$$

In case of quarters,

$$\text{Correction factor } (CF) = \frac{\text{NCR for first quarter - Old CR i.e., 100}}{4}$$

Now, make the adjustment by subtracting correction factor from each chain relative. These adjusted chain relatives are the required seasonal indices. If the seasons are months the sum of seasonal indices should be 1200 and in case of quarters the sum of seasonal indices should be 400. If the sum is not comes out to be as required, then we have to make an adjustment in the seasonal indices multiplying each of them by (1/sum of seasonal indices) × 1200 or (1/sum of seasonal indices) × 400 as the case may be. These adjusted seasonal indices are the required seasonal indices.

Example 10. Calculate the seasonal indices for the following data using link relative method:

Year	Quarterly			
	Q_1	Q_2	Q_3	Q_4
2004	58	52	51	53
2005	55	48	46	51
2006	58	53	53	57
2007	60	49	46	52
2008	50	45	41	55

Solution:

Year	Quarters			
	Q_1	Q_2	Q_3	Q_4
2004	–	89.66	98.08	103.92
2005	103.77	87.27	95.83	110.87
2006	113.73	91.38	100.00	107.55
2007	105.26	81.67	93.88	113.05
2008	96.15	90.00	91.11	134.15
Totals of link relatives	418.91	439.98	478.90	569.54
Average of link relatives	104.73	88.00	95.78	113.91
New chain relatives	100	$\dfrac{88 \times 100}{100} = 88$	$\dfrac{95.78 \times 88}{100} = 84.29$	$\dfrac{113.91 \times 84.29}{100} = 96.01$
Adjusted chain relatives	100	$88 - 0.1375$ $= 87.8625$	$84.29 - 2(0.1375)$ $= 84.015$	$96.01 - 3(0.1375)$ $= 95.598$
Seasonal indices	$\dfrac{100}{367.4755} \times 400$ $= 108.85$	$\dfrac{87.8625}{367.4755} \times 400$ $= 95.638$	$\dfrac{84.015}{367.4755} \times 400$ $= 91.45$	$\dfrac{95.598}{367.4755} \times 400$ $= 104.06$

9.10 MEASUREMENT OF CYCLICAL VARIATIONS

There are various methods available in literature to measure the cyclical variations. Some of these methods are Residual method, Direct method and Harmonic analysis method. The most commonly used method is residual method. In this method we remove the trend and seasonal variations from the series. In such a way the values obtained involve only cyclic and irregular variations in case of multiplicative model we have

$$C \times I = \frac{y}{T \times S}$$

and for additive model we have

$$C + I = y - T - S$$

Now, compute moving averages of the $C \times I$ values for the corresponding time periods to remove as much of the irregular variation I as possible. These moving averages are the required cyclic indices.

9.11 MEASUREMENT OF RANDOM OR IRREGULAR VARIATIONS

The term random variation is used to cover all types of variation other than trend, seasonal and cyclical variations. There is no mathematical method to measure them. It can be measured as the residual after regular (trend, seasonal and cyclical) factors have been removed from a series.

In case of multiplicative model it can be obtained as

$$I = \frac{y}{T \times S \times C}$$

and for additive model it can be obtained as

$$I = y - Y - S - C$$

There is little or no interest in such residuals in their own right and we are not devoting further attention to them.

EXERCISE 9.1
---

1. What is a time series and what are its components?
2. Briefly explain the applications of time series.
3. What do you understand by additive and multiplicative models of time series?
4. Describe the various methods for determining the trend of a time series.
5. Explain the different variations of time series.
6. Explain the simple average method and ratio to trend method for measuring the seasonal variations.
7. Fit a trend line to the following data by the free hand or graphical method:

Year	2000	2001	2002	2003	2004	2005	2006	2007
Sales	135	160	145	170	165	162	180	175

8. Fit a trend line to the following data by the free hand or graphical method:

Year	2001	2002	2003	2004	2005	2006	2007
Sales	280	290	292	283	294	299	292

9. Fit a trend line to the following data by the method of semi averages:

Year	1999	2000	2001	2002	2003	2004	2005	2006	2007	2008
Exports	270	280	300	240	260	260	280	290	300	320

10. Fit a trend line to the following data by the method of semi averages:

Year	2000	2001	2002	2003	2004	2005	2006	2007	2008
Sales	180	200	140	160	160	180	190	200	220

Time Series and Forecasting

11. Calculate trend by the three yearly moving averages of sales of a certain item from the following data:

Year	1999	2000	2001	2002	2003	2004	2005	2006	2007	2008
Sales	21	22	23	25	24	22	25	26	27	26

12. Calculate trend by the four yearly moving averages of profits of a commercial concern from the following data:

Year	1999	2000	2001	2002	2003	2004	2005	2006	2007
Profit	614	615	652	678	681	655	717	719	708

13. Calculate trend by the five yearly moving average of production of a certain item from the following data:

Year	1999	2000	2001	2002	2003	2004	2005	2006	2007	2008
Production	1011	1193	1125	1068	1119	1120	990	1099	1304	1676

14. Calculate trend by the five yearly moving average item from the following data:

Year	1998	1999	2000	2001	2002	2003	2004	2005	2006	2007
Value('000Rs)	123	140	110	98	104	133	95	105	150	135

15. Fit a trend line to the following data by the method of least squares:

Year	2001	2002	2003	2004	2005	2006	2007
Sales	80	90	92	83	94	99	92

16. Fit a straight line trend to the following data by the method of least squares and estimate the sales for 2008:

Year	2001	2002	2003	2004	2005
Sales (Rs. Lakhs)	70	74	80	86	90

17. Fit a straight line trend to the following data by the method of least squares and estimate the production for 2007:

Year	2001	2002	2003	2004	2005	2006
Production	10	12	15	16	18	19

18. Calculate the seasonal indices for the following data related to quarterly sales using simple average method:

Year	Quarterly Sales			
	Q_1	Q_2	Q_3	Q_4
2000	35	39	34	36
2001	35	41	37	48
2002	35	39	37	40
2003	40	46	38	45
2004	41	44	42	45
2006	42	46	43	47

19. Calculate the seasonal indices for the following data using simple average method:

Year	Quarters			
	Q_1	Q_2	Q_3	Q_4
2000	78	66	84	80
2001	76	74	82	78
2002	72	68	80	70
2003	74	70	84	74
2004	76	74	86	82

20. Calculate the seasonal indices for the following data using ratio to trend method:

Year	Quarters			
	Q_1	Q_2	Q_3	Q_4
2002	30	40	36	34
2003	34	52	50	44
2004	40	58	54	48
2005	54	76	68	62
2006	80	92	86	82

21. Calculate the seasonal indices for the following data using ratio to trend method:

Year	Quarters			
	Q_1	Q_2	Q_3	Q_4
2004	36	34	38	32
2005	38	48	52	42
2006	42	56	50	52
2007	56	74	68	62
2008	82	90	88	80

22. Calculate the seasonal indices for the following data using ratio to moving averages method:

Year	Quarterly			
	Q_1	Q_2	Q_3	Q_4
2002	68	62	61	63
2003	65	58	66	61
2004	68	63	63	67

23. Calculate the seasonal indices for the following data using ratio to moving averages method:

Year	Quarterly			
	Q_1	Q_2	Q_3	Q_4
2004	75	60	54	59
2005	86	65	63	80
2006	90	72	66	85
2007	100	78	72	93

24. Calculate the seasonal indices for the following data by link relative method:

Year	Quarterly			
	Q_1	Q_2	Q_3	Q_4
2005	37	38	37	40
2006	41	34	25	31
2007	35	37	35	41

25. Calculate the seasonal indices for the following data by link relative method:

Year	Quarterly			
	Q_1	Q_2	Q_3	Q_4
2003	45	54	72	60
2004	48	56	63	56
2005	49	63	70	65
2006	52	65	75	72
2007	60	70	83	86

ANSWERS

11. 22.00, 23.33, 24.00, 23.67, 23.67, 24.33, 26.00, 26.33
12. 648.125, 661.5, 674.625, 687.875, 696.375
13. 1103.2, 1125.0, 1084.4, 1079.2, 1126.4, 1237.8
14. 115, 117, 108, 107, 117.4, 123.6

15. $y = 90 + 2(x - 2004)$; 84, 86, 88, 90, 92, 94, 96
16. $y = 80 + 5.2(x - 2003)$; 69.6, 74.8, 80.0, 85.2, 90.4; 106
17. $y = 15 + 1.828(x - 2003)$; 10.43, 12.258, 14.086, 15.914, 17.742, 19.570; 23.226
18. 93.53, 104.60, 94.75, 107.06
19. 98.43, 92.15, 108.90, 100.52
20. 92.0, 117.4, 102.1, 88.4
21. 100.12, 109.55, 103.10, 87.23
22. 105.30, 95.21, 100.97, 98.52
23. 122.36, 92.43, 84.70, 100.51
24. 105.31, 102.26, 90.42, 102.01
25. 82.86, 98.45, 114.60, 104.08

Chapter – 10
Test of Significance and Analysis of Variance

10.1 INTRODUCTION

It is not easy to collect all the information about population and also it is not possible to study the characteristics of the entire population (finite or infinite) due to time factor, cost factor and other constraints. Thus we need sample. Sample is a finite subset of statistical individuals in a population and the number of individuals in a sample is called the sample size. Sampling is quite often used in our day-to-day practical life. For example in a shop we assess the quality of rice, wheat or any other commodity by taking a handful of it from the bag and then to decide to purchase it or not.

10.2 PARAMETER AND STATISTIC

The statistical constants of the population such as mean (μ), variance (σ^2), correlation coefficient (ρ) and population proportion (P) are called 'parameters'. Statistical constants computed from the samples corresponding to the parameters namely mean ($\bar{x}$), variance (s^2), sample correlation coefficient (r) and proportion (p) etc, are called statistic. Parameters are functions of the population values while statistics are functions of the sample observations. In general, population parameters are unknown and sample statistics are used as their estimates.

10.3 SAMPLING DISTRIBUTION:

The distribution of all possible values which can be assumed by some statistic measured from samples of same size 'n' randomly drawn from the same population of size N, is called as sampling distribution of the statistic. Consider a population with N values. Let us take a random sample of size n from this population, and then there are $^{N}C_{n}$ possible samples. From each of these $^{N}C_{n}$ samples if we compute a statistic (e.g., mean, variance, correlation coefficient, skewness etc) and then we form a frequency distribution for these ^NC_n values of a statistic. Such a distribution is called sampling distribution of that statistic.

10.4 STANDARD ERROR

The standard deviation of the sampling distribution of a statistic is known as its standard error. It is abbreviated as $S.E.$ For example, the standard deviation of the sampling distribution of the mean ($\bar{x}$) known as the standard error of the mean. The standard errors of some of the well known statistic for large samples are given below, where n is the sample size, (σ^2) is the population variance and P is the population proportion and $Q = 1 - P \cdot n_1$ and n_2 represent the sizes of two independent random samples respectively.

Statistic	Standard error
Sample mean ($\bar{x}$)	$\dfrac{\sigma}{\sqrt{n}}$
Observed sample proportion (p)	$\sqrt{\dfrac{PQ}{n}}$
Difference of two sample means ($\bar{x}_1 - \bar{x}_2$)	$\sqrt{\dfrac{\sigma_1^2}{n_1} + \dfrac{\sigma_2^2}{n_2}}$
Difference of two sample proportions ($p_1 - p_2$)	$\sqrt{\dfrac{P_1 Q_1}{n_1} + \dfrac{P_2 Q_2}{n_2}}$
Difference of two sample standard deviations ($s_1 - s_2$)	$\sqrt{\dfrac{\sigma_1^2}{2n_1} + \dfrac{\sigma_2^2}{2n_2}}$

10.5 USES OF STANDARD ERROR

1. Standard error plays a very important role in the large sample theory and forms the basis of the testing of hypothesis.
2. The magnitude of the standard error gives an index of the precision of the estimate of the parameter.
3. The reciprocal of the standard error is taken as the measure of reliability or precision of the sample, *i.e.*,

$$\text{Precision of a statistic} = \dfrac{1}{\text{S.E. of the statistic}}$$

4. Standard error enables us to determine the probable limits within which the population parameter may be expected to lie. The most probable limits for population mean can be obtained by adding and subtracting

three times standard error to the sample mean $(\bar{x})$, *i.e.*, $(\bar{x} \pm 3\sigma)$. The two limits are called 'confidence limits' and the gap between these two is called 'confidence interval' and the level at which it is attained is called 'confidence level'. Some of the frequently used confidence limits, confidence interval and confidence level are as follows:

Confidence limits	Confidence interval		Confidence level
	Minimum	Maximum	
$\bar{x} \pm \sigma_{\bar{x}}$	$\bar{x} - \sigma_{\bar{x}}$	$\bar{x} + \sigma_{\bar{x}}$	68.27%
$\bar{x} \pm 2\sigma_{\bar{x}}$	$\bar{x} - 2\sigma_{\bar{x}}$	$\bar{x} + 2\sigma_{\bar{x}}$	95.45%
$\bar{x} \pm 3\sigma_{\bar{x}}$	$\bar{x} - 3\sigma_{\bar{x}}$	$\bar{x} + 3\sigma_{\bar{x}}$	99.73%
$\bar{x} \pm 1.96\sigma_{\bar{x}}$	$\bar{x} - 1.96\sigma_{\bar{x}}$	$\bar{x} + 1.96\sigma_{\bar{x}}$	95%
$\bar{x} \pm 2.58\sigma_{\bar{x}}$	$\bar{x} - 2.58\sigma_{\bar{x}}$	$\bar{x} + 2.58\sigma_{\bar{x}}$	99%

10.6 HYPOTHESIS TESTING

A statistical hypothesis is a statement about a population parameter. The procedure which enable us to decide on the basis of sample result whether a hypothesis is true or not, is called test of hypothesis or test of significance. The conventional approach to hypothesis testing is not to construct a single hypothesis about the population parameter but rather to set up two different hypothesis. So that of one hypothesis is accepted, the other is rejected and vice versa. There are two types of statistical hypothesis, null hypothesis and alternative hypothesis

10.7 NULL HYPOTHESIS

Null hypothesis is the hypothesis which is tested for possible rejection under the assumption that it is true. Null hypothesis is denoted by H_0. It is very useful tool in test of significance. Null hypothesis asserts that there is no significant difference between the sample statistic and the population parameter and whatever difference is observed that is merely due to the fluctuations in sampling from the same population.

10.8 ALTERNATIVE HYPOTHESIS

Rejecting the null hypothesis implies that it is rejected in favour of some other hypothesis which is accepted. A hypothesis which is accepted when H_0 is rejected is called the alternative hypothesis and is denoted by H_1. In other words, alternative hypothesis is complementary to the null hypothesis.

For example, if we want to test the null hypothesis that the population has a specified mean μ_0 (say), *i.e.*,

Null hypothesis $H_0 : \mu = \mu_0$

then alternative hypothesis may be

1. $H_1 : \mu \neq \mu_0$ (*i.e.*, $\mu > \mu_0$ or $\mu < \mu_0$)
2. $H_1 : \mu > \mu_0$
3. $H_1 : \mu < \mu_0$

the alternative hypothesis in (1) is known as a 'two tailed alternative' and the alternative in (2) is known as right tailed (3) is known as left tailed alternative respectively. The settings of alternative hypothesis is very important since it enables us to decide whether we have to use a single tailed (right or left) or two tailed test.

10.9 LEVEL OF SIGNIFICANCE

In testing a given hypothesis, the maximum probability with which we would be willing to take risk is called level of significance of the test. This probability is denoted by α which is generally specified before samples are drawn. The levels of significance usually employed in testing of significance are 5% and 1%. If for example a 5% level of significance is chosen in deriving a test of hypothesis, then there are about 5 chances in 100 that we would reject the hypothesis when it should be accepted, i.e., we are about 95% confident that we made the right decision. In such a case we say that the hypothesis has been rejected at 5% level of significance which means that we could be wrong with probability 0.05.

10.10 CRITICAL REGION

A region in which null hypothesis H_0 is rejected is called the critical region or region of rejection.

10.11 CRITICAL VALUE

The value of test statistic which separates the critical (or rejection) region and the acceptance region is called the critical value or significant value.

The value of Z under the null hypothesis is known as test statistic. The critical value of the test statistic at the level of significance for a two tailed test is given by $Z_{\alpha/2}$ and for a one tailed test by Z_α.

10.12 ONE TAILED AND TWO TAILED TESTS

In any test, the critical region is represented by a portion of the area under the probability curve of the sampling distribution of the test statistic. A test of any statistical hypothesis where the alternative hypothesis is one tailed (right tailed or left tailed) is called a one tailed test.

For example, for testing the mean of a population $H_0 : \mu = \mu_0$, against the alternative hypothesis $H_1 : \mu > \mu_0$ (right tailed) or $H_1 : \mu < \mu_0$ (left tailed) is a single tailed test. In the right tailed test $H_1 : \mu > \mu_0$ the critical region lies entirely in right tail of the sampling distribution of x, while for the left tailed test $H_1 : \mu < \mu_0$ the critical region is entirely in the left of the distribution of x.

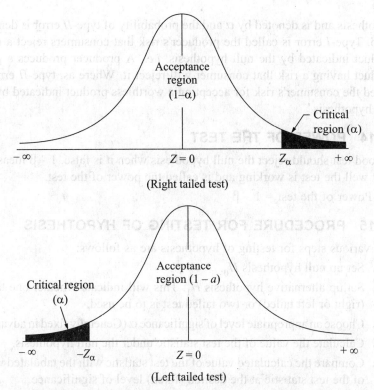

(Right tailed test)

(Left tailed test)

A test of statistical hypothesis where the alternative hypothesis is two tailed such as $H_1 : \mu = \mu_0$ against the alternative hypothesis $H_1 : \mu \neq \mu_0$ ($\mu > \mu_0$ or $\mu < \mu_0$) is known as two tailed test and in such a case the critical region is given by the portion of the area lying in both the tails of the probability curve of test of statistic.

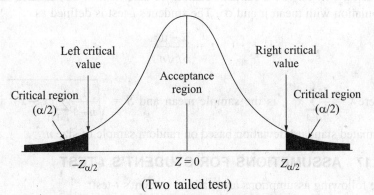

(Two tailed test)

10.13 TYPE-I ERROR AND TYPE-II ERROR

The rejection of the null hypothesis (H_0) when H_0 true is called the type-*I* error and the accepting H_0 when H_0 is false(H_0 is true) is called the type-*II* error. The probability of a type-*I* error is the level of significance of the test of hypothesis and is denoted by α and the probability of type-*II* error is denoted by β. Type-*I* error is called the producer's risk that consumers reject a good product indicated by the null hypothesis. *i.e.*, A producer produces a good product having a risk that consumer will reject it. Where as, type-*II* error is called the consumer's risk for accepting a worthless product indicated by the null hypothesis.

10.14 POWER OF THE TEST

A good test should reject the null hypothesis when it is false. $1 - \beta$ measures how well the test is working and is called the power of the test.

Power of the test = $1 - \beta$

10.15 PROCEDURE FOR TESTING OF HYPOTHESIS

The various steps for testing of hypothesis are as follows:
1. Set up null hypothesis H_0.
2. Set up alternative hypothesis H_1. This will indicate whether one tailed (right or left tailed) or two tailed test is to be used.
3. Choose an appropriate level of significance α (Generally fixed in advance).
4. Calculate the value of the test statistic under the null hypothesis.
5. Compare the calculated value of the test statistic with the tabulated value of the test statistic at the desired (fixed) level of significance.
6. If calculated value is greater or equal to the tabulated value then the null hypothesis H_0 is rejected otherwise H_0 is accepted.

10.16 STUDENT'S *t*-TEST

Let $x_1, x_2, \ldots, x_n$ be a random sample of size n ($n < 30$) from a normal population with mean μ and σ^2. The student's *t*-test is defined as

$$t = \frac{\overline{x} - \mu}{S/\sqrt{n}}$$

where $\overline{x} = \frac{1}{n}\sum_{i=1}^{n} x_i$, is the sample mean and $S = \sqrt{\frac{1}{n-1}\sum_{i=1}^{n}(x_i - \overline{x})^2}$ is the estimated standard deviation based on random sample of size n.

10.17 ASSUMPTIONS FOR STUDENT'S *t*-TEST

The following assumptions are made in student's *t*-test:
1. The parent population from which the sample drawn is normal.
2. The population standard deviation (σ) is unknown.
3. Sample size is less than 30.

10.18 t-TEST FOR SINGLE MEAN

Suppose we want to test
  (i) If a random sample x_i ($i = 1, 2, \ldots n$) of size n has been drawn from a normal population with a specified mean say μ or
  (ii) If the sample mean differs significantly from the hypothetical value μ of the population mean.

  Under null hypothesis H_0 :
  (i) The sample mean has been drawn from the population with mean μ
  (ii) There is no significant difference between the sample mean $\bar{x}$ and the population mean μ

The statistic
$$t = \frac{\bar{x} - \mu}{S/\sqrt{n}}$$

where $\bar{x} = \frac{1}{n}\sum_{i=1}^{n} x_i$ and $S = \sqrt{\frac{1}{n-1}\sum_{i=1}^{n}(x_i - \bar{x})^2}$

follows Student's t-distribution with $(n - 1)$ degrees of freedom.

We now compare the calculated value of t with the tabulated value of t at certain level of significance. If the absolute value of calculated t i.e., $|t| >$ tabulated t, H_0 is rejected otherwise H_0 may be accepted.

Example 1. Ten individuals are chosen at random from a normal population and the heights are found to be in inches 63, 63, 66, 67, 68, 69, 70, 70, 71 and 71. Test if the sample belongs to the population whose mean height is 66 inches. (Given $t_{0.05}$ = 2.26 for 9 degrees of freedom)

Solution.

x_i	$x_i - \bar{x}$	$(x_i - \bar{x})^2$
63	−4.8	23.04
63	−4.8	23.04
66	−1.8	3.24
67	−0.8	0.64
68	0.2	0.04
69	1.2	1.44
70	2.2	4.84
70	2.2	4.84
71	3.2	10.24
71	3.2	10.24
$\sum x_i = 678$		$\sum (x_i - \bar{x})^2 = 81.6$

Here, $n = 10$

$$\bar{x} = \frac{1}{n}\sum_{i=1}^{n} x_i = \frac{678}{10} = 67.8 \text{ inches}$$

$$S = \sqrt{\frac{1}{n-1}\sum_{i=1}^{n}(x_i - \bar{x})^2} = \sqrt{\frac{1}{9} \times 81.6} = \sqrt{9.0667} = 3.011$$

Null hypothesis $H_0 : \mu = 66$, i.e., population mean is 66 inches

Under null hypothesis H_0, the test statistic is

$$t = \frac{\bar{x} - \mu}{S/\sqrt{n}}$$

$$= \frac{67.8 - 66}{3.011/\sqrt{10}} = \frac{1.8 \times \sqrt{10}}{3.011} = \frac{5.692}{3.011}$$

$$= 1.8904$$

Degree of freedom $= n - 1 = 10 - 1 = 9$

$t_{0.05} = 2.26$ for 9 d.f.

As the calculated value of t is less than $t_{0.05} = 2.26$, the difference between $\bar{x}$ and μ may be due to fluctuations of random sampling, H_0 may be accepted. In other words, the data does not provide any significant evidence against the hypothesis that the population mean is 66 inches.

Example 2. A random sample of 16 values from a normal population showed a mean of 41.5 inches and the sum of squares of deviations from this mean equal to 135 square inches. Show that the assumption of a mean of 43.5 inches for the population is not reasonable.

(Given $t_{0.05} = 2.13$, $t_{0.01} = 2.95$ for 15 degrees of freedom)

Solution:

Here, $n = 16$, $\bar{x} = 41.5$ inches, $\sum(x_i - \bar{x})^2 = 135$ square inches

$$S = \sqrt{\frac{1}{n-1}\sum_{i=1}^{n}(x_i - \bar{x})^2} = \sqrt{\frac{1}{15} \times 135} = \sqrt{9} = 3$$

Null hypothesis $H_0 : \mu = 43.5$ inches, i.e., the data are consistent with the assumption that the mean height for population is 43.5 inches

Under null hypothesis H_0, the test statistic is

$$S = \frac{\bar{x} - \mu}{S/\sqrt{n}}$$

$$= \frac{41.5 - 43.5}{3/\sqrt{16}} = \frac{-2 \times 4}{3} = -2.667$$

Degree of freedom $= n - 1 = 16 - 1 = 15$

We are given that $t_{0.05} = 2.13$, $t_{0.01} = 2.95$ for 15 degrees of freedom As the calculated value of $|t| = 2.667$ is greater than $t_{0.05} = 2.13$, the null hypothesis H_0 is rejected at 5% level of significance and we conclude that the assumption of mean 43.5 inches for the population is not reasonable.

Remark: As the calculated value of $|t| = 2.667$ is less than $t_{0.01} = 2.95$, the null hypothesis H_0 may be accepted at 1% level of significance.

10.19 t-TEST FOR DIFFERENCE OF MEANS

Given two independent random samples x_i ($i = 1, 2,...,n_1$), y_j ($j = 1, 2,...,n_2$) of sizes n_1, n_2 with means $\bar{x}$, $\bar{y}$ and standard deviations S_1, S_2 respectively from normal populations with the same variance, we have to test the hypothesis that the population means are same. In other words, since a normal distribution is completely specified by its mean and variance, we have to test the hypothesis that the two independent samples come from the same normal population.

Under null hypothesis H_0:
The statistic is given by

$$t = \frac{\bar{x} - \bar{y}}{S\sqrt{\frac{1}{n_1} + \frac{1}{n_2}}}$$

Where $\bar{x} = \frac{1}{n_1}\sum_{i=1}^{n_1} x_i$, $\bar{y} = \frac{1}{n_2}\sum_{j=1}^{n_2} y_j$ and

$$S^2 = \frac{1}{(n_1 + n_2 - 2)}[(n_1 - 1)S_1^2 + (n_2 - 1)S_2^2]$$

or

$$S^2 = \frac{1}{(n_1 + n_2 - 2)}\left[\sum_{i=1}^{n_1}(x_i - \bar{x})^2 + \sum_{j=1}^{n_2}(y_j - \bar{y})^2\right]$$

follows Student's t-distribution with $(n_1 + n_2 - 2)$ degrees of freedom.

We now compare the calculated value of t with the tabulated value of t at certain level of significance. If the absolute value of calculated t i.e.,$|t| >$ tabulated t, H_0 is rejected i.e, the difference between the sample means is said to be significant otherwise H_0 may be accepted i.e., the data are said to be consistent with the hypothesis.

Example 3. The following data related to the heights (in cms.) of two different varieties of wheat plants:

Variety1	63	65	68	69	71	72	-	-	-	-
Variety2	61	62	65	65	69	66	70	70	72	73

Test the null hypothesis that the mean heights of plants of both varieties are same.

Solution. Here, $n_1 = 6$, $n_2 = 10$

Null hypothesis $\qquad\qquad\qquad H_0 : \mu_1 = \mu_2$

Alternative hypothesis $\qquad\quad H_1 : \mu_1 > \mu_2$ (right tail)

Under null hypothesis H_0, the test statistic is given by

$$t = \frac{\bar{x} - \bar{y}}{S\sqrt{\dfrac{1}{n_1} + \dfrac{1}{n_2}}}$$

Variety1			Variety2		
x_i	$x_i - \bar{x}$	$(x_i - \bar{x})^2$	y_j	$y_j - \bar{y}$	$(y_j - \bar{y})^2$
63	–5	25	61	–6	36
65	–3	9	62	–5	25
68	0	0	65	–2	4
69	1	1	65	–2	4
71	3	9	66	–1	1
72	4	16	66	–1	1
			70	3	9
			70	3	9
			72	5	25
			73	6	36
$\Sigma x_i = 48$		$\Sigma(x_i - \bar{x})^2 = 60$	$\Sigma y_j = 670$		$\Sigma(y_j - \bar{y})^2 = 150$

$$\bar{x} = \frac{1}{n_1}\sum_{i=1}^{n_1} x_i = \frac{408}{6} = 68, \quad \bar{y} = \frac{1}{n_2}\sum_{j=1}^{n_2} y_j = \frac{670}{10} = 67$$

$$S^2 = \frac{1}{(n_1 + n_2 - 2)}\left[\sum_{i=1}^{n_1}(x_i - \bar{x})^2 + \sum_{j=1}^{n_2}(y_j - \bar{y})^2\right]$$

$$= \frac{1}{6+10-2}[60+150] = \frac{210}{14} = 15 \Rightarrow S = 3.8673$$

So $\quad t = \dfrac{\bar{x} - \bar{y}}{S\sqrt{\dfrac{1}{n_1} + \dfrac{1}{n_2}}} = \dfrac{68 - 67}{3.873\sqrt{\dfrac{1}{6} + \dfrac{1}{10}}} = \dfrac{1}{3.873 \times 0.5164} = 0.499$

Tabulated $t_{0.05} = 1.76$ for 14 degrees of freedom in case of one tailed test. As the calculated value of $t = 0.499$ is less than $t_{0.05} = 1.76$, it not significant at 5% level of significance, the null hypothesis H_0 may be accepted and we conclude that the heights of the plants are not different at 5% level of significance.

Example 4. The mean values of birth weight with standard deviations and sample sizes are given below by socio-economic status. Is the mean difference in birth weight significant between socio-economic groups?

	High socio-economic group	Low socio-economic group
Sample size	$n_1 = 15$	$n_2 = 10$
Birth weight (k.g)	$\bar{x} = 2.91$	$\bar{y} = 2.26$
Standard deviation	$S_1 = 0.27$	$S_2 = 0.22$

Solution. Here, $n_1 = 15$, $n_2 = 10$, $\bar{x} = 2.91$, $\bar{y} = 2.26$, $S_1 = 0.27$ and $S_2 = 0.22$

Null hypothesis $\quad H_0 : \mu_1 = \mu_2$

Alternative hypothesis $\quad H_1 : \mu_1 > \mu_2$ (right tail) *i.e.*, high socio-economic group is superior to low socio-economic group.

Under null hypothesis H_0, the test statistic is given by

$$t = \frac{\bar{x} - \bar{y}}{S\sqrt{\dfrac{1}{n_1} + \dfrac{1}{n_2}}}$$

$$S^2 = \frac{1}{(n_1 + n_2 - 2)}[(n_1 - 1)S_1^2 + (n_2 - 1)S_2^2]$$

$$= \frac{1}{15 + 10 - 2}[(15 - 1) \times (0.27)^2 + (10 - 1) \times (0.22)^2]$$

$$= \frac{1.0206 + 0.4356}{23} = \frac{1.4562}{23} = 0.063$$

$\Rightarrow \quad S = 0.25$

So $\quad t = \dfrac{\bar{x} - \bar{y}}{S\sqrt{\dfrac{1}{n_1} + \dfrac{1}{n_2}}} = \dfrac{2.91 - 2.26}{0.25\sqrt{\dfrac{1}{15} + \dfrac{1}{10}}} = \dfrac{0.65 \times 2.45}{0.25} = 6.37$

Tabulated $t_{0.05} = 1.71$ for 23 degrees of freedom in case of one tailed test. As the calculated value of $t = 6.37$ is greater than $t_{0.05} = 1.71$, it highly significant at 5% level of significance, the null hypothesis H_0 is rejected and we conclude that the mean of high group is greater than low group at 5% level of significance.

Example 5. In a test examination given to two groups of students, the marks obtained were as follows:

Group1	25	32	30	34	24	14	32	24	30	31	35	25	
Group2	44	34	22	10	47	31	40	30	32	35	18	21	35	29	22

Examine the significance of difference between the arithmetic average of marks secured by students of the two groups.

Solution. Here, $\quad n_1 = 12, n_2 = 15$

Null hypothesis $\quad H_0 : \mu_1 = \mu_2$

Alternative hypothesis (two tailed) $H_1 : \mu_1 \neq \mu$ (two tailed)

Group 1			Group 2		
x_i	$x_i - \bar{x}$	$(x_i - \bar{x})^2$	y_j	$y_j - \bar{y}$	$(y_j - \bar{y})^2$
25	−3	9	44	14	196
32	4	16	34	4	16
30	2	4	22	−8	64
34	6	36	10	−20	400
24	−4	16	47	17	289
14	−14	196	31	1	1
32	4	16	40	10	100
24	−4	16	30	0	0
30	2	4	32	2	4
31	3	9	35	5	25
35	7	49	18	−12	144
25	−3	9	21	−9	81
			35	5	25
			29	−1	1
			22	−8	64
$\sum x_i = 336$		$\sum (x_i - \bar{x})^2 = 380$	$\sum y_j = 450$		$\sum (y_j - \bar{y})^2 = 1410$

$$\bar{x} = \frac{1}{n_1}\sum_{i=1}^{n_1} x_i = \frac{336}{12} = 28, \bar{y} = \frac{1}{n_2}\sum_{j=1}^{n_2} y_j = \frac{450}{15} = 30$$

$$S^2 = \frac{1}{(n_1 + n_2 - 2)}\left[\sum_{i=1}^{n_1}(x_i - \bar{x})^2 + \sum_{j=1}^{n_2}(y_j - \bar{y})^2\right]$$

$$= \frac{1}{12 + 15 - 2}[380 + 1410] = \frac{1790}{25} = 71.16 \Rightarrow S = 8.46$$

Under null hypothesis H_0, the test statistic is given by

$$t = \frac{\bar{x} \sim \bar{y}}{S\sqrt{\frac{1}{n_1} + \frac{1}{n_2}}}$$

So
$$t = \frac{30 - 28}{8.46\sqrt{\frac{1}{12} + \frac{1}{15}}} = \frac{2}{8.46 \times 0.387} = 0.61$$

Tabulated $t_{0.05} = 2.06$ for 25 degrees of freedom.

As the calculated value of $t = 0.61$ is less than $t_{0.05} = 2.06$ at 5% level of significance, the null hypothesis H_0 may be accepted and we conclude that two averages do not differ significantly.

10.20 PAIRED t-TEST FOR DIFFERENCE OF MEANS

If the size of the two samples is the same, say equal to n, and the data are paired, *i.e.*, (x_i, y_i), $(i = 1, 2,..., n)$ corresponds to the same i^{th} sample unit. The problem is to test if the sample means differ significantly or not.

Here, we consider the increments, $d_i = x_i - y_i$, $(i = 1, 2,..., n)$.

Under null hypothesis H_0: increments are due to fluctuations of sampling

The statistic is given by

$$t = \frac{\bar{d}}{S/\sqrt{n}}$$

Where $\bar{d} = \frac{1}{n}\sum_{i=1}^{n} d_i$ and $S^2 = \frac{1}{(n-1)}\sum_{i=1}^{n}(d_i - \bar{d})^2$

follows Student's t-distribution with $(n-1)$ degrees of freedom. If $\sum d_i$ is negative, we may consider $|\bar{d}|$. This test is generally one tailed test. Therefore, the alternative hypothesis is $H_1 : \mu_1 > \mu_2$ or $H_1 : \mu_1 < \mu_2$.

We now compare the calculated value of t with the tabulated value of t at certain level of significance. If the absolute value of calculated t *i.e.*, $|t| >$ tabulated t, H_0 is rejected otherwise H_0 may be accepted.

Example 6. Memory capacity of 8 students was tested before and after training. State at 5% level of significance whether the training was effective from the following scores:

Student	1	2	3	4	5	6	7	8	Total
Before training	49	53	51	52	47	50	52	53	407
After training	52	55	52	53	50	54	54	53	423

Solution. Let x denotes the scores before training and y denotes the score after training.

Null hypothesis $H_0 : \mu_1 = \mu_2$, *i.e.*, there is no significant difference in the scores before and after the training. In other words, the given increments are just by chance (fluctuations of sampling).

Alternative hypothesis $H_1 : \mu_1 < \mu_2$ (left tail) *i.e.*, training has been effective

Student	Scores before training (x)	Scores after training (y)	$d = x - y$	d^2
1	49	52	–3	9
2	53	55	–2	4
3	51	52	–1	1
4	52	53	–1	1
5	47	50	–3	9

Student	Scores before training (x)	Scores after training (y)	$d = x - y$	d^2
6	50	54	−4	16
7	52	54	−2	4
8	53	53	0	0
			$\sum d = -16$	$\sum d^2 = 44$

Under H_0 the test statistic is given by

$$t = \frac{\bar{d}}{S/\sqrt{n}}$$

$$\bar{d} = \frac{1}{n}\sum_{i=1}^{n} d_i = \frac{-16}{8} = -2$$

$$S^2 = \frac{1}{(n-1)}\sum_{i=1}^{n}(d_i - \bar{d})^2$$

$$= \frac{1}{(n-1)}[\sum_{i=1}^{n} d_i^2 - n(\bar{d})^2]$$

$$= \frac{1}{7}[44 - 8\times(-2)^2] = \frac{44-32}{7} = \frac{12}{7}$$

$$= 1.714$$

$\Rightarrow \quad S = 1.31$

So $\quad t = \dfrac{\bar{d}}{S/\sqrt{n}} = \dfrac{-2}{1.31/\sqrt{8}} = \dfrac{-2\times 2.83}{1.31} = -4.32$

Tabulated $t_{0.05} = 1.90$ for $(8 - 1) = 7$ degrees of freedom in case of one tailed test.

As the calculated value of $|t| = 4.32$ is greater than $t_{0.05} = 1.90$ at 5% level of significance, the null hypothesis H_0 is rejected and we conclude that the scores differ significantly before and after the training *i.e.*, training were effective.

Example 7. A certain drug administered to 10 patients showed the following additional hours of sleep:

−1.0, 0.5, 2.7, −0.6, 1.2, 1.8, 1.6, 3.5, 0.2, −1.7

Cab it be concluded that the drug does produce additional hours of sleep?

Solution. Here, d_i are given as

$d_i = x_i - y_i = -1.0, 0.5, 2.7, -0.6, 1.2, 1.8, 1.6, 3.5, 0.2, -1.7$

$n = 10$

$$\bar{d} = \frac{1}{n}\sum_{i=1}^{n} d_i \quad \frac{-1.0 + 0.5 + 2.7 - 0.6 + 1.2 + 1.8 + 1.6 + 3.5 + 0.2 - 1.7}{10}$$

$$= \frac{8.2}{10} = 0.82$$

$$\sum d_i^2 = 1 + 0.25 + 7.29 + 0.36 + 1.44 + 3.24 + 2.56 + 12.25 + 0.04 + 2.89$$
$$= 31.32$$

Null hypothesis $H_0 : \mu_1 = \mu_2$, i.e., the drug does not produce any additional hours of sleep.

Alternative hypothesis $H_1 : \mu_1 < \mu_2$ (left tail) i.e., drug is effective

Under H_0 the test statistic is given by

$$t = \frac{\bar{d}}{S/\sqrt{n}}$$

$$S^2 = \frac{1}{(n-1)} \sum_{i=1}^{n} (d_i - \bar{d})^2 = \frac{1}{(n-1)} [\sum_{i=1}^{n} d_i^2 - n(\bar{d})^2]$$

$$= \frac{1}{10-1}[31.32 - 8 \times (0.82)^2] = \frac{1}{9}[31.32 - 6.724] = 2.733$$

$\Rightarrow \quad S = 1.653$

So $\quad t = \dfrac{\bar{d}}{S/\sqrt{n}} = \dfrac{0.82}{1.653/\sqrt{10}} = \dfrac{2.593}{1.653} = 1.57$

Tabulated $t_{0.05} = 1.833$ for $(10-1) = 9$ degrees of freedom in case of one tailed test.

As the calculated value of $t = 1.57$ is less than $t_{0.05} = 1.833$ at 5% level of significance, the null hypothesis H_0 is accepted and we conclude that the drug do not produce additional hours of sleep.

EXERCISE

1. A brand of matches is sold in boxes on which it is claimed that the average contents are 40 matches. A check on a pack of 5 boxes gives the following results:

   41, 39, 37, 40, 38

   (i) Test the manufacturer's claim keeping the interests of both the manufacturer and the customer in mind.

   (ii) As a customer test the manufacturer's claim.

2. A sample of size 10 drawn from a normal population has a mean 31 and a variance 2.25. Is it reasonable to assume that the mean of the population is 30? (Use 1% level of significance).

3. A random sample of size 10 from a normal population with mean μ gives a sample mean of 40 and sample standard deviation of 6. Test the hypothesis that $\mu = 44$ against $\mu \neq 44$ at 5% level of significance.

4. A new drug manufacturer wants to market a new drug only if he could be quite sure that the mean temperature of a healthy person taking the drug could not rise above $98.6°F$ otherwise he will withhold the drug. The drug is administered to a random sample of 17 healthy persons. The mean temperature was found to be $98.4°F$ with a standard deviation of $0.6°F$. Assuming that the distribution of the temperature is normal and $\alpha = 0.01$, what should the manufacturer do?

5. The marks of students in two groups were obtained as

Group1	18	20	36	50	49	36	34	49	41
Group2	29	28	26	35	30	44	46	–	–

Test whether the groups were identical. (Use 5% level of significance)

6. Two different types of drug A and B were tried on certain patients for increasing weight. Five person were given drug A and seven persons were given drug B. The increase in weight in pounds is given below:

Drug A	8	12	13	9	3	–	–
Drug B	10	8	12	15	6	8	11

Do the two drugs differ significantly with regard to their effect in increasing weight? (Use 5% level of significance)

7. The mean life of a sample of 10 electric bulbs was found to be 1456 hours with standard deviation of 423 hours. A second sample of 17 bulbs chosen from a different batch showed a mean life of 1280 hours with standard deviation of 398 hours. Is there a significant difference between the means of the two batches at 5% level of significance?

8. To verify whether a course in Statistics improved performance, a similar test was given 12 participants both before and after the course. The original marks recorded in alphabetical order of the participants were: 44, 40, 61, 52, 32, 44, 70, 41, 67, 72, 53 and 72. After the course, the marks were in the same order 53, 38, 69, 57, 46, 39, 73, 48, 73, 74, 60 and 78. Was the course useful? (Use 5% level of significance)

9. A certain medicine given to each of the 9 patients resulted in the following increase of blood pressure:

$$7, 3, -1, 4, -3, 5, 6, -4, -1$$

Can it be concluded that the medicine will in general be accompanied by an increase in blood pressure. (Use 5% level of significance)

ANSWERS

1. (*i*) Accept manufacturer's claim (*ii*) Manufacturer's claim is justified
2. Yes
3. Accept null hypothesis
4. The manufacturer should market the drug

5. Two groups are identical
6. No
7. No
8. Yes
9. No

10.21 Z-TEST

Z-test is applied with the following assumptions:
1. The sample size must be large ($n > 30$)
2. The variable is assumed to follow normal distribution
3. The data must be quantitative

Z-test is used to test the significance for the attributes and variables.

10.22 TEST OF SIGNIFICANCE FOR ATTRIBUTES

The sampling of attributes may be regarded as the drawing of samples from a population whose members consist of presence or absence of a particular characteristic. The various types of test are as follows:

10.23 TEST FOR NUMBER OF SUCCESSES

The sampling distribution of the number of successes follows a binomial distribution. Let X be the number of successes in n independent trials with constant probability P of success for each trial.

We have in case of binomial distribution $E(X) = nP$ and $V(X) = nPQ$, where $Q = 1 - P$ probability of failure in each trial

Now, $\qquad \text{S.E.}(X) = \sqrt{nPQ}$

The normal test for the number of successes is given by

$$Z = \frac{X - E(X)}{\text{S.E.}(X)} = \frac{X - nP}{\sqrt{nPQ}} \sim N(0,1)$$

Example 8. A coin is tossed 324 times and the head turned up 175 times. Test the hypothesis that the coin is unbiased.

Solution. Here, $n = 324$, X = number of heads = 175

If the coin is unbiased then,

$$P = \text{Probability of getting a head in a toss} = \frac{1}{2}$$

$$Q = 1 - P = 1 - \frac{1}{2} = \frac{1}{2}$$

Null hypothesis H_0 the coin is unbiased *i.e.*, $P = \frac{1}{2}$

Under H_0 the test statistic is given by

$$Z = \frac{X - E(X)}{S.E.(X)} = \frac{X - nP}{\sqrt{nPQ}} = \frac{175 - 324 \times \frac{1}{2}}{\sqrt{324 \times \frac{1}{2} \times \frac{1}{2}}} = \frac{13}{9} = 1.44$$

As calculated value of $|Z| = 1.44$ is less than the tabulated value of $Z = 1.96$ at 5% level of significance, H_0 is accepted and we conclude that the coin is unbiased.

Example 9. A die is thrown 1000 times and a throw of 5 or 6 was obtained 420 times. On the assumption of random throwing do the data indicate an unbiased die?

Solution. Here, $n = 1000$, $X =$ number of success $= 420$

If the die is unbiased then,

$$P = \text{Probability of getting 5 or 6} = \frac{1}{6} + \frac{1}{6} = \frac{1}{3}$$

$$Q = 1 - P = 1 - \frac{1}{3} = \frac{2}{3}$$

Null hypothesis H_0 : the die is unbiased

Under H_0 the test statistic is given by

$$Z = \frac{X - E(X)}{S.E.(X)} = \frac{X - nP}{\sqrt{nPQ}} = \frac{420 - 1000 \times \frac{1}{3}}{\sqrt{1000 \times \frac{1}{3} \times \frac{2}{3}}} = \frac{86.67}{14.91} = 5.813$$

As calculated value of $|Z| = 5.813$ is greater than the tabulated value of $Z = 1.96$ at 5% level of significance, H_0 is rejected and we conclude that the die is biased.

10.24 TEST FOR SINGLE PROPORTION

This test is used to test the significant difference between proportion of the sample and the population.

Let X be the number of successes in n independent trials with constant probability P of success for each trial.

We have $E(X) = nP$ and $V(X) = nPQ$, where $Q = 1 - P =$ probability of failure in each trial

Now, $\quad p = \dfrac{X}{n}$ (p = observed proportion of success)

$$E(p) = E\left(\frac{X}{n}\right) = \frac{1}{n} E(X) = \frac{nP}{n} = P$$

$$V(p) = V\left(\frac{X}{n}\right) = \frac{1}{n^2} V(X) = \frac{nPQ}{n^2} = \frac{PQ}{n}$$

$$S.E.(p) = \sqrt{\frac{PQ}{n}}$$

The normal test for the proportion of successes is given by

$$Z = \frac{p - E(p)}{S.E.(p)} = \frac{p - P}{\sqrt{\frac{PQ}{n}}} \sim N(0,1)$$

Z is called a test statistic which is used to test the significant difference of the sample and the population proportion.

Example 9. A manufacturer claims that only 4% of his products supplied by him are defective. A random sample of 600 products contained 36 defectives. Test the claim of the manufacturer.

Solution. Here, $n = 600$

$$p = \text{sample proportion of defectives} = \frac{36}{600} = 0.06$$

$$P = \text{proportion of defectives in the population} = \frac{4}{100} = 0.04$$

$$Q = 1 - P = 1 - 0.04 = 0.96$$

Null hypothesis $H_0 : P = 0.04$ is true *i.e.*, the claim of the manufacturer is right

Under H_0 the test statistic is given by

$$Z = \frac{p - P}{\sqrt{\frac{PQ}{n}}} = \frac{0.06 - 0.04}{\sqrt{\frac{0.04 \times 0.96}{600}}} = \frac{0.02}{0.008} = 2.5$$

As calculated value of $|Z| = 2.5$ is greater than the tabulated value of $Z = 1.96$ at 5% level of significance, H_0 is rejected and we conclude that the manufacturer claim is wrong.

Note: If we set the alternative hypothesis $H_1 : P \neq 0.04$ we apply two tailed test and if we set the alternative hypothesis $H_1 : P > 0.04$ we apply right (one) tailed test.

Example 10. 500 oranges are taken at random from a large basket and 65 are found to be bad. Find the *S.E.* of the proportion of bad ones in a sample of this size and assign limits within which the percentage of bad apples most probably lies.

Solution. Here, $n = 500$

$$X = \text{number of bad oranges in the sample} = 65$$

$$p = \text{proportion of bad oranges in the sample} = \frac{65}{500} = 0.3$$

$$q = 1 - p = 1 - 0.13 = 0.87$$

Since the proportion of bad oranges P in the population is not known, we can take

$$P = p = 0.13 \text{ and } Q = q = 0.87$$

$$\text{S.E. of proportion} = \sqrt{\frac{PQ}{n}} = \sqrt{\frac{0.13 \times 0.87}{500}} = 0.015$$

Limits for proportions of bad oranges in the population are

$$P \pm 3\sqrt{\frac{PQ}{n}} = 0.13 \pm 3\sqrt{\frac{0.13 \times 0.87}{500}} = 0.13 \pm 0.045 = 0.175 \text{ and } 0.085$$

$\Rightarrow$ 17.5% and 8.5%.

Example 11. A manufacturer claimed that at least 95% of the equipment which he supplied to a factory conformed to specifications. An examination of a sample of 300 equipments revealed that 27 are faulty. Test the claim at significance level of (i) 5% and (ii) 1%.

Solution. Here, $n = 300$

$X =$ number of equipments conforming to specifications in the samples
$= 300 - 27 = 273$

$p =$ sample proportion conforming to specifications $= \frac{273}{300} = 0.91$

$P =$ proportion of equipments conforming to specifications in the population
$= 95\% = 0.95$

$Q = 1 - P = 1 - 0.95 = 0.05$

Null hypothesis $H_0 : P = 0.95$ i.e., the proportion of equipments conforming to specifications in the population is 95%.

Alternative hypothesis $H_0 : P < 0.95$ i.e., at least 95% conformed to specifications. (left tailed test)

Under H_0 the test statistic is given by

$$Z = \frac{p - P}{\sqrt{\frac{PQ}{n}}} = \frac{0.91 - 0.95}{\sqrt{\frac{0.95 \times 0.05}{300}}} = -\frac{0.04}{0.0126} = -3.175$$

(i) As calculated value of $|Z|$ is greater than the tabulated value of $Z = 1.645$ at 5% level of significance, H_0 is rejected and we conclude that the manufacturer claim is not acceptable.

(ii) As calculated value of $|Z| = 3.175$ is greater than the tabulated value of $Z = 2.326$ at 1% level of significance, H_0 is rejected and we conclude that the manufacturer claim is not acceptable.

10.25 TEST FOR DIFFERENCE OF PROPORTIONS

This test is used to test the significant difference between the sample proportions.

Let two samples X_1 and X_2 of sizes n_1 and n_2 respectively taken from two different populations, then $p_1 = \dfrac{X_1}{n_1}$ and $p_2 = \dfrac{X_2}{n_2}$.

To test the significant difference between the sample proportions p_1 and p_2 we set the null hypothesis H_0 : there is no significant difference between the sample proportions.

Under the null hypothesis H_0 the test statistic is given by

$$Z = \frac{p_1 - p_2}{\sqrt{PQ\left(\dfrac{1}{n_1} + \dfrac{1}{n_2}\right)}}$$

where $\qquad P = \dfrac{n_1 p_1 + n_2 p_2}{n_1 + n_2}$ and $Q = 1 - P$

If sample proportions are not given, we set the null hypothesis $H_0 : p_1 = p_2$
Under the null hypothesis H_0 the test statistic is given by

$$Z = \frac{P_1 - P_2}{\sqrt{\dfrac{P_1 Q_1}{n_1} + \dfrac{P_2 Q_2}{n_2}}}$$

where $\qquad Q_1 = 1 - P_1$ and $Q_2 = 1 - P_2$

Note:

1. The probable limits for the observed proportion of success are

   $E(p) \pm Z_\alpha \sqrt{V(P)}$ i.e., $P \pm Z_\alpha \sqrt{\dfrac{PQ}{n}}$, where Z_α is the significant value

   at the level of significance α.

2. If P is not known then the probable limits for the proportion in the

   population are $p \pm Z_\alpha \sqrt{\dfrac{pq}{n}}$.

3. If α is not given, then we can use 3σ limits. Hence, probable limits for

   the observed proportion of success are $P \pm 3\sqrt{\dfrac{PQ}{n}}$ and probable limits

   for the proportion in the population are $p \pm 3\sqrt{\dfrac{pq}{n}}$.

4. A set of three selected values is commonly used for α. Each α and corresponding Z_α and $Z_{\alpha/2}$ values are given in the following table:

Level of significance	For two tailed test	For one tailed test
α	$\lvert Z_{\alpha/2} \rvert$	$\lvert Z_\alpha \rvert$
10%	1.645	1.282
5%	1.960	1.645
1%	2.576	2.326

Example 11. Before an increase in excise duty on tea, 400 people out of a sample of 500 persons were found to be tea drinkers. After an increase in the excise duty, 400 persons were known to be tea drinkers in a sample of 600 people. Do you think that there has been a significant decrease in the consumption of tea after the increase in the excise duty?

Solution. Here, $n_1 = 500$, $X_1 = 400$, $n_2 = 600$ and $X_2 = 400$

$$p_1 = \text{Proportion of drinkers in first sample} = \frac{X_1}{n_1} = \frac{400}{500} = 0.8$$

$$p_2 = \text{Proportion of drinkers in second sample} = \frac{X_2}{n_2} = \frac{400}{600} = 0.67$$

Since proportion P of the population is not given, it can be estimated by using

$$P = \frac{n_1 p_1 + n_2 p_2}{n_1 + n_2} = \frac{400 + 400}{500 + 600} = \frac{800}{1100} = \frac{8}{11} \text{ and } Q = 1 - \frac{8}{11} = \frac{3}{11}$$

Null hypothesis $H_0 : P_1 = P_2$ i.e., there is no significant difference in the consumption of tea before and after increase of excise duty.

Alternative hypothesis $H_0 : P_1 > P_2$ (right tailed test)

Under H_0 the test statistic is given by

$$Z = \frac{p_1 - p_2}{\sqrt{PQ\left(\frac{1}{n_1} + \frac{1}{n_2}\right)}} = \frac{0.8 - 0.67}{\sqrt{\frac{8}{11} \times \frac{3}{11}\left(\frac{1}{500} + \frac{1}{600}\right)}} = \frac{0.13}{0.027} = 4.815$$

As calculated value of $\lvert Z \rvert = 4.815$ is greater than the tabulated value of $Z = 1.645$ at 5% level of significance, H_0 is rejected and we conclude that there is a significant decrease in the consumption of tea due to increase in excise duty.

Example 12. During a country wide investigation the incidence of a chronic decease was found to be 1%. In a village of 400 strength 5 were reported to be affected whereas in another village of 1200 strength 10 were reported to be affected. Does this indicate any significant difference?

Solution. Here, $n_1 = 400$, $X_1 = 5$, $n_2 = 1200$ and $X_2 = 10$

$$p_1 = \frac{X_1}{n_1} = \frac{5}{400} = 0.0125, p_2 = \frac{X_2}{n_2} = \frac{10}{1200} = 0.0083$$

$P = 1\% = 0.01$ and $Q = 1 - 0.01 = 0.99$

Null hypothesis $H_0 : P_1 = P_2$ i.e., there is no significant difference.
Alternative hypothesis $H_0 : P_1 \neq P_2$ (two tailed test)
Under H_0 the test statistic is given by

$$Z = \frac{p_1 - p_2}{\sqrt{PQ\left(\frac{1}{n_1} + \frac{1}{n_2}\right)}} = \frac{0.0125 - 0.0083}{\sqrt{0.01 \times 0.99\left(\frac{1}{400} + \frac{1}{1200}\right)}} = \frac{0.0042}{0.00574} = 0.732$$

As calculated value of $|Z| = 0.732$ is less than the tabulated value of $Z = 1.96$ at 5% level of significance, H_0 is accepted and we conclude that there is no significant difference.

Example 13. 500 articles from a factory are examined and found to be 2% defective. 800 similar articles from a second factory are found to have only 1.5% defectives. Can it reasonably conclude that the products of the first factory are inferior to those of second?

Solution. Here, $n_1 = 500$, $n_2 = 800$

p_1 = Proportion of defectives from first factory = $\frac{2}{100} = 0.02$

p_2 = Proportion of defectives from second factory = $\frac{1.5}{100} = 0.015$

Since proportion P of the population is not given, it can be estimated by using

$$P = \frac{n_1 p_1 + n_2 p_2}{n_1 + n_2} = \frac{10 + 12}{500 + 800} = \frac{22}{1300} = 0.017$$

and $Q = 1 - P = 1 - 0.017 = 0.983$

Null hypothesis $H_0 : P_1 = P_2$ i.e., there is no significant difference between the products of two factories.

Alternative hypothesis $H_0 : P_1 \neq P_2$ (two tailed test)
Under H_0 the test statistic is given by

$$Z = \frac{p_1 - p_2}{\sqrt{PQ\left(\frac{1}{n_1} + \frac{1}{n_2}\right)}} = \frac{0.02 - 0.015}{\sqrt{0.017 \times 0.983\left(\frac{1}{500} + \frac{1}{800}\right)}} = \frac{0.005}{0.00737} = 0.678$$

As calculated value of $|Z| = 0.678$ is less than the tabulated value of $Z = 1.96$ at 5% level of significance, H_0 is accepted and we conclude that there is no significant difference between the products of first and second factory i.e., the products of the first factory are not inferior to those of second.

Example 14. A manufacturing firm claims that its brand A product outsells its brand B product by 8%. If it is found that 84 out of a sample of 400 persons prefer brand A and 36 out of another sample of 200 persons prefer brand B. Test whether the 8% difference is valid claim.

Solution. Here, $n_1 = 400$, $n_2 = 200$

$$p_1 = \text{Proportion of preference of brand } A = \frac{84}{400} = 0.21$$

$$p_2 = \text{Proportion of preference of brand } B = \frac{36}{200} = 0.18$$

Since proportion P of the population is not given, it can be estimated by using

$$P = \frac{n_1 p_1 + n_2 p_2}{n_1 + n_2} = \frac{84 + 36}{400 + 200} = \frac{120}{600} = 0.2$$

and $Q = 1 - P = 1 - 0.2 = 0.8$

Null hypothesis $H_0 : P_1 - P_2 = 0.08$ i.e., 8% difference is there in the sales of brand A and brand B

Alternative hypothesis $H_0 : P_1 - P_2 \ne 0.08$ (two tailed test)

Under H_0 the test statistic is given by

$$Z = \frac{(p_1 - p_2) - (P_1 - P_2)}{\sqrt{PQ\left(\dfrac{1}{n_1} + \dfrac{1}{n_2}\right)}} = \frac{(0.21 - 0.18) - 0.08}{\sqrt{0.2 \times 0.8 \left(\dfrac{1}{500} + \dfrac{1}{800}\right)}} = -\frac{0.05}{0.0346} = -1.44$$

As calculated value of $|Z| = 1.44$ is less than the tabulated value of $Z = 1.96$ at 5% level of significance, H_0 is accepted and we conclude that the claim of 8% difference in the sales of brand A and brand B is valid.

Exampl 15. In two large populations there are 30% and 25% respectively of fair haired people. Is this difference likely to be hidden in samples of 1400 and 1000 respectively from the two populations?

Solution. Here, $n_1 = 1400$, $n_2 = 1000$

$$P_1 = \text{Proportion of fair haired in the first population} = \frac{30}{100} = 0.3$$

$$P_2 = \text{Proportion of fair haired in the second population} = \frac{25}{100} = 0.25$$

$Q_1 = 1 - P_1 = 1 - 0.3 = 0.7$, $Q_2 = 1 - P_2 = 1 - 0.25 = 0.75$

Null hypothesis $H_0 : p_1 = p_2$ i.e., (Sample proportions are equal) the difference in population proportions is likely to be hidden in sampling.

Alternative hypothesis $H_1 : p_1 \ne p_2$ (two tailed test)

Under H_0 the test statistic is given by

$$Z = \frac{P_1 - P_2}{\sqrt{\dfrac{P_1 Q_1}{n_1} + \dfrac{P_2 Q_2}{n_2}}} = \frac{0.30 - 0.25}{\sqrt{\dfrac{0.3 \times 0.7}{1400} + \dfrac{0.25 \times 0.75}{1000}}} = \frac{0.05}{0.01837} = 2.72$$

As calculated value of $|Z| = 2.72$ is greater than the tabulated value of $Z = 1.96$ at 5% level of significance, H_0 is rejected and we conclude that these samples will exhibit the difference in the population proportions.

EXERCISE

1. A coin was tossed 400 times and the head turned up 216 times. Test the hypothesis that the coin is unbiased at 5% level of significance.
2. In a hospital 525 female and 475 male babies were born in a month. Do these figures confirm the hypothesis that females and males are born in equal number? Use 5% level of significance.
3. A die is thrown 10000 times and a throw of 3 or 4 was obtained 4200 times. On the assumption of random throwing do the data indicate an unbiased die?
4. Given that on the average 4% of insured men of age 65 die within a year and that of 60 of a particular group of 1000 such men (age 65) died within a year. Can this group be regarded as a representative sample?
5. 325 men out of 600 men chosen from a big city were found to be smokers. Does this information support the conclusion that the majority of men in the city are smokers?
6. A random sample of 400 apples is taken from a large basket and 40 are found to be bad. Estimate the proportion of bad apples in the basket and assign limits within which the percentage most probably lies.
7. A manufacturer claimed that at least 95% of the equipments which he supplied to a factory conformed to specifications. An examination of a sample of 200 pieces equipments revealed that 18 were faulty. Test the manufacturer's claim at a level of significance (*i*) 5% and (*ii*) 1%.
8. 1000 articles from a factory are examined and found to be 2.5% defective. 1500 similar articles from the second factory are found to have only 2% defectives. Can it reasonably conclude that the products of the first factory are inferior to those of second?
9. A manufacturing firm claims that its brand *A* product outsells its brand *B* product by 8%. If it is found that 42 out of a sample of 200 persons prefer brand *A* and 18 out of another sample of 100 persons prefer brand *B*. Test whether the 8% difference is valid claim.
10. In a survey on a particular matter in a college, 850 males and 560 females voted. 500 males and 320 females voted yes. Does this indicate a significant difference of opinion between male and female on this matter at 1% level of significance?
11. Two samples of sizes 1200 and 900 respectively drawn from two large populations. In the two large populations there are 30% and 25% respectively of fair haired people. Test whether these two samples will reveal the difference in the population proportions.

12. Before an increase in excise duty on tea 800 persons out of a sample of 1000 persons were found to be tea drinkers. After an increase in excise duty 800 people were tea drinkers in a sample of 1200 people. Test whether there is a significant decrease in the consumption of tea after the increase in excise duty.

ANSWERS

1. H_0 is accepted
2. Yes, H_0 is accepted
3. H_0 is rejected
4. H_0 is rejected
5. H_0 is rejected at 5% level of significance
6. 8.5; 11.5
7. Using left tailed test, H_0 is rejected at both 5% and 1% level of significance
8. No, H_0 is accepted
9. H_0 is accepted
10. H_0 is accepted
11. H_0 is rejected at 5% level of significance
12. H_0 is rejected

10.26 TEST OF SIGNIFICANCE FOR VARIABLES

The sampling of variables may be regarded as the drawing of samples from a population whose members consist of quantitative measurements. The various types of test for variables are as follows:

10.27 TEST OF SIGNIFICANCE FOR SINGLE MEAN

This test is used to test the significant difference between sample mean and population mean. Let $x_1, x_2, \ldots, x_n$ be a random sample of size n from a normal population with mean μ and variance σ^2. The standard error (S.E.) of mean of a random sample of size n from a population is given by

$$S.E.(\bar{x}) = \frac{\sigma}{\sqrt{n}},$$ where σ is the standard deviation of the population.

We set the null hypothesis H_0 that the sample has been drawn from a large population with mean μ and variance σ^2, i.e., there is no significant difference between the sample mean ($\bar{x}$) and population mean (μ).

Under the null hypothesis H_0 the test statistics is

$$Z = \frac{\bar{x} - \mu}{\sigma/\sqrt{n}}$$

If standard deviation of the population (σ) is not known, we use the test statistic given as

$$Z = \frac{\bar{x} - \mu}{s/\sqrt{n}},$$ where s is the standard deviation of the sample.

Note: The limits of the population mean μ are given by $\bar{x} \pm Z_\alpha \dfrac{\sigma}{\sqrt{n}}$ i.e.,

$$\bar{x} - Z_\alpha \frac{\sigma}{\sqrt{n}} \leq \mu \leq \bar{x} + Z_\alpha \frac{\sigma}{\sqrt{n}}$$

These limits are called the confidence limits for μ.

Example 16. A normal population has a mean as 6.8 and standard deviation 1.5. A random sample of 400 members has a mean as 6.75. Is there any significant difference between sample and population mean?

Solution. Here, $\mu = 6.8, \bar{x} = 6.75, \sigma = 1.5$ and $n = 400$

Null hypothesis $H_0 : \bar{x} = \mu$ (there is no significant difference between $\bar{x}$ and μ)

Alternative hypothesis H_1 : there is a significant difference between $\bar{x}$ and μ

Under H_0 the test statistic is given by

$$Z = \frac{\bar{x} - \mu}{\sigma/\sqrt{n}} = \frac{6.75 - 6.8}{1.5/\sqrt{400}} = -\frac{0.05}{0.075} = -0.67$$

As calculated value of $|Z| = 0.67$ is less than the tabulated value of $Z = 1.96$ at 5% level of significance, H_0 is accepted and we conclude that there is no significant difference between sample mean $\bar{x}$ and population mean μ.

Example 17. A random sample of 400 members has a mean 99. Can it be reasonably regarded as a sample from a large population with mean 100 and standard deviation 8 at 5% level of significance?

Solution. Here, $\mu = 100, \bar{x} = 99, \sigma = 8$ and $n = 400$

Null hypothesis $H_0 : \mu = 100$

Alternative hypothesis $H_1 : \mu \neq 100$ (two tailed test)

Under H_0 the test statistic is given by

$$Z = \frac{\bar{x} - \mu}{\sigma/\sqrt{n}} = \frac{99 - 100}{8/\sqrt{400}} = -\frac{1}{0.4} = -2.5$$

As calculated value of $|Z| = 2.5$ is greater than the tabulated value of $Z = 1.96$ at 5% level of significance, H_0 is rejected and we conclude that there is a significant difference between sample mean $\bar{x}$ and population mean μ i.e., it can not be regarded as a sample from a large population.

Example 18. The management of a company claims that the average weekly income of their employees is Rs. 900. The trade union disputes this claim stressing that it is rather less. An independent sample of 150 randomly employees estimated the average to be Rs. 856 with standard deviation of Rs. 354. Would you accept the view of the management?

Solution. Here, $\mu = 900$, $\bar{x} = 854$, $s = 354$ and $n = 150$

Null hypothesis H_0 : there is no significant difference between $\bar{x}$ and μ i.e., the view of the management is correct.

Alternative hypothesis H_1 : $\mu \neq 900$ (two tailed test)

Under H_0 the test statistic is given by

$$Z = \frac{\bar{x}-\mu}{s/\sqrt{n}} = \frac{854-900}{354/\sqrt{150}} = -\frac{46}{28.904} = -1.59$$

As calculated value of $|Z| = 1.59$ is less than the tabulated value of $Z = 1.96$ at 5% level of significance, H_0 is accepted and we conclude that there is a significant difference between sample mean $\bar{x}$ and population mean μ i.e., the view of the management is correct.

Example 19. In a population with a standard deviation of 14.8, what sample size is needed to estimate the mean of population within ± 1.2 with 95% confidence?

Solution. Let the required sample size be n.

Here, $\bar{x} - \mu = \pm 1.2$, $\sigma = 14.8$ and $Z = 1.96$

We know that

$$Z = \frac{\bar{x}-\mu}{\sigma/\sqrt{n}}$$

Using the given data, we have

$$1.96 = \frac{\pm 1.2}{14.8/\sqrt{n}} = \frac{\pm 1.2\sqrt{n}}{14.8}$$

On squaring both the sides we have,

$$(1.96)^2 = \left(\frac{\pm 1.2}{14.8}\right)^2 \times n$$

or

$$n = \left(\frac{1.96 \times 14.8}{\pm 1.2}\right)^2 = 584.35 \approx 584$$

Example 20. A random sample of 900 measurements from a large population gave a mean value of 64. If this sample has been drawn from a normal population with standard deviation of 20, find the 95% and 99% confidence limits for the mean in the population.

Solution. Here, $n = 900$, $\bar{x} = 64$ and $\sigma = 20$

At 95% confidence $Z = 1.96$

At 99% confidence $Z = 2.58$

The confidence limits for 95% confidence are

$$64 \pm 1.96 \times \frac{20}{\sqrt{900}} = 64 \pm 1.307 = 62.693 \text{ and } 65.307$$

The confidence limits for 99% confidence are

$$64 \pm 2.58 \times \frac{20}{\sqrt{900}} = 64 \pm 1.72 = 62.28 \text{ and } 65.72$$

10.28 TEST OF SIGNIFICANCE FOR DIFFERENCE OF MEANS

This test is used to test the significant difference between the means of two large samples. Let $\bar{x}_1$ be the mean of a sample of size n_1 from a population with mean μ_1 and variance σ_1^2 and let $\bar{x}_2$ be the mean of a sample of size n_2 from another population with mean μ_2 and variance σ_2^2.

We set the null hypothesis that there is no significant difference between the sample means i.e., $\mu_1 = \mu_2$.

Under the null hypothesis H_0 the test statistic is

$$Z = \frac{\bar{x}_1 - \bar{x}_2}{\sqrt{\frac{\sigma_1^2}{n_1} + \frac{\sigma_2^2}{n_2}}}$$

If the samples are drawn from the same population with common standard deviation, then under the null hypothesis H_0 the test statistic is

$$Z = \frac{\bar{x}_1 - \bar{x}_2}{\sigma\sqrt{\frac{1}{n_1} + \frac{1}{n_2}}} \qquad (\because \sigma_1 = \sigma_2 = \sigma)$$

Note: 1. If σ_1 and σ_2 are not known, the test statistic is

$$Z = \frac{\bar{x}_1 - \bar{x}_2}{\sqrt{\frac{s_1^2}{n_1} + \frac{s_2^2}{n_2}}}$$

2. If common standard deviation (σ) is not known and $\sigma_1 = \sigma_2$ then σ can be obtained by using

$$\sigma = \sqrt{\frac{n_1 s_1^2 + n_2 s_2^2}{n_1 + n_2}}$$

The test statistic is

$$Z = \frac{\bar{x}_1 - \bar{x}_2}{\sqrt{\frac{n_1 s_1^2 + n_2 s_2^2}{n_1 + n_2}\left(\frac{1}{n_1} + \frac{1}{n_2}\right)}}$$

Example 21. Examine whether there is any significant difference between the two samples for the following data:

Sample	Size	Mean
1	50	140
2	60	150

Standard deviation of the population = 10.

Solution. Here, $n_1 = 50$, $n_2 = 60$, $\bar{x}_1 = 140$, $\bar{x}_2 = 150$ and $\sigma = 10$

Null hypothesis $H_0 : \mu_1 = \mu_2$ i.e., samples are drawn from the same normal population.

Alternative hypothesis $H_1 : \mu_1 \neq \mu_2$ (two tailed test)

Under the null hypothesis H_0 the test statistic is

$$Z = \frac{\bar{x}_1 - \bar{x}_2}{\sigma\sqrt{\frac{1}{n_1} + \frac{1}{n_2}}} = \frac{140 - 150}{10\sqrt{\frac{1}{50} + \frac{1}{60}}} = -\frac{10}{1.915} = -5.22$$

As calculated value of $|Z| = 5.22$ is greater than the tabulated value of $Z = 1.96$ at 5% level of significance, H_0 is rejected and we conclude that the samples are not drawn form the same normal population.

Example 22. Intelligence test on two groups of boys and girls gave the following results:

	Size	Mean	S.D.
Girls	70	70	10
Boy	100	75	11

Examine whether the difference between mean scores is significant.

Solution. Here, $n_1 = 70$, $n_2 = 100$, $\bar{x}_1 = 70$, $\bar{x}_2 = 75$, $s_1 = 10$ and $s_2 = 11$

Null hypothesis $H_0 : \bar{x}_1 = \bar{x}_2$ i.e., there is no significant difference between mean scores

Alternative hypothesis $H_1 : \bar{x}_1 \neq \bar{x}_2$ (two tailed test)

Under the null hypothesis H_0 the test statistic is

$$Z = \frac{\bar{x}_1 - \bar{x}_2}{\sqrt{\frac{s_1^2}{n_1} + \frac{s_2^2}{n_2}}} = \frac{70 - 75}{\sqrt{\frac{10^2}{70} + \frac{11^2}{100}}} = -\frac{5}{2.639} = -1.895$$

As calculated value of $|Z| = 1.895$ is less than the tabulated value of $Z = 1.96$ at 5% level of significance, H_0 is accepted and we conclude that there is no significant difference between mean scores.

Example 23. Two samples were taken from two normal populations. The following information was available on these samples regarding the expenditure in rupees per month per family:

Sample 1: $n_1 = 42, \bar{x}_1 = 744.85, \sigma_1^2 = 158165.43$

Sample 2: $n_2 = 32, \bar{x}_2 = 516.78, \sigma_2^2 = 26413.61$

Test whether the average expenditure per month per family is equal.

Solution:

Here, $n_1 = 42, \bar{x}_1 = 744.85, \sigma_1^2 = 158165.43, n_2 = 32, \bar{x}_2 = 516.78$ and $\sigma_2^2 = 26413.61$

Null hypothesis $H_0 : \mu_1 = \mu_2$ i.e., the average expenditure per month per family is equal

Alternative hypothesis $H_1 : \mu_1 \neq \mu_2$ (two tailed test)

Under the null hypothesis H_0 the test statistic is

$$Z = \frac{\bar{x}_1 - \bar{x}_2}{\sqrt{\frac{\sigma_1^2}{n_1} + \frac{\sigma_2^2}{n_2}}} = \frac{744.85 - 516.78}{\sqrt{\frac{158165.43}{42} + \frac{26413.61}{32}}} = \frac{228.07}{67.76} = 3.37$$

As calculated value of $Z = 3.37$ is greater than the tabulated value of $Z = 1.96$ at 5% level of significance, H_0 is rejected and we conclude that the average expenditure per month per family is not equal.

Example 24. The means of two large samples of 1000 and 2000 members are 168.75 cm and 170 cm respectively. Can the samples be regarded as drawn from the same population of standard deviation 6.25 cm?

Solution. Here, $n_1 = 1000, n_2 = 2000, \bar{x}_1 = 168.75, \bar{x}_2 = 170$ and $\sigma = 6.25$

Null hypothesis $H_0 : \mu_1 = \mu_2$ i.e., samples are drawn from the same population.

Alternative hypothesis $H_1 : \mu_1 \neq \mu_2$ (two tailed test)

Under the null hypothesis H_0 the test statistic is

$$Z = \frac{\bar{x}_1 - \bar{x}_2}{\sigma \sqrt{\frac{1}{n_1} + \frac{1}{n_2}}} = \frac{168.75 - 170}{6.25 \sqrt{\frac{1}{1000} + \frac{1}{2000}}} = -\frac{1.25}{0.242} = -5.165$$

As calculated value of $|Z| = 5.165$ is greater than the tabulated value of $Z = 1.96$ at 5% level of significance, H_0 is rejected and we conclude that the samples are not drawn form the same population.

Example 25. Two random samples of sizes 1000 and 2000 farms gave an average yield of 2000 kg and 2050 kg respectively. The variance of wheat farms in the country may be taken as 100 kg. Examine whether the two samples differ significantly in yield.

Solution. Here, $n_1 = 1000$, $n_2 = 2000$, $\bar{x}_1 = 2000$, $\bar{x}_2 = 2050$ and $\sigma^2 = 100 \Rightarrow \sigma = 10$

Null hypothesis $H_0 : \mu_1 = \mu_2$ i.e., samples are drawn from the same population.

Alternative hypothesis $H_1 : \mu_1 \neq \mu_2$ (two tailed test)

Under the null hypothesis H_0 the test statistic is

$$Z = \frac{\bar{x}_1 - \bar{x}_2}{\sigma\sqrt{\frac{1}{n_1} + \frac{1}{n_2}}} = \frac{2000 - 2050}{10\sqrt{\frac{1}{1000} + \frac{1}{2000}}} = -\frac{50}{0.387} = -129.20$$

As calculated value of $|Z| = 129.20$ is greater than the tabulated value of $Z = 1.96$ at 5% level of significance, H_0 is rejected and we conclude that the samples are not drawn form the same population.

10.29 TEST OF SIGNIFICANCE FOR DIFFERENCE OF STANDARD DEVIATIONS

This test is used to test the significant difference between the standard deviations of two populations. Let two independent random samples of sizes n_1 and n_2 having standard deviations σ_1 and σ_2 respectively.

We set the null hypothesis H_0 that the sample standard deviations do not differ significantly i.e., $\sigma_1 = \sigma_2$.

Under the null hypothesis H_0 the test statistic is

$$Z = \frac{s_1 - s_2}{\sqrt{\frac{\sigma_1^2}{2n_1} + \frac{\sigma_2^2}{2n_2}}}$$

If σ_1 and σ_2 are not known then the test statistic is

$$Z = \frac{s_1 - s_2}{\sqrt{\frac{s_1^2}{2n_1} + \frac{s_2^2}{2n_2}}}$$

Example 26. The standard deviation of weight of all students in a college was found to be 4 kgs. Two random samples are drawn. The standard deviations of the weight of 100 undergraduate students is 3.5 kgs. and 50 post graduate students is 3 kgs. Test the significance of the difference in standard deviations of the samples at 5% level of significance.

Solution. Here, $n_1 = 100$, $n_2 = 50$, $s_1 = 3.5$, $s_2 = 3$ and $\sigma = 4$

Null hypothesis $H_0 : \sigma_1 = \sigma_2$ i.e., sample standard deviations do not differ significantly

Alternative hypothesis $H_1 : \sigma_1 \neq \sigma_2$ (two tailed test)

Under the null hypothesis H_0 the test statistic is

$$Z = \frac{s_1 - s_2}{\sigma\sqrt{\frac{1}{2n_1} + \frac{1}{2n_2}}} = \frac{3.5 - 3}{4\sqrt{\frac{1}{200} + \frac{1}{100}}} = \frac{0.5}{0.49} = 1.02$$

As calculated value of $|Z| = 1.02$ is less then the tabulated value of $Z = 1.96$ at 5% level of significance, H_0 is accepted and we conclude that the sample standard deviations do not differ significantly.

Example 27. Random samples drawn from two large cities gave the following information relating to the heights of adult males:

	Mean height (in inches)	Standard deviation	No. in samples
City 1	67.42	2.58	1000
City 2	67.25	2.50	1200

Test the significance difference in standard deviations of the samples at 5% level of significance.

Solution. Here, $n_1 = 1000$, $n_2 = 1200$, $\bar{x}_1 = 67.42$, $\bar{x}_2 = 67.25$, $s_1 = 2.58$, $s_2 = 2.50$ and σ is not known.

Null hypothesis $H_0 : \sigma_1 = \sigma_2$ i.e., sample standard deviations do not differ significantly

Alternative hypothesis $H_1 : \sigma_1 \neq \sigma_2$ (two tailed test)

Under the null hypothesis H_0 the test statistic is

$$Z = \frac{s_1 - s_2}{\sqrt{\frac{s_1^2}{2n_1} + \frac{s_2^2}{2n_2}}} = \frac{2.58 - 2.50}{\sqrt{\frac{(2.58)^2}{2000} + \frac{(2.50)^2}{2400}}} = \frac{0.08}{0.077} = 1.039$$

As calculated value of $|Z| = 1.039$ is less then the tabulated value of $Z = 1.96$ at 5% level of significance, H_0 is accepted and we conclude that the sample standard deviations do not differ significantly.

Example 28. In a survey of incomes of two classes of workers of two random samples gave the following data:

	Mean annual income (in Rs.)	Standard deviation	Size of sample
Sample 1	582	24	100
Sample 2	546	28	100

Examine whether the difference between
(i) Mean and
(ii) The standard deviations significant.

Solution. Here, $n_1 = 100$, $n_2 = 100$, $\bar{x}_1 = 582$, $\bar{x}_2 = 546$, $s_1 = 24$, $s_2 = 28$ and σ is not known.

(i) Null hypothesis $H_0 : \bar{x}_1 = \bar{x}_2$ i.e., there is no significant difference between sample means

Alternative hypothesis $H_1 : \bar{x}_1 \neq \bar{x}_2$ (two tailed test)

Under the null hypothesis H_0 the test statistic is

$$Z = \frac{\bar{x}_1 - \bar{x}_2}{\sqrt{\frac{s_1^2}{n_1} + \frac{s_2^2}{n_2}}} = \frac{582 - 546}{\sqrt{\frac{(24)^2}{100} + \frac{(28)^2}{100}}} = \frac{36}{3.6878} = 9.762$$

As calculated value of $|Z| = 9.762$ is greater than the tabulated value of $Z = 1.96$ at 5% level of significance, H_0 is rejected and we conclude that sample means differ significantly.

(ii) Null hypothesis $H_0 : \sigma_1 = \sigma_2$ i.e., sample standard deviations do not differ significantly

Alternative hypothesis $H_1 : \sigma_1 \neq \sigma_2$ (two tailed test)

Under the null hypothesis H_0 the test statistic is

$$Z = \frac{s_1 - s_2}{\sqrt{\frac{s_1^2}{2n_1} + \frac{s_2^2}{2n_2}}} = \frac{24 - 28}{\sqrt{\frac{(24)^2}{200} + \frac{(28)^2}{200}}} = -\frac{4}{2.6077} = -1.53$$

As calculated value of $|Z| = 1.53$ is less then the tabulated value of $Z = 1.96$ at 5% level of significance, H_0 is accepted and we conclude that the sample standard deviations do not differ significantly.

EXERCISE

1. A random sample of 900 members has a mean 3.4 cm Can it be reasonably regarded as a sample from a large population of mean 3.2 cm and standard deviation 2.3 cm?

2. A random sample of 400 male students is found to have a mean height of 160 cm. Can it be reasonably regarded as a sample from a large population of mean height 162.5 cm and standard deviation 4.5 cm?

3. A random sample of 200 measurements from a large population gave a mean value of 50 and standard deviation of 9. Determine 95% confidence interval for the mean of population.

4. A random sample of 400 measurements from a large population gave a mean value of 82 and standard deviation of 18. Determine 95% confidence interval for the mean of population.
5. A company manufacturing electric bulbs claims that the average life of its bulbs is 1600 hours. The average life and standard deviation of random sample of 100 such bulbs were 1570 hours and 120 hours respectively. Should we accept the claim of the company?
6. An insurance agent has claimed that the average age of policy holders who insure through him is less than the average for all the agents which is 30.5 years. A random sample of 100 policy holders who had insured through him reveals that the mean and standard deviation are 28.8 years and 6.35 years respectively. Test his claim at 5% level of significance.
7. The guaranteed average life of a certain type of bulbs is 1000 hours with a standard deviation of 125 hours. It is decided to sample the output so as to ensure that 90% of the bulbs do not fall short of the guaranteed average by more than 2.5%. What must be the minimum size of the sample?
8. The number of accidents per day was studied for 144 days in city A and for 100 days in city B. The mean numbers of accidents and standard deviations were respectively 4.5 and 1.2 for city A and 5.4 and 1.5 for city B. Is city A more prone to accidents than city B.
9. The mean yields of a crop from two places in a district were 210 kgs and 220 kgs per acre from 100 acres and 150 acres respectively. Can it be regarded that the sample were drawn from the same district which has the standard deviation of 11 kgs per acre?
10. Given the following data:

	No. of cases	Mean wages (in Rs.)	Standard deviation of wages (in Rs.)
Sample 1	582	24	100
Sample 2	546	28	100

Examine whether the two mean wages differ significantly.

11. Two random samples relating to the heights of the soldiers (Sample1) and sailors (Sample2) gave the following information:

	No. of cases	Mean (in inches)	Standard deviation (in inches)
Sample 1	6400	67.85	2.562
Sample 2	1600	68.55	2.52

Do the data indicate that the sailors are on the average taller than soldiers?

380 *Numerical and Statistical Techniques*

12. Intelligence tests on two groups of boys and girls gave the following results:

	Size	Mean	Standard deviation
Boys	100	73	10
Girls	60	75	8

    Examine whether the two mean scores is significant.

13. The yield of a crop in a random sample of 1000 farms in a certain area has a standard deviation of 192 kgs. Another random sample of 1000 farms gives a standard deviation of 224 kgs. Are the standard deviations significantly different?

14. The standard deviation of a random sample of 900 members is 4.6 and that of another random sample of 1600 is 4.8. Examine if the standard deviations are significantly different.

15. The mean yield of two sets of plots and their variability are as follows:

	Set of 40 plots	Set of 60 plots
Mean yield per plot.	1258 Kgs.	1243 Kgs.
S.D. per plot	34	28

    Examine whether
    (i) the difference in the variability in yields is significant,
    (ii) the difference in the mean yields is significant.

ANSWERS

1. Yes, H_0 is accepted
2. Yes, H_0 is accepted
3. 48.8 and 51.2
4. 80.24 and 83.76
5. No, H_0 is rejected at 5% level of significance
6. Claim is valid
7. $n = 4$
8. No
9. No
10. Yes, highly significant
11. Highly significant
12. Not significant at 5%
13. Yes
14. Not significant
15. (i) Not significant (ii) Significant

10.30 F-TEST

This test uses the variance ratio to test the significance of difference between two sampled variances. F-test which is based on F-distribution is called so in honor of a great statistician Prof. R. A. Fisher.

Let $x_1, x_2, \ldots, x_{n1}$ and $y_1, y_2, \ldots, y_{n2}$ be the values of two independent random samples drawn from the normal populations with the same variance σ^2.

The null hypothesis $H_0 : \sigma_1^2 = \sigma_2^2 = \sigma^2$ i.e., the population variances are equal

Under the null hypothesis H_0 the test statistic is given by

$$F = \frac{S_1^2}{S_2^2} ; S_1 > S_2$$

where
$$S_1^2 = \frac{1}{n_1 - 1} \sum_{i=1}^{n_1} (x_i - \bar{x})^2$$

$$S_2^2 = \frac{1}{n_2 - 1} \sum_{j=1}^{n_2} (y_j - \bar{y})^2$$

are unbiased estimate of the common population variance σ^2 obtained from two independent samples and it follows F-distribution with $v_1 = n_1 - 1$ and $v_2 = n_2 - 1$ degrees of freedom.

Note: The greater variance among the variances is to be taken in the numerator and n_1 corresponds to the greater variance.

10.31 PROCEDURE OF F-TEST

The various steps involved in F-test are as follows:

1. Set up the null hypothesis $H_0 : \sigma_1^2 = \sigma_2^2 = \sigma^2$ i.e., the independent estimates of the common population variance do not differ significantly.
2. Find the degrees of freedom v_1 and v_2 given by $v_1 = n_1 - 1$ and $v_2 = n_2 - 1$ respectively.
3. Calculate the variances of two samples and then calculate F.
4. From F-distribution table note the value of F for v_1 and v_2 degrees of freedom at the desired level of significance.
5. Compare the calculated value of F with tabulated value of F at the desired level of significance. If the calculated value of F is less than the tabulated value, then the difference is not significant and we conclude that the same could have come from two populations with the same variance i.e., accept H_0 otherwise reject H_0.

10.32 ASSUMPTIONS FOR F-TEST

The following assumptions are made in F-test:

1. Both the populations are normal.
2. Samples have been drawn on random basis
3. Observations are independent.
4. No measurement error is there.

10.33 CRITICAL VALUES OF F-DISTRIBUTION

The available F-table gives the critical values of F for the right tailed test, *i.e.* the critical region is determined by the right tail areas. Thus, the significance value $F_\alpha(v_1, v_2)$ at the level of significance and (v_1, v_2) degrees of freedom is determined by

$$P[F > F_\alpha(v_1, v_2)] = \alpha, \text{ as shown below:}$$

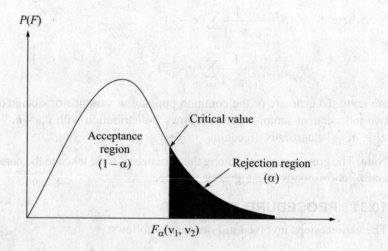

Example 29. In one sample of size 8 the sum of squares of deviations of the sample values from the sample mean is 84.4 and in the other sample of size 10 it is 102.6. Test whether this difference is significant at 5% level of significance. Given that for $v_1 = 7$ and $v_3 = 9$; $F_{0.05} = 3.29$.

Solution. Here, $n_1 = 8$, $n_2 = 10$, $\sum(x-\bar{x})^2 = 84.4$ and $\sum(y-\bar{y})^2 = 102.6$

$$S_1^2 = \frac{1}{n_1-1}\sum_{i=1}^{n_1}(x_i-\bar{x})^2 = \frac{1}{7} \times 84.4 = 12.057$$

$$S_2^2 = \frac{1}{n_2-1}\sum_{j=1}^{n_2}(y_j-\bar{y})^2 = \frac{1}{9} \times 102.6 = 11.4$$

Null hypothesis $H_0 : \sigma_1^2 = \sigma_2^2 = \sigma^2$, *i.e.*, the estimates of σ^2 given by the samples are homogeneous.

Under H_0 the test statistic is given by

$$F = \frac{S_1^2}{S_2^2} = \frac{12.057}{11.4} = 1.057$$

For $v_1 = 7$ and $v_2 = 9$, we have $F_{0.05} = 3.29$. As the calculated value of $F = 1.057$ is less than $F_{0.05} = 3.29$ at 5% level of significance, the null hypothesis H_0 is accepted.

Example 30. Two random samples gave the following information:

Sample	Size	Sample mean	Sum of squares of deviations from the mean
1	10	15	90
2	12	14	108

Test whether the samples have been drawn from the same normal population. Given that for $v_1 = 9$ and $v_2 = 11$; $F_{0.05} = 2.90$.

Solution. Here, $n_1 = 10$, $n_2 = 12$, $\bar{x} = 15$, $\bar{y} = 14$, $\sum (x-\bar{x})^2 = 90$ and $\sum (y-\bar{y})^2 = 108$

$$S_1^2 = \frac{1}{n_1 - 1} \sum_{i=1}^{n_1} (x_i - \bar{x})^2 = \frac{1}{9} \times 90 = 10$$

$$S_2^2 = \frac{1}{n_2 - 1} \sum_{j=1}^{n_2} (y_j - \bar{y})^2 = \frac{1}{11} \times 108 = 9.82$$

Null hypothesis $H_0 : \sigma_1^2 = \sigma_2^2 = \sigma^2$, i.e., two samples have been drawn from the same normal population.

Under H_0 the test statistic is given by

$$F = \frac{S_1^2}{S_2^2} = \frac{10}{9.82} = 1.018$$

For $v_1 = 9$ and $v_2 = 11$, we have $F_{0.05} = 2.90$. As the calculated value of $F = 1.018$ is less than $F_{0.05} = 2.90$ at 5% level of significance, the null hypothesis H_0 is accepted.

Example 31. The samples of sizes 9 and 8 give the sum of squares of deviations from their respective means as 160 and 91 square units respectively. Test whether the samples have been drawn from the same normal population. Given that for $v_1 = 8$ and $v_2 = 7$; $F_{0.05} = 3.73$.

Solution. Here, $n_1 = 9$, $n_2 = 8$, $\sum (x-\bar{x})^2 = 160$ and $\sum (y-\bar{y})^2 = 91$

$$S_1^2 = \frac{1}{n_1 - 1} \sum_{i=1}^{n_1} (x_i - \bar{x})^2 = \frac{1}{8} \times 160 = 20$$

$$S_2^2 = \frac{1}{n_2-1}\sum_{j=1}^{n_2}(y_j-\bar{y})^2 = \frac{1}{7}\times 91 = 13$$

Null hypothesis $H_0 : \sigma_1^2 = \sigma_2^2 = \sigma^2$, i.e., two samples have been drawn from the same normal population.

Under H_0 the test statistic is given by

$$F = \frac{S_1^2}{S_2^2} = \frac{20}{13} = 1.54$$

For $v_1 = 8$ and $v_2 = 7$, we have $F_{0.05} = 3.73$. As the calculated value of $F = 1.54$ is less than $F_{0.05} = 3.73$ at 5% level of significance, the null hypothesis H_0 is accepted.

Example 32. Two samples are drawn from two normal populations. From the following data test whether the two samples have the same variances at 5% level of significance.

Sample1	60	65	71	74	76	82	85	87	–	–
Sample2	61	66	67	85	78	88	86	85	63	91

Solution. Here, $n_1 = 8$, $n_2 = 10$

Null hypothesis $H_0 : S_1^2 = S_2^2$, i.e., two samples have the same variance

Alternative hypothesis $H_1 : S_1^2 \neq S_2^2$,

Sample 1			Sample 2		
x_i	$x_i - \bar{x}$	$(x_i - \bar{x})^2$	y_j	$y_j - \bar{y}$	$(y_j - \bar{y})^2$
60	–15	225	61	–16	256
65	–10	100	66	–11	121
71	–4	16	67	–10	100
74	–1	1	85	8	64
76	1	1	78	1	1
82	7	49	88	11	121
85	10	100	86	9	81
87	12	144	85	8	64
			63	–14	196
			91	14	196
$\sum x_i = 600$		$\sum(x_i-\bar{x})^2 = 636$	$\sum y_j = 770$		$\sum(y_j-\bar{y})^2 = 1200$

$$\bar{x} = \frac{\sum x}{n_1} = \frac{600}{8} = 75, \quad \bar{y} = \frac{\sum y}{n_2} = \frac{770}{10} = 77$$

$$S_1^2 = \frac{1}{n_1-1}\sum_{i=1}^{n_1}(x_i-\bar{x})^2 = \frac{1}{7}\times 636 = 90.857$$

$$S_2^2 = \frac{1}{n_2-1}\sum_{j=1}^{n_2}(y_j-\bar{y})^2 = \frac{1}{9}\times 1200 = 133.33$$

Under H_0 the test statistic is given by

$$F = \frac{S_2^2}{S_1^2} = \frac{133.33}{90.857} = 1.467$$

$(\because S_2^2 > S_1^2)$

For $v_1 = 9$ and $v_2 = 7$, we have $F_{0.05} = 3.68$. As the calculated value of $F = 1.467$ is less than $F_{0.05} = 3.68$ at 5% level of significance, the null hypothesis H_0 is accepted and we conclude that the sample1 and sample 2 have the same variance.

EXERCISE

1. In a sample of 8 observations, the sum of squared deviations of items from the mean was 94.5. In another sample of 10 observations, the value was found to be 101.7. Test whether the difference is significant at 5% level.

2. The following are the values in thousands of an inch obtained by two engineers in 10 successive measurements with the same micrometer. Is one engineer significantly more consistent than the other?

Engineer A	503	505	497	505	495	502	499	493	510	501
Engineer B	502	497	492	498	499	495	497	496	498	–

3. The nicotine content (in milligrams) of two samples of tobacco were found to be as follows:

Sample A	24	27	26	21	25	–
Sample B	27	30	28	31	22	36

Can it be said that the two samples come from the same normal population?

4. The daily wages in Rs. of skilled workers in two cities are as follows:

City	Size of sample of workers	S.D. of wages in the sample
A	16	25
B	13	32

Test at 5% level of significance the equality of variances of the wage distribution in the two cities.

5. The time taken by workers in performing a job by method1 and method2 is as follows:

Method 1	20	16	26	27	23	22	–
Method 2	27	33	42	35	32	34	38

Do the data show that the variances of time distribution of population from which these samples are drawn do not differ significantly?

6. Two random samples drawn from two normal populations are given as follows:

Sample 1	63	65	68	69	71	72	–	–	–	–
Sample 2	63	62	65	66	69	69	70	71	72	73

Test whether the two populations have the same variance at 5% level of significance?

ANSWERS
---

1. No
2. Not significant
3. Yes
4. Accepted
5. Not significant
6. Yes

10.34 CHI-SQUARE TEST

In test of hypothesis of parameters, it is usually assumed that the random variable follows a particular distribution. To confirm whether our assumption is right, Chi-square test is used which measures the discrepancy between the observed (actual) frequencies and theoretical (expected) frequencies, on the basis of outcomes of a trial or observational data. Chi-square is a letter of Greek alphabet and is denoted by χ^2. It is a continuous distribution which assumes only positive values.

10.35 CHI-SQUARE TEST TO TEST THE GOODNESS OF FIT

The value of χ^2 is used to test whether the deviations of the observed (actual) frequencies from the theoretical (expected) frequencies are significant or not. Chi-square test is also used to test whether a set of observations fit a given distribution or not. Therefore, Chi-square provides a test of goodness of fit.

If $O_1, O_2, \ldots, O_n$ is a set of observed (actual) frequencies $E_1, E_2, \ldots, E_n$ is the corresponding set of theoretical (expected) frequencies, then the statistic χ^2 is given by

$$\chi^2 = \sum_{i=1}^{n} \frac{(O_i - E_i)^2}{E_i}$$

is distributed with $(n - 1)$ degrees of freedom.

Here, we test the null hypothesis H_0 : there is no significant difference between the observed (actual) values and the corresponding theoretical (expected) values.

Alternative hypothesis H_1 : H_0 is not true.

If $\chi^2_{cal} \geq \chi^2_{tab}$ (or $\chi^2_{\alpha, n-1}$) then H_0 is rejected otherwise H_0 is accepted.

Note: If the null hypothesis H_0 is true, the test statistic χ^2 follows Chi-square distribution with $(n-1)$ degrees of freedom, where

$$\sum_{i=1}^{n} O_i = \sum_{i=1}^{n} E_i; \text{ i.e., } \sum_{i=1}^{n} (O_i - E_i) = 0.$$

Example 33. The following table gives the number of accidents that took place in an industry during various days of the week. Test whether the accidents are uniformly distributed over the week:

Days	Mon	Tue.	Wed.	Thu.	Fri.	Sat.
No. of accidents	16	20	14	13	17	16

Solution. Here, $n = 6$, total number of accidents $= 96$

Null hypothesis H_0 : the accidents are uniformly distributed over the week

Under H_0, the expected number of accidents of each of these days

$$= \frac{\text{Total number of accidents}}{\text{Number of days}} = \frac{96}{6} = 16$$

The observed and expected numbers of accidents are as follows:

O_i	16	20	14	13	17	16
E_i	16	16	16	16	16	16
$(O_i - E_i)^2$	0	16	4	9	1	0

Now, $$\chi^2 = \sum_{i=1}^{6} \frac{(O_i - E_i)^2}{E_i} = \frac{0 + 16 + 4 + 9 + 1 + 0}{16} = \frac{30}{16} = 1.875$$

Tabulated value of χ^2 for 5 (6-1 = 5) degrees of freedom at 5% level of significance is 11.07.

As calculated value $\chi^2 = 1.875$ of is less than the tabulated value of $\chi^2 = 11.07$ at 5% level of significance, H_0 is accepted and we conclude that the accidents are uniformly distributed over the week.

Example 34. A die is thrown 120 times and the results of these throws are given as:

No. appeared on the die	1	2	3	4	5	6
Frequency	16	30	22	18	14	20

Test whether the die is biased or not.

Solution. Here, $n = 6$, total frequency $= 120$

Null hypothesis H_0 : the die is unbiased

Under H_0, the expected frequencies for each digit $= \dfrac{120}{6} = 20$

The observed and expected frequencies are as follows:

O_i	16	30	22	18	14	20
E_i	20	20	20	20	20	20
$(O_i - E_i)^2$	16	100	4	4	36	0

Now, $\chi^2 = \sum_{i=1}^{6} \dfrac{(O_i - E_i)^2}{E_i} = \dfrac{16+100+4+4+36+0}{20} = \dfrac{160}{20} = 8$

Tabulated value of χ^2 for 5 $(6 - 1 = 5)$ degrees of freedom at 5% level of significance is 11.07.

As calculated value $\chi^2 = 8$ is less than the tabulated value of $\chi^2 = 11.07$ at 5% level of significance, H_0 is accepted and we conclude that the die is unbiased.

Example 35. The following table shows the distribution of digits in numbers chosen at random from a telephone directory:

Digits	0	1	2	3	4	5	6	7	8	9
Frequency	1026	1107	997	966	1075	933	1107	972	964	853

Test at 5% level whether the digits may be taken to occur equally frequently in the directory.

Solution. Here, $n = 10$, total frequency $= 10,000$

Null hypothesis H_0 : all the digits occur equally frequently in the directory

Under H_0, the expected frequencies for each digit $= \dfrac{10,000}{10} = 1000$

The observed and expected frequencies are as follows:

O_i	1026	1107	997	966	1075	933	1107	972	964	853
E_i	1000	1000	1000	1000	1000	1000	1000	1000	1000	1000
$(O_i - E_i)^2$	676	11449	9	1156	5625	4489	11449	784	1296	21609

Now, $\chi^2 = \sum_{i=1}^{10} \dfrac{(O_i - E_i)^2}{E_i} = \dfrac{676 + 11449 + \ldots + 21609}{1000}$

$= \dfrac{58542}{1000} = 58.542$

Tabulated value of χ^2 for 9 ($10 - 1 = 9$) sdegrees of freedom at 5% level of significance is 16.92.

As calculated value of $\chi^2 = 58.542$ is greater than the tabulated value of $\chi^2 = 16.92$ at 5% level of significance, H_0 is rejected and we conclude that all the digits in the numbers in the telephone directory do not occur equally frequently.

Example 36. Survey of 320 families of 5 children each revealed the following information:

No. of male births	5	4	3	2	1	0
No. of female births	0	1	2	3	4	5
No. of families	14	56	110	88	40	12

Test whether the data are consistent with the hypothesis that Binomial law holds and the chance of male and female births are equally probable.

Solution. Null hypothesis H_0: The male and female births are equally probable

i.e. $p = q = \dfrac{1}{2}$, where p is the probability of female birth and q is the probability of male birth.

The expected frequencies are calculated by using Binomial distribution as:

$E(r) = N \times P(X = r)$, where $r = 0, 1, 2, 3, 4, 5$; where N is the total frequency and $E(r)$ is the number of families with r female children.

$P(X = r) = {}^N C_r p^r q^{n-r}$; n is the number of children

$E(0)$ = No. of families with 0 female children

$$= 320 \times {}^5C_0 \left(\frac{1}{2}\right)^0 \left(\frac{1}{2}\right)^{5-0} = 320 \times \frac{1}{32} = 10$$

$E(1)$ = No. of families with 1 female children

$$= 320 \times {}^5C_1 \left(\frac{1}{2}\right)^1 \left(\frac{1}{2}\right)^{5-1} = 320 \times \frac{5}{32} = 50$$

$E(2)$ = No. of families with 2 female children

$$= 320 \times {}^5C_2 \left(\frac{1}{2}\right)^2 \left(\frac{1}{2}\right)^{5-2} = 320 \times \frac{10}{32} = 100$$

$E(3)$ = No. of families with 3 female children

$$= 320 \times {}^5C_3 \left(\frac{1}{2}\right)^3 \left(\frac{1}{2}\right)^{5-3} = 320 \times \frac{10}{32} = 100$$

$E(4)$ = No. of families with 4 female children

$$= 320 \times {}^5C_4 \left(\frac{1}{2}\right)^4 \left(\frac{1}{2}\right)^{5-4} = 320 \times \frac{5}{32} = 50$$

$E(5)$ = No. of families with 5 female children

$$= 320 \times {}^5C_5 \left(\frac{1}{2}\right)^5 \left(\frac{1}{2}\right)^{5-5} = 320 \times \frac{1}{32} = 10$$

The observed and expected frequencies are as follows:

O_i	14	56	110	88	40	12
E_i	10	50	100	100	50	10
$(O_i - E_i)^2$	16	36	100	144	100	4

Now, $\chi^2 = \sum_{i=1}^{10} \frac{(O_i - E_i)^2}{E_i} = \frac{16}{10} + \frac{36}{50} + \frac{100}{100} + \frac{144}{100} + \frac{100}{50} + \frac{4}{10}$

$= 1.60 + 0.72 + 1.00 + 1.44 + 2.00 + 0.40 = 7.16$

Tabulated value of χ^2 for 5 (6 – 1 = 5) degrees of freedom at 5% level of significance is 11.07.

As calculated value of $\chi^2 = 7.16$ is less than the tabulated value of $\chi^2 = 11.07$ at 5% level of significance, H_0 is accepted and we conclude that male and female births are equally probable.

Example 36. Fit a Poisson distribution for the following data and test the goodness of fit.

No. of defects (x)	0	1	2	3	4	5
Frequency	6	13	13	8	4	3

Solution. Null hypothesis H_0 : Poisson distribution is a good fit to the data.

We first find the Poisson distribution for the above data.

Mean of given distribution $= \frac{\sum f_i x_i}{\sum f_i} = \frac{94}{47} = 2$

Here, $\lambda = 2$ (For a Poisson distribution mean $= \lambda$)

$$N = \sum f_i = 47$$

The expected frequencies of the Poisson distribution are given by:

$$E(r) = N \times P(X = r) = N \times e^{-\lambda} \frac{\lambda^r}{r!}, \text{ where } r = 0, 1, 2, 3, 4, 5$$

The expected frequencies are as follows:

$$E(0) = 47 \times e^{-2} \cdot \frac{2^0}{0!} = 6.36 \approx 6 \qquad (e^{-2} = 0.1353)$$

$$E(1) = 47 \times e^{-2} \cdot \frac{2^1}{1!} = 12.72 \approx 13$$

$$E(2) = 47 \times e^{-2} \cdot \frac{2^2}{2!} = 12.72 \approx 13$$

$$E(3) = 47 \times e^{-2} \cdot \frac{2^3}{3!} = 8.48 \approx 9$$

$$E(4) = 47 \times e^{-2} \cdot \frac{2^4}{4!} = 4.24 \approx 4$$

$$E(5) = 47 \times e^{-2} \cdot \frac{2^5}{5!} = 1.696 \approx 2$$

The observed and expected frequencies are as follows:

x	0	1	2	3	4	5
O_i	6	13	13	8	4	3
E_i	6.36	12.72	12.72	8.48	4.24	1.696
$(O_i - E_i)^2$	0.1296	0.0784	0.0784	0.2304	0.0576	1.7004

Now, $\quad \chi^2 = \sum_{i=0}^{5} \frac{(O_i - E_i)^2}{E_i} = \frac{0.1296}{6.36} + \frac{0.0784}{12.72} + \frac{0.0784}{12.72}$

$$+ \frac{0.2304}{8.48} + \frac{0.0576}{4.24} + \frac{1.7004}{1.696}$$

$= 0.02038 + 0.00616 + 0.00616 + 0.02717 + 0.01358 + 1.0026 = 1.07605$

Tabulated value of χ^2 for 4 (6 – 1 – 1 = 4) degrees of freedom at 5% level of significance is 9.488.
As calculated value of $\chi^2 = 1.07605$ is less than the tabulated value of $\chi^2 = 9.488$ at 5% level of significance, H_0 is accepted and we conclude that the Poisson distribution is a good fit to the data.

Example 37. The theory predicts that the proportion of beans in the four groups $A, B, C,$ and D should be in the ratio 11 : 4 : 3 : 2. In an experiment with 2000 beans the number of four groups $A, B, C,$ and D are 1070, 430, 330 and 170 respectively. Does the experimental result support the theory?

Solution. Null hypothesis H_0: the experimental result support the theory, *i.e.*, there is no significant difference between observed and theoretical frequencies.

Under H_0 the expected (theoretical) frequencies can be calculated as:
Total number of beans = 1070 + 430 + 330 + 170 = 2000
Sum of the ratios = 11 + 4 + 3 + 2 = 20

$$E(A) = 2000 \times \frac{11}{20} = 1100$$

$$E(B) = 2000 \times \frac{4}{20} = 400$$

$$E(C) = 2000 \times \frac{3}{20} = 300$$

$$E(D) = 2000 \times \frac{2}{20} = 200$$

The observed and expected frequencies are as follows:

O_i	1070	430	330	170
E_i	1100	400	300	200
$(O_i - E_i)^2$	900	900	900	900

Now, $\chi^2 = \sum_{i=1}^{4} \frac{(O_i - E_i)^2}{E_i} = \frac{900}{1100} + \frac{900}{400} + \frac{900}{300} + \frac{900}{200}$

$= 0.8182 + 2.250 + 3.000 + 4.500 = 10.5682$

Tabulated value of χ^2 for 3 (4 – 1 = 3) degrees of freedom at 5% level of significance is 7.815.

As calculated value of $\chi^2 = 10.5682$ is greater than the tabulated value of $\chi^2 = 7.815$ at 5% level of significance, H_0 is rejected and we conclude that the experimental result does not support the theory.

10.36 CHI-SQUARE TEST TO TEST THE INDEPENDENCE OF ATTRIBUTES

The value of χ^2 is used to test whether two attributes are associated or not, *i.e.*, independence of attributes. To test the attributes contingency table is used.

A contingency table is a two way table in which rows are classified according to one attribute or criterion and columns are classified according to the other attribute or criterion. Each cell contains that number of items O_{ij} possessing the qualities of the *i*-th row and *j*-th column, where, $i = 1, 2, ... ,r$ and $j = 1, 2, ..., s$. In such a case contingency table is said to be of order $(r \times s)$. Each row or column total is known as marginal total.

Also we have the sum of row totals $\sum_{i=1}^{r} R_i$ is equal to the sum of columns totals $\sum_{j=1}^{s} C_j$, *i.e.*, $\sum_{i=1}^{r} R_i = \sum_{j=1}^{s} C_j = N$, where N is the total frequency.

Let us consider the two attributes A and B, where A is divided into r classes $A_1, A_2, ..., A_r$ and B is divided into s classes $B_1, B_2, ... , B_s$. If R_i

represents the number of persons possessing the attributes A_i ; C_j represents the number of persons possessing the attributes B_j and O_{ij} represent the number of persons possessing attributes A_i and B_j respectively. The contingency table of order $(r \times s)$ is shown in the following table:

Rows	Columns				Total
	B_1	B_2	...	B_s	
A_1	O_{11}	O_{12}	...	O_{1s}	R_1
A_2	O_{21}	O_{22}	...	O_{2s}	R_2
.	.	.	...	.	.
.	.	.	...	.	.
.	.	.	...	.	.
A_r	O_{r1}	O_{r2}	...	O_{rs}	R_r
Total	C_1	C_2		C_s	N

Corresponding to each O_{ij} the expected frequency E_{ij} in a contingency table is calculated by

$$E_{ij} = \frac{R_i \times C_j}{N} = \frac{\text{Row total} \times \text{Column total}}{\text{Grand total}}$$

Here, we test the null hypothesis H_0 : there is no association between the attributes under study, *i.e.*, attributes A and B are independent.

Alternative hypothesis H_1 : attributes are associated, *i.e.*, attributes A and B are dependent.

Null hypothesis H_0 can be tested by the statistic

$$\chi^2 = \sum_{i=1}^{r} \sum_{j=1}^{s} \frac{(O_{ij} - E_{ij})^2}{E_{ij}}$$

is distributed with $(r-1)(s-1)$ degrees of freedom.

If $\chi^2_{cal} \geq \chi^2_{tab}$ (or $\chi^2_{\alpha;(r-1)(s-1)}$), then H_0 is rejected otherwise H_0 is accepted.

NOTE:
1. For a contingency table with r rows and s columns, the degrees of freedom $= (r-1)(s-1)$.
2. For a 2×2 contingency table

a	b
c	d

We use the following formula to calculate the value of statistic χ^2 as

$$\chi^2 = \frac{N(ad-bc)^2}{(a+b)(b+d)(a+c)(c+d)}, \text{ where } N = a+b+c+d$$

χ^2 has $(2-1)(2-1) = 1$ degree of freedom.

3. Yate's correction: In a 2 × 2 contingency table, if any of cell frequency is less than 5, we make a correction to make χ^2 continuous. Decrease by 1/2 those cell frequencies which are greater than expected frequencies and increase by 1/2 those cell frequencies which are less than expected frequencies. This will affect the marginal totals. This correction is known as a Yate's correction.

After applying the Yate's correction, the corrected value of χ^2 is given by

$$\chi^2 = \frac{N\left(|ad-bc|-\dfrac{N}{2}\right)^2}{(a+b)(b+d)(a+c)(c+d)}$$

10.37 CONDITIONS FOR χ^2 TEST

1. The number of observations collected must be large, i.e., $n \geq 30$.
2. No theoretical frequency should be very small.
3. The sample observations should be independent.
4. N, the total of frequencies should be reasonably large, say, greater than 50.

10.38 USES OF χ^2 TEST

1. To test the goodness of fit.
2. To test the discrepancies between observed and expected frequencies.
3. To determine the association between attributes.

Example 39. Find the expected frequencies of 2 × 2 contingency table

a	b
c	d

Solution.

Attributes	B_1	B_2	Total
A_1	a	b	$a+b$
A_2	c	d	$c+d$
Total	$a+c$	$b+d$	$N = a+b+c+d$

The expected frequencies are:

$$E(a) = E(A_1, B_1) = \frac{(a+b)(a+c)}{a+b+c+d}$$

$$E(b) = E(A_1, B_2) = \frac{(a+b)(b+d)}{a+b+c+d}$$

$$E(c) = E(A_2, B_1) = \frac{(c+d)(a+c)}{a+b+c+d}$$

$$E(d) = E(A_2, B_2) = \frac{(c+d)(b+d)}{a+b+c+d}$$

Example 40. The following data is collected on two characters:

	Smokers	Non smokers
Literate	83	57
Illiterate	45	68

From this information find out whether there is any relation between literacy and the smoking.

Solution. Null hypothesis H_0 : there is no relation between literacy and the smoking. i.e., they are independent.

	Smokers	Non smokers	Total
Literate	83	57	140 ($R1$)
Illiterate	45	68	113 ($R2$)
Total	128 ($C1$)	125 ($C2$)	$N = 253$

Under the null hypothesis, expected frequencies can be calculated by using

$$Eij = \frac{R_i \times C_j}{N} (i = 1, 2\ ; j = 1, 2)$$

Expected frequencies are

	Smokers	Non smokers	Total
Literate	$\frac{140 \times 128}{253} = 70.83$	$\frac{140 \times 125}{253} = 69.17$	140
Illiterate	$\frac{113 \times 128}{253} = 57.17$	$\frac{113 \times 125}{253} = 55.83$	113
Total	128	125	$N = 253$

Now,

$$\chi^2 = \sum_{i=1}^{2} \sum_{j=1}^{2} \frac{(O_{ij} - E_{ij})^2}{E_{ij}}$$

$$= \frac{(83 - 70.83)^2}{70.83} + \frac{(57 - 69.17)^2}{69.17} + \frac{(45 - 57.17)^2}{57.17} + \frac{(68 - 55.83)^2}{55.83}$$

Tabulated value of χ^2 for 1 [(2 – 1) (2 – 1) = 1] degree of freedom at 5% level of significance is 3.841.

As calculated value of $\chi^2 = 9.475$ is greater than the tabulated value of $\chi^2 = 3.841$ at 5% level of significance, H_0 is rejected and we conclude that there is a relation between literacy and smoking or they are not independent.

Example 51. In a locality 100 persons were selected randomly and asked about their educational achievements. The results are given below:

Sex	Education		
	Middle	High School	College
Male	10	15	25
Female	25	10	15

Based on this information can you say the education depends on sex?

Solution. Null hypothesis H_0 Education is independent of sex.

Under the null hypothesis, expected frequencies can be calculated by using

$$E_{ij} = \frac{R_i \times C_j}{N} \quad (i=1,2; j=1,2,3)$$

Sex	Education			Total
	Middle	High School	College	
Male	10	15	25	50 ($R1$)
Female	25	10	15	50 ($R2$)
Total	35 ($C1$)	25 ($C2$)	40 ($C3$)	$N = 100$

Expected frequencies are

Sex	Education			Total
	Middle	HighSchool	College	
Male	$\frac{50 \times 35}{100} = 17.5$	$\frac{50 \times 25}{100} = 12.5$	$\frac{50 \times 40}{100} = 20$	50
Female	$\frac{50 \times 35}{100} = 17.5$	$\frac{50 \times 25}{100} = 12.5$	$\frac{50 \times 40}{100} = 20$	50
Total	35	25	40	$N = 100$

Now,

$$\chi^2 = \sum_{i=1}^{2}\sum_{j=1}^{3} \frac{(O_{ij} - E_{ij})^2}{E_{ij}}$$

$$= \frac{(10-17.5)^2}{17.5} + \frac{(15-12.5)^2}{12.5} + \frac{(25-20)^2}{20} + \frac{(25-17.5)^2}{17.5}$$

$$+ \frac{(10-12.5)^2}{12.5} + \frac{(15-20)^2}{20} = 3.214 + 0.5 + 1.25 + 3.214 + 0.5 + 1.25$$

$$= 9.928$$

Tabulated value χ^2 of for 2 [(2 – 1) (3 – 1) =2] degrees of freedom at 5% level of significance is 5.991.

As calculated value of $\chi^2 = 9.928$ is greater than the tabulated value of $\chi^2 = 5.991$ at 5% level of significance, H_0 is rejected and we conclude that education is not independent of sex or there is a relation between education and sex.

Example 52. From the following table regarding the colour of eyes of father and son, test whether the colour of the son's eye is associated with that of father's:

Eye colour of father	Eye colour of son		Total
	Light	Not light	
Light	471	151	622
Not light	148	230	378
Total	619	381	1000

Solution. Null hypothesis H_0: the colour of son's eye is not associated with that of father, *i.e.*, they are independent.

Under the null hypothesis, expected frequencies can be calculated by using

$$E_{ij} = \frac{R_i \times C_j}{N} \quad (i = 1, 2; j = 1, 2)$$

Expected frequencies are

Eye colour of father	Eye colour of son		Total
	Light	Not light	
Light	$\frac{622 \times 619}{1000} = 385.018$	$\frac{622 \times 381}{1000} = 236.982$	622
Not light	$\frac{378 \times 619}{1000} = 233.982$	$\frac{378 \times 381}{1000} = 144.018$	378
Total	619	381	1000

Now,

$$\chi^2 = \sum_{i=1}^{2} \sum_{j=1}^{2} \frac{(O_{ij} - E_{ij})^2}{E_{ij}}$$

$$= \frac{(471 - 385.018)^2}{385.018} + \frac{(151 - 236.982)^2}{236.982} + \frac{(148 - 233.982)^2}{233.982}$$

$$+ \frac{(230 - 144.018)^2}{144.018}$$

$$= 19.201 + 31.196 + 31.596 + 51.333 = 133.326$$

Tabulated value of χ^2 for 1 [(2 − 1) (2 − 1) =1] degree of freedom at 5% level of significance is 3.841.

As calculasted value of $\chi^2 = 133.326$ is greater than the tabulated value of $\chi^2 = 3.841$ at 5% level of significance, H_0 is rejected and we conclude that the colour of son's eye is associated with that of father or they are dependent.

Example 53. The following table gives the number of good and bad parts produced by each of three shifts in a factory:

	Good parts	Bad parts	Total
Day shift	960	40	1000
Evening shift	940	50	990
Night shift	950	45	995
Total	2850	135	2985

Test whether the production of bad parts is independent of the shifts on which they were produced.

Solution. Null hypothesis H_0 : the production of bad parts is independent of the shifts on which they were produced, *i.e.*, production and shifts are independent.

Under the null hypothesis, expected frequencies can be calculated by using

$$E_{ij} = \frac{R_i \times C_j}{N} \quad (i = 1, 2, 3; \ = 1, 2)$$

Expected frequencies are

	Good parts	Bad parts	Total
Day shift	$\frac{1000 \times 2850}{2985} = 954.772$	$\frac{1000 \times 135}{2985} = 45.226$	1000
Evening shift	$\frac{990 \times 2850}{2985} = 945.226$	$\frac{990 \times 135}{2985} = 44.774$	990
Night shift	$\frac{995 \times 2850}{2985} = 950.000$	$\frac{995 \times 135}{2985} = 45.000$	995
Total	2850	135	2985

Now,

$$\chi^2 = \sum_{i=1}^{3} \sum_{j=1}^{2} \frac{(O_{ij} - E_{ij})^2}{E_{ij}}$$

$$= \frac{(960 - 954.774)^2}{954.774} + \frac{(40 - 45.226)^2}{45.226} + \frac{(940 - 945.226)^2}{945.226}$$

$$+ \frac{(50 - 44.774)^2}{44.774} + \frac{(950 - 950)^2}{950} + \frac{(45 - 45)^2}{45}$$

$$= 0.0286 + 0.6039 + 0.0289 + 0.6099 + 0 + 0 = 1.2713$$

Tabulated value of χ^2 for 2 [$(3-1)(2-1) = 2$] degrees of freedom at 5% level of significance is 5.991.

As calculated value of $\chi^2 = 1.2713$ is less than the tabulated value of $\chi^2 = 5.991$ at 5% level of significance, H_0 is accepted and we conclude that the production of bad parts is independent of the shifts on which they were produced or production and shifts are independent.

EXERCISE

1. The frequency distribution of the digits on asset of random numbers was observed to be:

Digits	0	1	2	3	4	5	6	7	8	9
Frequency	18	19	23	21	16	25	22	20	21	15

   Test the hypothesis that the digits are uniformly distributed.

2. The sales in a supermarket during a week are given below:

Days	Mon.	Tue.	Wed.	Thu.	Fri.	Sat.
Sales (,000 Rs.)	65	54	60	56	71	84

   Test the hypothesis that the sales do not depend on the day of the week, using a 5% level of significance.

3. The following table gives the number of accidents that took place in an industry during various days of the week:

Days	Mon.	Tue.	Wed.	Thu.	Fri.	Sat.
No. of accidents	14	18	12	11	15	14

   Test if the accidents are uniformly distributed over the week.

4. A die is thrown 276 times and the results of these throws are given below:

No. appeared on the die	1	2	3	4	5	6
Frequency	40	32	29	59	57	59

   Test whether the die is biased or not.

5. A sample analysis of examination results of 500 students was made. It was found that 220 had failed; 170 had secured a third class; 90 were placed in second class; 20 got first class. Are these results commensurable with the general examination result which is in the ratio of 4 : 3 : 2 : 1 for the above said categories respectively.

6. Four dice were thrown 112 times and the number of times 1, 3 or 5 was thrown as follows:

No. of dice throwing 1, 3 or 5	0	1	2	3	4
Frequency	10	25	40	30	7

   Test the hypothesis that all dice were fair.

7. Fit a Poisson distribution for the following data and test the goodness of fit:

No. of defects (x)	0	1	2	3	4
Frequency	109	65	22	3	1

8. The following table gives the classification of 100 workers according to sex and nature of work:

Sex	Nature of work	
	Skilled	Unskilled
Male	40	20
Female	10	30

Using χ^2 test examine whether the nature of work is independent of the sex of the worker.

9. For the data given in the following table use χ^2 test to test the effectiveness of inoculation in preventing the attack of smallpox:

	Attacked	Not attacked
Inoculated	25	220
Not inoculated	90	160

10. Two investigators draw samples from the same town in order to estimate the number of persons falling in the income groups 'poor', 'middle class' and 'well to do'. Their results are as follows:

	Income groups		
Day shift	Poor	Middle class	Well to do
A	140	100	15
B	140	50	20

Test whether the sampling techniques of the two investigators are significantly dependent of the income groups of the people.

ANSWERS

1. Yes 2. Accepted 3. Yes
4. Biased 5. No 6. Yes
7. Poisson distribution is a good fit to the data
8. No
9. Inoculation against smallpox is a preventive measure
10. Sampling techniques are dependent of the income groups

10.39 ANALYSIS OF VARIANCE

We have discussed the test of significance of the difference of two means by t-test *i.e.,* whether two samples differ significantly with respect to some normal population means, we use the technique of analysis of variance (ANOVA). This technique was developed by *R.A.* Fisher and was mainly used in agricultural research. The analysis of variance is an extremely useful technique in other disciplines like as biology, economics, sociology, psychology etc. The analysis of variance technique is important in the context of all those situations where we want to compare more than two populations. The analysis of variance is a procedure for testing the difference among different groups of data for homogeneity.

In analysis of variance, the total variation in a set of data is splitted in two types, that amount which can be attributed to chance (non-assignable causes) and that amount which can be attributed to specified (assignable) causes. It determines how much of variation is due to one group of causes and how much of it is due to other group of causes. Generally, the total variation is split up into the two components *i.e.,*

Total variation = variation between samples + variation within samples.

It is noted that if the variation between the sample means is significantly greater than the variation within samples, it is assumed that the samples are drawn from different populations.

10.40 ASSUMPTIONS IN THE ANALYSIS OF VARIANCE

The analysis of variance is based on certain assumptions. The following assumptions must be satisfied for the validity of the analysis of variance test.

1. The samples are independently drawn.
2. The effects of various components are additive.
3. They occur at random and independent of each other in the groups.
4. The population for each sample must be normally distributed with identical mean and variance.

10.41 TECHNIQUE OF ANALYSIS OF VARIANCE

The technique of analysis of variance in case of one factor and two factors is similar. However, in case of one factor analysis, the total variation is divided into two parts only *i.e.,* (*i*) variation between the samples (*ii*) variation within the samples is the residual variation.

In case of two factor analysis, the total variation is divided into three parts *i.e.,* (*i*) variation due to factor one (rows) (*ii*) variation due to factor two (columns) and (*iii*) residual reradiation.

10.42 THE BASIC PRINCIPLE OF ANALYSIS OF VARIANCE

The basic principle of analysis of variance is to test for differences among the means of the populations by examining the amount of variation within each of these samples. For testing these differences among means, we find two estimates of the population variance *i.e.*, one based on between sample variance and the other based on within sample variance. Then the two estimates of the population are compared with F-test, by

$$F = \frac{\text{Estimate of population variane based on between samples variance}}{\text{Estimate of population variance based on within sample variance}}$$

If calculated value of F is greater than the tabulated value of F for given degrees of freedom at certain level of significance, we reject the null hypothesis and we conclude that there are significant differences among the sample means.

10.43 ANALYSIS OF VARIANCE IN ONE WAY CLASSIFICATION

When data are classified according to one criterion only it is called one-way classified data. For example, different feeds given to animals, yields of different varieties of a plant, different types of gasoline used in engine sets etc. Here, we have to determine if there are differences with in that factor. In other words, when a set of observations is distributed over the different levels of a factor they form one-way classified data. Let us take a factor, say, at k levels. Let there be n_i observations denoted by x_{ij} ; $i = 1, 2, \ldots k$ and $j = 1, 2, \ldots n_i$ against the ith level. Then the observations x_{ij} classified in k groups according to the k levels of a factor are said to form one-way classified data.

Let there be N observations, classified into k classes; $A_1, A_2, \ldots, A_k$; the number of observations in the ith class being n_i. Let x_{ij} be the jth observations in the ith class. The scheme of classification is given as follows:

Classes			
A_1	A_2	.....	A_k
x_{11}	x_{21}	.....	x_{k1}
x_{12}	x_{22}	.....	x_{k2}
$\vdots$	$\vdots$	.....	$\vdots$
x_{1n_1}	x_{2n_2}	.....	x_{kn_k}

The various steps for the analysis of one-way classified data are as follows:

1. Calculate total for each class *i.e.*, $T_1, T_2, \ldots T_k$

   where $T_i = \sum_{j=1}^{n_i} x_{ij}$

2. Calculate the grand total

$$G = \sum_{i=1}^{k} \sum_{j}^{n_i} x_{ij} = \sum_{i=1}^{k} T_i$$

3. Calculate the correction factor (C.F.)

$$C.F. = \frac{G^2}{N}$$

where N is the total number of observations in all the samples.

4. Calculate total sum of squares (TSS)

$$TSS = \sum \sum x_{ij}^2 - C.F$$

5. Calculate sum of squares due to treatment (between samples)

$$SST = \sum_{i=1}^{k} \frac{T_i^2}{n_i} - C.F.$$

6. Calculate sum of squares due to error (within sample)
$$SSE = TSS - SST$$

7. Set up the table of analysis of variance and calculate F.

Analysis table for one-way classification

Source of variation	Sum of squares (SS)	Degrees of freedom (d.f.)	Mean sum of squares $MSS = \frac{SS}{d.f}$	Variance ratio F
Between samples	SST	$k-1$	$MSST = \frac{SST}{k-1}$	$F = \frac{MSST}{MSSE}$
Within samples (Error)	SSE	$N-k$	$MSSE = \frac{SSE}{N-k}$	
Total	TSS	$N-1$		

8. Compare the calculated value of F with the tabulated value of F for the $(k-1)$, $(N-k)$ degrees of freedom at a certain level of significance. Generally, we take 5% level of significance. If calculated value of F is greater than the tabulated value of F at certain level of significance, null hypothesis H_0 is rejected, otherwise H_0 is accepted.

Example 1. The following table gives the yields on 20 sample plots under four varieties of seeds:

Variety			
A	B	C	D
8	12	18	13
10	11	12	9
12	9	16	12
8	14	6	16
7	4	8	15

Prepare analysis of variance table and test if the varieties differ significantly among themselves.

Solution. Here, $k = 4$, $N = 4 \times 5 = 20$

Null hypothesis H_0 : The mean yield of varieties of seeds do not differ significantly.

Variety			
A	B	C	D
8	12	18	13
10	11	12	9
12	9	16	12
8	14	6	16
7	4	8	15
$T_1 = 45$	$T_2 = 50$	$T_3 = 60$	$T_4 = 65$

Grand total $\quad (G) = T_1 + T_2 + T_3 + T_4 = 45 + 50 + 60 + 65 = 220$

Correction factor $(C.F.) = \dfrac{G^2}{N} = \dfrac{(220)^2}{20} = 2420$

Total sum of squares $(TSS) = \sum\sum x_{ij}^2 - C.F.$

$\qquad = [(8)^2 + (10)^2 + \ldots + (16)^2 + (15)^2] - 2420$

$\qquad = 2678 - 2420 = 258$

Sum of squares between the samples $(SST) = \sum_{i=1}^{4} \dfrac{T_i^2}{n_i} - C.F.$

$\qquad = \left[\dfrac{(45)^2}{5} + \dfrac{(50)^2}{5} + \dfrac{(60)^2}{5} + \dfrac{(65)^2}{5}\right] - 2420$

$\qquad = 2470 - 2420 = 50$

sum of squares within samples (error) $(SSE) = TSS - SST$

$\qquad = 258 - 50 = 208$

Now, we prepare tube of analysis of variance as follows:

Source of variation	Sum of squares (SS)	Degrees of freedom (d.f.)	Mean sum of squares (MSS)	Variance ratio F
Between samples	50	$4-1=3$	$\frac{50}{3} = 16.7$	$F = \frac{16.7}{13.0} = 1.28$
Within samples	208	$20-4=16$	$\frac{208}{16} = 13.0$	
Total	258	$20-1=19$		

Tabulated value of $F_{0.05}$ for (3, 16) degrees of freedom at 5% level of significance is 3.24.

As the calculated value of $F = 1.28$ is less then the tabulated value of F, null hypothesis H_0 is accepted at 5% level of significance and we conclude that mean yield of varieties of seeds do not differ significantly.

Example 2. To assess the significance of possible variation in performance of an antibiotics after its administration to *TB* patients of different hospitals of a city, a common test was performed on a number of patients taken at random from each of the four hospitals. The results are as follows:

Hospital			
A	B	C	D
20	25	24	23
19	23	20	20
21	21	22	20

Make an analysis of variance of data.

Solution. Heres, $k = 4$, $N = 3 \times 4 = 12$

Null hypothesis H_0 : There is no significant difference in the means effects of antibiotics on *TB* patients in different hospitals.

A	B	C	D
20	25	24	23
19	23	20	20
21	21	22	20
$T_1 = 60$	$T_2 = 69$	$T_3 = 66$	$T_4 = 63$

Grand total $(G) = T_1 + T_2 + T_3 + T_4 = 60 + 69 + 66 + 63 = 258$

Correction factor $(C.F.) = \frac{G^2}{N} = \frac{(258)^2}{12} = \frac{66564}{12} = 5547$

Total sum of squares $(TSS) = \sum\sum x_{ij}^2 - C.F.$
$= [(20)^2 + (19)^2 + \ldots + (20)^2 + (20)^2] - 5547$
$= 5586 - 5547 = 39$

Sum of squares between samples $(SST) = \sum_{i=1}^{4} \frac{T_i^2}{n_i} - C.F.$

$$= \frac{(60)^2}{3} + \frac{(69)^2}{3} + \frac{(66)^2}{3} + \frac{(63)^2}{3} - 5547$$
$$= 5562 - 5547 = 15$$

Sum of squares within samples $(SSE) = TSS - SST$
$$= 39 - 15 = 24$$

Now, we prepare table of analysis of variance as follows:

Source of variation	Sum of squares (SS)	Degrees of freedom (d.f)	Mean sum of sqares (MSS)	Variance ratio F
Between samples	15	$4 - 1 = 3$	$\frac{15}{3} = 5$	$F = \frac{5}{3} = 1.67$
Within samples	24	$12 - 4 = 8$	$\frac{24}{8} = 3$	
Total	39	$12 - 1 = 11$		

Tabulated value of $F_{0.05}$ for (3, 8) degrees of freedorm at 5% level of significance is 8.83.

As the calculated value of $F = 1.67$ is less than the tabulated value of $F = 8.83$, null hypothesis H_0 is accepted and we conclude that the mean effect of antibiotics on TB patients in different hospitals of the city is insignificant

Example 3. The following data give the production in kg. of three varieties of wheat sown in 12 plots:

Variety of wheat		
1	2	3
14	14	18
16	13	16
18	15	19
	22	19
		20

Is there any significance difference in the production of three varieties? (Use 5% level of significance)

Solution Here, $k = 3$, $N = 12$

Null hypothesis H_0 : There is no significant difference in the mean yield of three varieties of wheat.

Variety of wheat		
1	2	3
14	14	18
16	13	16
18	15	19
	22	19
		20
$T_1 = 48$	$T_2 = 64$	$T_3 = 92$

Grand total $(G) = T_1 + T_2 + T_3 = 48 + 64 + 92 = 204$

Correction factor $(C.F.) = \dfrac{G^2}{N} = \dfrac{(204)^2}{12} = \dfrac{41616}{12} = 3468$

Total sum of squares $(TSS) = \sum\sum x_{ij}^2 - C.F.$
$= [(14)^2 + (16)^2 + \ldots + (19)^2 + (20)^2] - 3468$
$= 3552 - 3468 = 84$

Sum of squares between the samples $(SST) = \sum_{i=1}^{3} \dfrac{T_i^2}{n_i} - C.F.$

$= \left[\dfrac{(48)^2}{3} + \dfrac{(64)^2}{4} + \dfrac{(92)^2}{5}\right] - 3468$

$= 3484.80 - 3468 = 16.8$

Sum of squares within samples $(SSE) = TSS - SST$
$= 84 - 16.8 = 67.20$

Now, we prepare table of analysis of variance as follows:

Source of varatioon	Sum of squares (SS)	Degrees of freedom (d.f.)	Mean sum of squares (MSS)	Variance ratio F
Between samples	16.8	3 − 1 = 2	$\dfrac{16.8}{2} = 8.40$	$F = \dfrac{8.40}{7.467} = 1.125$
Within samples	67.20	12 − 3 = 9	$\dfrac{67.20}{9} = 7.467$	
Total	84	11		

Tabulated value of $F_{0.05}$ for (2, 9) degrees of freedom at 5% level of significance is 4.26.

As the calculated value of $F = 1.125$ is less than the tabulated value of $F = 4.26$, null hypothesis H_0 is accepted at 5% level of significance and we conclude that there is no significant difference in the mean yield of three varieties of wheat.

Example 4. The following data gives the yields on 12 plots of land in three samples under three varieties of fertilizers:

A	B	C
25	20	24
22	17	26
24	16	30
21	19	20

Is there any significant difference in the average yields of land under the three varieties of fertilizers?

Solution: Here, $k = 3$, $N = 3 \times 4 = 12$

Null hypothesis H_0 : There is no significant difference in the average yields under the three varieties.

A	B	C
25	20	24
22	17	26
24	16	30
21	19	20
$T_1 = 92$	$T_2 = 72$	$T_3 = 100$

Grand total $(G) = T_1 + T_2 + T_3 = 92 + 72 + 100 = 264$

Correction factor $(C.F.) = \dfrac{G^2}{N} = \dfrac{(264)^2}{12} = 5808$

Total sum of squares $(TSS) = \sum\sum x_{ij}^2 - C.F.$
$= [(25)^2 + (22)^2 + ... + (30)^2 + (20)^2] - 5808$
$= 5984 - 5808 = 176$

Sum of squares between the samples $(SST) = \sum\limits_{i=1}^{3} \dfrac{T_i^2}{n_i} - C.F.$

$= \left[\dfrac{(92)^2}{4} + \dfrac{(72)^2}{4} + \dfrac{(100)^2}{4}\right] - 5808$
$= 5912 - 5808$
$= 104$

Sum of squares within samples $(SSE) = TSS - SST$
$= 176 - 104 = 72$

Now, we prepare table of analysis of variance as follows:

Source of variation	Sum of squares (SS)	Degrees of freedom (d.f.)	Mean sum of squares (MSS)	Variance ratio F
Between samples	104	$3 - 1 = 2$	$\dfrac{104}{2} = 52$	$F = \dfrac{52}{8} = 6.5$
Within samples	72	$12 - 3 = 9$	$\dfrac{72}{9} = 8$	
Total	176	$12 - 1 = 11$		

Tabulated value of $F_{0.05}$ for (2, 9) degrees of freedom at 5% level of significance is 4.26.

As the calculated value of $F = 6.5$ is greater than the tabulated value of $F = 4.26$ null hypothesis H_0 is rejected at 5% level of significance and we conclude that the difference in average yields under the three varieties is significant.

10.44 ANALYSIS OF VARIANCE IN TWO WAY CLASSIFICATION

We can plan an experiment in such a way as to study the effects of two factors in the same experiment. For each factor there will be a number of classes or levels. When data is classified according to two factors then it is called two-way classified data. For example, if two drugs are manufactured by a company or the same drug is manufactured by different companies and it is required to find the efficacy of the drug as well as the side effects of the drug, we require a two-way analysis, other examples, the agricultural output may be classified on the basis of different varieties of seeds and also on the basis of different varieties of fertilizers used, the production of a number of units may depend upon the machine type and workmen etc. In case of two-way classified data, total variation is broken into three parts:

1. Variation between the columns
2. Variation between the rows
3. Variation due to error or residual variance *i.e.*,

   Total SS = SS due to columns + SS due to rows + SS due to error

or $TSS = SSC + SSR + SSE$

Let the two factors A and B with the levels $A_1, A_2, ..., A_p$ and $B_1, B_2, ..., B_q$ respectively. Let x_{ij} be the observation under the ith level A and the jth level of B. The scheme of classification is given as follows:

	B_1	B_2	...	B_q	Total
A_1	x_{11}	x_{12}	...	x_{1q}	TR_1
A_2	x_{21}	x_{22}	...	x_{2q}	TR_2
⋮	⋮	⋮	...	⋮	⋮
A_p	x_{p1}	x_{p2}	...	x_{pq}	TR_p
Total	TC_1	TC_2	...	TC_q	G

The various steps for the analysis of two-way classified data are as follows:

1. Calculate total for each of the A-class (or rows)

   $TR_1, TR_2, ... , TR_p$

2. Calculate total for each of the B-class (or columns)

   $TC_1, TC_2, ... , TC_q$

3. Calculate the grand total

$$G = \sum_{i=1}^{p} TR_i = \sum_{j=1}^{q} TC_j = \sum_{i=1}^{p} \sum_{j=1}^{q} x_{ij}$$

4. Calculate the correction factor (C.F.)

$$C.F. = \frac{G^2}{N} \quad \text{where } N = p.q$$

$p \to$ Number of rows; $q \to$ Number of columns

5. Calculate total sum of squares due to rows (TSS)

$$TSS = \sum\sum x_{ij}^2 - C.F.$$

6. Calculate sum of squares due to rows (SSR)

$$SSR = \frac{\sum_{i=1}^{p} TR_i^2}{q} - C.F.$$

7. Calculate sum of squares due to columns (SSC)

$$SSC = \frac{\sum_{j=1}^{q} TC_j^2}{p} - C.F.$$

8. Calculate sum of squares due to error (SSE)

$$SSE = TSS - SSR - SSC$$

9. Set up the table of analysis of variance and calculate F.

Analysis table for two-way classification

Source of variation	Sum of squares (SS)	Degrees of freedom (d.f.)	Mean sum of squares $MSS = \frac{SS}{d.f.}$	Variance ratio F
Between the rows (level A)	SSR	$p - 1$	$MSSR = \frac{SSR}{p-1}$	$F = \frac{MSSR}{MSSE}$
Between the columns (level B)	SSC	$q - 1$	$MSSC = \frac{SSR}{q-1}$	$F = \frac{MSSC}{MSSE}$
Error	SSE	$(p-1)(q-1)$	$MSSE = \frac{SSE}{(p-1)(q-1)}$	
Total	TSS	$pq - 1$		

Compare the calculated values of F with the tabulated values of F for the $[(p - 1), (p - 1)(q - 1)]$ and $[(q - 1), (p - 1)(q - 1)]$ degrees of freedoms respectively at a certain level of significance. If calculated value of F is greater than the tabulated value of F at certain level of significance, null hypothesis H_0 is rejected, otherwise H_0 is accepted.

Example 5. A farmer applies three types of fertilizers on four separate plots. The figures on yield per area are tabulated as follows :

Fertilizers	Plots				Total
	A	B	C	D	
Nitrogen	6	4	8	6	24
Potash	7	6	6	9	28
Phosphate	8	5	10	9	32
Total	21	15	24	24	84

Find out if the plots are materially different and also if the fertilizers make any material difference.

Solution. Here, $p = 3$, $q = 4$, $N = 12$

Null hypothesis

(i) H_0: Plots do not differ materially.

(ii) H_0: The fertilizers do not differ materially.

$$\text{Grand total } (G) = 84$$

$$\text{Correction factor } (C.F) = \frac{G^2}{N} = \frac{(84)^2}{12} = 588$$

$$\text{Total sum of squares } (TSS) = \sum\sum x_{ij}^2 - C.F.$$
$$= [(6)^2 + (4)^2 + \ldots + (10)^2 + (9)^2] - 588$$
$$= 624 - 588 = 36$$

$$\text{Sum of squares due to rows } (SSR) = \frac{\sum TR_i^2}{q} - C.F.$$

$$= \left[\frac{(24)^2}{4} + \frac{(28)^2}{4} + \frac{(32)^2}{4}\right] - 588$$
$$= 596 - 588 = 8$$

$$\text{Sum of squares due to columns } (SSC) = \frac{\sum TC_j^2}{p} - C.F.$$

$$= \left[\frac{(21)^2}{3} + \frac{(15)^2}{3} + \frac{(24)^2}{3} + \frac{(24)^2}{3}\right] - 588$$
$$= 606 - 588 = 18$$

Sum of squares due to error $(SSE) = TSS - SSR - SSC$
$$= 36 - 8 - 18 = 10$$

Now, we prepare table of analysis of variance as follows :

Source of variation	Sum of squares (SS)	Degrees of freedom (d.f.)	Mean sum of squares (MSS)	Variance ratio F
Between rows (fertilizers)	8	3 − 1 = 2	$\dfrac{8}{2} = 4$	$F(2, 6) = \dfrac{4}{1.67} = 2.4$
Between columns (Plots)	18	4 − 1 = 3	$\dfrac{18}{3} = 6$	$F(3, 6) = \dfrac{6}{1.67} = 3.6$
Error	10	(3 − 1)(4 − 1) = 6	$\dfrac{10}{6} = 1.67$	
Total	36	12 − 1 = 11		

Tabulated values : $F_{(2, 6)} = 5.14$ and $F_{(3, 6)} = 4.76$ at 5% level of significance.

As both the calculated values of F are less than the corresponding tabulated values of F, both the null hypothesis are accepted at 5% level of significance and we conclude that the plots and fertilizers do not matter significantly.

Example 6. The following table gives the figures of monthly drop in acidity level and chlorine concentration in a lake water. Apply two-way classification of analysis of variance and interpret your results.

Chlorine concentration	Acidity level			
	Low	Medium	High	Very high
Low	22	19	9	7
Medium	11	11	8	4
High	9	10	6	4

Solution Here, $p = 3$, $q = 4$, $N = 12$

Null hypothesis

(i) H_0: There is no significant difference in acidity level

(ii) H_0: There is no significant difference in different concentrations of chlorine.

Chlorine concentration	Acidity level				Total
	Low	Medium	High	Very high	
Low	22	19	9	7	57
Medium	11	11	8	4	34
High	9	10	6	4	29
Total	42	40	23	15	120

Grand total $(G) = 120$

Correction factor $(C.F.) = \dfrac{G^2}{N} = \dfrac{(120)^2}{12} = 1200$

Total sum of squares $(TSS) = \sum\sum x_{ij}^2 - C.F.$
$= [(22)^2 + (19)^2 + ... + (6)^2 + (4)^2] - 1200$
$= 1530 - 1200 = 330$

Sum of squares due to rows $(SSR) = \dfrac{\sum TR_i^2}{q} - C.F.$

$= \left[\dfrac{(57)^2}{4} + \dfrac{(34)^2}{4} + \dfrac{(29)^2}{4}\right] - 1200$

$= 1311.50 - 1200 = 111.50$

Sum of squares due to columns $(SSC) = \dfrac{\sum TC_j^2}{p} - C.F.$

$= \left[\dfrac{(42)^2}{3} + \dfrac{(40)^2}{3} + \dfrac{(23)^2}{3} + \dfrac{(15)^2}{3}\right] - 1200$

$= 1372.66 - 1200 = 172.66$

Sum of squares due to error $(SSE) = TSS - SSR - SSC$
$= 330 - 111.50 - 172.66 = 45.84$

Now, we prepare table of analysis of variance as follows:

Source of variation	Sum of squares (SS)	Degrees of freedom (d.f.)	Mean sum of squares (MSS)	Variance ratio F
Between rows (chlorine concentration)	111.50	$3 - 1 = 2$	$\dfrac{111.50}{2} = 55.75$	$F_{(2,6)} = \dfrac{55.75}{7.64} = 7.30$
Between columns (Acidity level)	172.66	$4 - 1 = 3$	$\dfrac{172.66}{3} = 57.55$	$F_{(3,6)} = \dfrac{57.55}{7.64} = 7.53$
Error	45.84	$(3-1)(4-1) = 6$	$\dfrac{45.84}{6} = 7.64$	
Total	330	$12 - 1 = 11$		

Tabulated values: $F_{(2, 6)} = 5.14$ and $F_{(3, 6)} = 4.76$ at 5% level of significance.

As both the calculated values of F are greater than the corresponding tabulated values of F, both the null hypothesis are rejected at 5% level of significance and we conclude that

(i) there is a significant difference in acidity level

(ii) there is a significant difference in different concentrations of chlorine.

Example 7. Set up the analysis of variance table for the following two-way classified data:

Per acre production data for wheat (in metric tonnes)

Variety of fertilizers	Variety of seeds		
	A	B	C
W	6	5	5
X	7	5	4
Y	3	3	3
Z	8	7	4

State whether variety differences are significant at 5% level of significance.

Solution. Here, $p = 4$, $q = 3$, $N = 12$

Null hypothesis

(i) H_0: There is no significant difference among the variety of seeds.

(ii) H_0: There is no significant difference among the variety of fertilizers.

Variety of fertilzers	Variety of seeds			Total
	A	B	C	
W	6	5	5	16
X	7	5	4	16
Y	3	3	3	9
Z	8	7	4	19
Total	24	20	16	60

Grand total $(G) = 60$

Correction factor $(C.F.) = \dfrac{G^2}{N} = \dfrac{(60)^2}{12} = 300$

Total sum of squares $(TSS) = \sum \sum x_{ij}^2 - C.F.$

$= [(6)^2 + (5)^2 + \ldots + (7)^2 + (4)^2] - 300$

$= 332 - 300 = 32$

Sum of squares due to rows $(SSR) = \dfrac{\sum TR_i^2}{q} - C.F.$

$$= \left[\dfrac{(16)^2}{3} + \dfrac{(16)^2}{3} + \dfrac{(9)^2}{3} + \dfrac{(19)^2}{3}\right] - 300$$

$= 318 - 300 = 18$

Sum of squares due to columns $(SSC) = \dfrac{\sum TC_j^2}{p} - C.F.$

$$= \left[\dfrac{(24)^2}{4} + \dfrac{(20)^2}{4} + \dfrac{(16)^2}{4}\right] - 300 = 308 - 300 = 8$$

Sum of squares due to error $(SSE) = TSS - SSR - SSC$

$= 32 - 18 - 8 = 6$

Now, we prepare table of analysis of variance as follows :

Source of variation	Sum of squares (SS)	Degrees of freedom (d.f.)	Mean sum of squares (MSS)	Variance ratio F
Between rows (varieties of fertilizers)	18	4 – 1 = 3	$\dfrac{18}{3} = 6$	$F_{(3, 6)} = \dfrac{6}{1} = 6$
Between columns (varieties of seeds)	8	3 – 1 = 2	$\dfrac{8}{2} = 4$	$F_{(2, 6)} = \dfrac{4}{1} = 4$
Error	6	(4 – 1)(3 – 1) = 6	$\dfrac{6}{6} = 1$	
Total	32	12 – 1 = 11		

Tabulated values: $F_{(2, 6)} = 5.14$ and $F_{(3, 6)} = 4.76$ at 5% level of significance.

As the calculated value of F for variety of fertilizers is greater than the tabulated value of F, null hypothesis H_0 is rejected at 5% level of significance and we conclude that the differences among variety of fertilizers is significant. As the calculated value of F for variety of seeds is less than the tabulated value of F, null hypothesis H_0 is accepted at 5% level of significance and we conclude that the differences among variety of seeds is not significant.

EXERCISE

1. What do you under stand by analysis of variance?
2. What are the assumptions that are made in the analysis of variance?
3. Discuss the technique of analysis of variance for one-way classification.
4. Discuss the technique of analysis of variance for two-way classification.
5. The following data relates to the production in kg. of three varieties of wheat - A, B and C used on 15 plots:

Wheat variety	Yields (in kg)				
A	5	6	8	9	7
B	8	10	11	12	4
C	7	3	5	4	1

Test whether there is any significant difference in the production of three varieties. (Use 5% level of significance)

6. The three samples below have been obtained from normal populations with equal variances. Test the hypothesis at 5% level of significance that the population means are equal.

A	B	C
5	7	12
10	5	9
7	10	13
14	9	12
11	9	14

7. A manufacturing company purchased three new machines of different models and wishes to determine whether one of them is faster then the others in producing a certain output. Five hourly production figures are observed at random from each machine and results are given below :

Observations	Machine		
	I	II	III
1	25	31	24
2	30	39	30
3	36	38	28
4	38	42	25
5	31	35	28

Use analysis of variance and determine whether the machines are significantly different in their mean speed at 5% level of significance.

8. The three samples below have been obtained from normal populations with equal variances.

A	B	C
8	7	12
10	5	9
7	10	13
14	9	12
11	9	14

Test the hypothesis at 5% level of significance that the population means are equal.

9. Following are the weekly sale records (*in Rs.*) of three salesmen A, B and C of a company during 13 sale calls :

A	B	C
300	600	700
400	300	300
300	300	400
500	400	600
–	–	500

Test whether the sales of three salesmen are different at 5% level of significance.

10. To study the performance of three detergents and three different water temperatures; the following whiteness readings were obtained with specially designed equipment.

Water termperature	Detergents		
	A	B	C
Cold water	57	55	67
Warm water	49	52	68
Hot water	54	46	58

Perform a two-way analysis of variance using 5% level of significance.

11. A tea company appoints four salesmen A, B, C and D and observes their sales in three seasons – summer, winter and monsoon. The figure (in lacs.) are given in the following table :

Seasons	Salesmen				Seasons total
	A	B	C	D	
Summer	35	36	21	35	128
Winter	28	29	31	32	120
Monsoon	26	28	29	29	112
Salesmen total	90	93	81	96	360

Carry out an analysis of variance at 5% level of significance.

12. Four different drugs have been developed for the cure of a certain disease. There drugs are tried on patients of three different hospitals. The number of cases of recovery from the disease per people are given below.

Hospital	Drugs			
	A	B	C	D
X	24	20	24	17
Y	20	25	30	9
Z	13	18	31	13

Carry out an analysis of variance and interpret your results.

ANSWERS

5. Significant
6. Significant
7. Significant
8. Significant
9. Significant
10. For detergents, significant: For water temp, insignificant
11. For seasons, insignificant
    For salesman, insignificant
12. For hospital, insignificant
    For drugs, significant

Appendix

Values of e^x and e^{-x}

x	e^x	e^{-x}	x	e^x	e^{-x}
0.00	1.000	1.000	3.00	20.086	0.0497
0.10	1.105	0.904	3.10	22.198	0.0450
0.20	1.221	0.818	3.20	24.533	0.0407
0.30	1.349	0.740	3.30	27.113	0.0368
0.40	1.491	0.670	3.40	29.964	0.0333
0.50	1.648	0.606	3.50	33.115	0.0301
0.60	1.822	0.548	3.60	36.598	0.0273
0.70	2.013	0.496	3.70	40.447	0.0247
0.80	2.225	0.449	3.80	44.701	0.0223
0.90	2.459	0.406	3.90	49.402	0.0202
1.00	2.718	0.367	4.00	54.598	0.0183
1.10	3.004	0.332	4.10	60.340	0.0165
1.20	3.320	0.301	4.20	66.686	0.0149
1.30	3.669	0.272	4.30	73.700	0.0135
1.40	4.055	0.246	4.40	81.451	0.0122
1.50	4.481	0.223	4.50	90.017	0.0111
1.60	4.953	0.201	4.60	99.484	0.0100
1.70	5.473	0.182	4.70	109.95	0.0090
1.80	6.049	0.165	4.80	121.51	0.0082
1.90	6.685	0.149	4.90	134.29	0.0074
2.00	7.389	0.135	5.00	148.41	0.0067
2.10	8.166	0.122	5.10	164.02	0.0060
2.20	9.025	0.110	5.20	81.27	0.0055
2.30	9.974	0.100	5.30	200.34	0.0049
2.40	11.023	0.090	5.40	221.41	0.0045
2.50	12.182	0.082	5.50	244.69	0.0040
2.60	13.464	0.074	5.60	270.43	0.0036
2.70	14.880	0.067	5.70	298.87	0.0033
2.80	16.445	0.060	5.80	330.30	0.0030
2.90	18.174	0.055	5.90	365.04	0.0027
3.00	20.086	0.497	6.00	403.43	0.0024

Poisson Distribution

$$P(x=r) = \frac{\lambda^r e^{-\lambda}}{r!}, \quad \lambda = np > 0; r = 0, 1, 2, \ldots$$

$\lambda = np$ \ r	0.1	0.2	0.3	0.4	0.5	0.6	0.7	0.8	0.9	1.0
0	.9048	.8187	.7408	.6703	.6065	.5488	.4966	.4493	.4066	.3679
1	.0905	.1637	.2222	.2681	.3033	.3293	.3476	.3595	.3659	.3679
2	.0045	.0164	.0333	.0536	.0758	.0988	.1217	.1438	.1647	.1839
3	.0002	.0011	.0033	.0072	.0126	.0198	.0284	.0383	.0494	.0613
4	.0000	.0001	.0002	.0007	.0016	.0030	.0050	.0077	.0111	.0153
5	.0000	.0000	.0000	.0001	.0002	.0004	.0007	.0012	.0020	.0031
6	.0000	.0000	.0000	.0000	.0000	.0000	.0001	.0002	.0003	.0005
7	.0000	.0000	.0000	.0000	.0000	.0000	.0000	.0000	.0000	.0001

$\lambda = np$ \ r	1.1	1.2	1.3	1.4	1.5	1.6	1.7	1.8	1.9	2.0
0	.3329	.3012	.2725	.2466	.2231	.2019	.1827	.1653	.1496	.1353
1	.3662	.3614	.3543	.3452	.3347	.3230	.3106	.2975	.2842	.2707
2	.2014	.2169	.2303	.2417	.2510	.2584	.2640	.2678	.2700	.2707
3	.0738	.0867	.0998	.1128	.1255	.1378	.1496	.1607	.1710	.1804
4	.0203	.0260	.0324	.0395	.0471	.0551	.0636	.0723	.0812	.0902
5	.0045	.0062	.0084	.0111	.0141	.0176	.0216	.0260	.0309	.0361
6	.0008	.0012	.0018	.0026	.0035	.0047	.0061	.0078	.0098	.0120
7	.0001	.0002	.0013	.0005	.0008	.0011	.0015	.0020	.0027	.0034
8	.0000	.0000	.0001	.0001	.0001	.0002	.0003	.0005	.0006	.0009
9	.0000	.0000	.0000	.0000	.0000	.0000	.0001	.0001	.0001	.0002

$\lambda = np$ \ r	2.1	2.2	2.3	2.4	2.5	2.6	2.7	2.8	2.9	3.0
0	.1225	.1108	.1003	.0907	.0821	.0743	.0672	.0608	.0550	.0498
1	.2572	.2438	.2306	.2177	.2052	.1931	.1815	.1703	.1596	.1494
2	.2700	.2681	.2652	.2613	.2565	.2510	.2450	.2384	.2314	.2240
3	.1890	.1966	.2033	.2090	.2138	.2176	.2205	.2225	.2237	.2240
4	.0992	.1082	.1169	.1254	.1336	.1414	.1488	.1557	.1622	.1680
5	.0417	.0476	.0538	.0602	.0668	.0735	.0804	.0872	.0940	.1008
6	.0146	.0174	.0206	.0241	.0278	.0319	.0362	.0407	.0455	.0504
7	.0044	.0055	.0068	.0083	.0099	.0118	.0139	.0163	.0188	.0216
8	.0011	.0015	.0019	.0025	.0031	.0038	.0047	.0057	.0068	.0081
9	.0003	.0004	.0005	.0007	.0009	.0011	.0014	.0018	.0022	.0027
10	.0001	.0001	.0001	.0002	.0002	.0003	.0004	.0005	.0006	.0008
11	.0000	.0000	.0000	.0000	.0000	.0001	.0001	.0001	.0002	.0002
12	.0000	.0000	.0000	.0000	.0000	.0000	.0000	.0000	.0000	.0001

Appendix 421

$\lambda = np$ / r	3.1	3.2	3.3	3.4	3.5	3.6	3.7	3.8	3.9	4.0
0	.0450	.0408	.0369	.0334	.0302	.0273	.0247	.0224	.0202	.0183
1	.1397	.1304	.1217	.1135	.1057	.0084	.0915	.0850	.0789	.0733
2	.2165	.2087	.2008	.1929	.1850	.1771	.1692	.1615	.1539	.1465
3	.2237	.3226	.2209	.2186	.2158	.2125	.2087	.2046	.2001	.1954
4	.1734	.1781	.1823	.1858	.1888	.1912	.1931	.1944	.1951	.1954
5	.1075	.1140	.1203	.1264	.1322	.1377	.1429	.1477	.1522	.1563
6	.0555	.0608	.0662	.0716	.0771	.0826	.0881	.0936	.0989	.1042
7	.0246	.0278	.0312	.0348	.0385	.0425	.0466	.0508	.0551	.0595
8	.0095	.0111	.0129	.0148	.0169	.0191	.0215	.0241	.0269	.0298
9	.0033	.0049	.0047	.0056	.0066	.0076	.0089	.0102	.0116	.0132
10	.0010	.0013	.0016	.0019	.0023	.0028	.0033	.0039	.0045	.0053
11	.0003	.0004	.0005	.0006	.0007	.0009	.0011	.0013	.0016	.0019
12	.0001	.0001	.0001	.0002	.0002	.0003	.0003	.0004	.0005	.0006
13	.0000	.0000	.0000	.0000	.0001	.0001	.0001	.0001	.0002	.0002
14	.0000	.0000	.0000	.0000	.0000	.0000	.0000	.0000	.0000	.0001
	4.1	4.2	4.3	4.4	4.5	4.6	4.7	4.8	4.9	5.0
0	.0166	.0150	.0136	.0123	.0111	.0101	.0091	.0082	.0074	.0067
1	.0679	.0630	.0583	.0540	.0500	.0462	.0427	.0395	.0365	.0337
2	.1393	.1323	.1254	.1188	.1125	.1063	.1005	.0948	.0894	.0842
3	.1904	.1852	.1798	.1743	.1687	.1631	.1517	.1517	.1460	.1404
4	.1951	.1944	.1933	.1917	.1898	.1875	.1849	.1820	.1789	.1755
5	.1600	.1633	.1662	.1687	.1708	.1725	.1738	.1747	.1753	.1755
6	.1093	.1143	.1191	.1237	.1281	.1323	.1362	.1432	.1432	.1462
7	.0640	.0686	.0732	.0778	.0824	.0869	.0914	.0959	.1002	.1044
8	.0328	.0360	.0393	.0428	.0463	.0500	.0537	.0575	.0614	.0653
9	.0150	.0168	.0188	.0209	.0232	.0255	.0280	.0307	.0334	.0363
10	.0061	.0071	.0081	.0092	.0104	.0118	.0132	.0147	.0164	.0181
11	.0023	.0027	.0032	.0037	.0043	.0049	.0056	.0064	.0073	.0082
12	.0008	.0009	.0011	.0014	.0016	.0019	.0022	.0026	.0030	.0034
13	.0002	.0003	.0004	.0005	.0006	.0007	.0008	.0009	.0011	.0013
14	.0001	.0001	.0001	.0001	.0002	.0002	.0003	.0003	.0004	.0005
15	.0000	.0000	.0000	.0000	.0001	.0001	.0001	.0001	.0001	.0001

Normal Distribution

An entry in the table is the areas under the entire normal curve between normal variate $Z = 0$ and a positive value of Z. Areas for negative value of Z are obtained by symmetry.

Z to First Decimal	Second Decimal									
	.00	.01	.02	.03	.04	.05	.06	.07	.08	.09
0.0	.0000	.0040	.0080	.0120	.0160	.0199	.0239	.0279	.0319	.0359
0.1	.0398	.0438	.0478	.0517	.0557	.0596	.0636	.0675	.0714	.0753
0.2	.0793	.0832	.0871	.0910	.0948	.0987	.1026	.1064	.1103	.1141
0.3	.1179	.1217	.1255	.1293	.1331	.1368	.1406	.1443	.1480	.1517
0.4	.1554	.1591	.1628	.1664	.1700	.1736	.1772	.1808	.1844	.1879
0.5	.1915	.1950	.1985	.2019	.2054	.2088	.2123	.2157	.2190	.2224
0.6	.2257	.2291	.2324	.2357	.2389	.2422	.2454	.2486	.2518	.2549
0.7	.2580	.2611	.2642	.2674	.2704	.2734	.2764	.2794	.2823	.2852
0.8	.2881	.2910	.2939	.2967	.2995	.3023	.3051	.3078	.3106	.3133
0.9	.3159	.3186	.3212	.3238	.3264	.3289	.3315	.3340	.3365	.3389
1.0	.3413	.3438	.3461	.3485	.3508	.3531	.3554	.3577	.3599	.3621
1.1	.3643	.3665	.3686	.3708	.3729	.3749	.3770	.3790	.3810	.3830
1.2	.3849	.3869	.3888	.3907	.3925	.3944	.3962	.3980	.3997	.4015
1.3	.4032	.4049	.4066	.4082	.4099	.4115	.4131	.4147	.4162	.4177
1.4	.4192	.4207	.4222	.4236	.4251	.4265	.4279	.4292	.4306	.4319
1.5	.4332	.4345	.4357	.4370	.4382	.4394	.4406	.4418	.4429	.4441
1.6	.4452	.4463	.4474	.4484	.4495	.4505	.4515	.4525	.4535	.4545
1.7	.4554	.4564	.4573	.4582	.4591	.4599	.4608	.4616	.4625	.4633
1.8	.4641	.4649	.4656	.4664	.4671	.4678	.4686	.4693	.4699	.4706
1.9	.4713	.4719	.4726	.4732	.4738	.4744	.4750	.4756	.4761	.4767
2.0	.4772	.4778	.4783	.4788	.4793	.4798	.4803	.4808	.4812	.4817
2.1	.4821	.4826	.4830	.4834	.4838	.4842	.4846	.4850	.4854	.4857
2.2	.4861	.4865	.4868	.4871	.4874	.4878	.4881	.4884	.4887	.4890
2.3	.4893	.4896	.4898	.4901	.4904	.4906	.4909	.4911	.4913	.4916
2.4	.4918	.4920	.4922	.4925	.4927	.4929	.4931	.4932	.4934	.4936
2.5	.4938	.4940	.4941	.4943	.4945	.4946	.4948	.4949	.4951	.4952
2.6	.4953	.4955	.4956	.4957	.4959	.4960	.4961	.4962	.4963	.4964
2.7	.4965	.4966	.4967	.4968	.4969	.4970	.4971	.4972	.4973	.4974
2.8	.4974	.4975	.4976	.4977	.4977	.4978	.4979	.4979	.4980	.4981
2.9	.4981	.4982	.4982	.4983	.4984	.4984	.4985	.4985	.4986	.4986
3.0	.4986	.4987	.4987	.4988	.4988	.4989	.4989	.4989	.4990	.4990

Values of t_α

ν	$\alpha = 0.10$	$\alpha = 0.05$	$\alpha = 0.025$	$\alpha = 0.01$	$\alpha = 0.005$
1	3.078	6.314	12.706	31.821	63.657
2	1.886	2.920	4.303	6.965	9.925
3	1.638	2.353	3.182	4.541	5.841
4	1.533	2.132	2.776	3.747	4.604
5	1.476	2.015	2.571	3.365	4.032
6	1.440	1.943	2.447	3.143	3.707
7	1.415	1.895	2.365	2.998	3.499
8	1.397	1.860	2.306	2.896	3.355
9	1.383	1.833	2.262	2.821	3.250
10	1.372	1.812	2.228	2.764	3.169
11	1.363	1.796	2.201	2.718	3.106
12	1.356	1.782	2.179	2.681	3.055
13	1.350	1.771	2.160	2.650	3.012
14	1.345	1.761	2.145	2.624	2.977
15	1.341	1.753	2.131	2.602	2.947
16	1.337	1.746	2.120	2.583	2.921
17	1.333	1.740	2.110	2.567	2.898
18	1.330	1.734	2.101	2.552	2.878
19	1.328	1.729	2.093	2.539	2.861
20	1.325	1.725	2.086	2.528	2.845
21	1.323	1.721	2.080	2.518	2.831
22	1.321	1.717	2.074	2.508	2.819
23	1.319	1.714	2.069	2.500	2.807
24	1.318	1.711	2.064	2.492	2.797
25	1.316	1.708	2.060	2.485	2.787
26	1.315	1.706	2.056	2.479	2.779
27	1.314	1.703	2.052	2.473	2.771
28	1.313	1.701	2.048	2.467	2.763
29	1.311	1.699	2.045	2.462	2.756
inf.	1.282	1.645	1.960	2.326	2.576

(a) Value for one tailed-test is $t_\nu(\alpha)$.
(b) Value for two tailed-test is $t_\nu(\alpha/2)$.

Values of $F_{0.05}$

v_1 = Degrees of freedom for numerator

v_2 = Degrees of freedom for denominator

v_2 \ v_1	1	2	3	4	5	6	7	8	9	10	12	15	20	24	30	40	60	120	∞
1	161	200	216	225	230	237	239	241	242	244	246	248	249	250	251	252	253	254	
2	18.50	19.00	19.20	19.20	19.30	19.30	19.40	19.40	19.40	19.40	19.40	19.40	19.50	19.50	19.50	19.50	19.50	19.50	19.50
3	10.10	9.55	9.28	9.12	9.01	8.94	8.89	8.85	8.81	8.79	8.74	8.70	8.66	8.64	8.62	8.59	8.57	8.55	8.53
4	7.71	6.94	6.59	6.39	6.26	6.16	6.09	6.04	6.00	5.96	5.91	5.86	5.80	5.77	5.75	5.72	5.69	5.66	5.63
5	6.61	5.79	5.41	5.19	5.05	4.95	4.88	4.82	4.77	4.74	4.68	4.62	4.56	4.53	4.50	4.46	4.43	4.40	4.37
6	5.99	5.14	4.76	4.53	4.39	4.28	4.21	4.15	4.10	4.06	4.00	3.94	3.87	3.84	3.81	3.77	3.74	3.70	3.67
7	5.59	4.74	4.35	4.12	3.97	3.87	3.79	3.73	3.68	3.64	3.57	3.51	3.44	3.41	3.38	3.34	3.30	3.27	3.23
8	5.32	4.46	4.07	3.84	3.69	3.58	3.50	3.44	3.39	3.35	3.28	3.22	3.15	3.12	3.08	3.04	3.01	2.97	2.93
9	5.12	4.26	3.86	3.63	3.48	3.37	3.29	3.23	3.18	3.14	3.07	3.01	2.94	2.90	2.86	2.83	2.79	2.75	2.71
10	4.96	4.10	3.71	3.48	3.33	3.22	3.14	3.07	3.02	2.98	2.91	2.85	2.77	2.74	2.70	2.66	2.62	2.58	2.54
11	4.84	3.98	3.59	3.36	3.20	3.09	3.01	2.95	2.90	2.85	2.79	2.72	2.65	2.61	2.57	2.53	2.49	2.45	2.40
12	4.75	3.89	3.49	3.26	3.11	3.00	2.91	2.85	2.80	2.75	2.69	2.62	2.54	2.51	2.47	2.38	2.38	2.30	2.30
13	4.67	3.81	3.41	3.18	3.03	2.92	2.83	2.77	2.71	2.67	2.60	2.53	2.46	2.42	2.38	2.34	2.30	2.25	2.21
14	4.60	3.74	3.34	3.11	2.96	2.85	2.76	2.70	2.65	2.60	2.53	2.46	2.39	2.35	2.31	2.27	2.22	2.18	2.13
15	4.54	3.68	3.29	3.06	2.90	2.79	2.71	2.64	2.59	2.54	2.48	2.40	2.33	2.29	2.25	2.20	2.16	2.11	2.07
16	4.49	3.63	3.24	3.01	2.85	2.74	2.66	2.59	2.54	2.49	2.42	2.35	2.28	2.24	2.19	2.15	2.11	2.06	2.01
17	3.45	3.59	3.20	2.96	2.81	2.70	2.61	2.55	2.49	2.45	2.38	2.31	2.23	2.19	2.15	2.10	2.06	2.01	1.96
18	4.41	3.55	3.16	2.93	2.77	2.66	2.58	2.51	2.46	2.41	2.34	2.27	2.19	2.15	2.11	2.06	2.02	1.97	1.93
19	4.38	3.52	3.13	2.90	2.74	2.63	2.54	2.48	2.42	2.38	2.31	2.23	2.16	2.11	2.07	2.03	1.98	1.93	1.88

20	4.35	3.49	3.10	2.87	2.71	2.60	2.51	2.45	2.39	2.35	2.28	2.20	2.12	2.08	2.04	1.99	1.95	1.90	1.84
21	4.32	3.47	3.07	2.84	2.68	2.57	2.49	2.42	2.37	2.32	2.25	2.18	2.10	2.05	2.01	1.96	1.92	1.87	1.81
22	4.30	3.44	3.05	2.82	2.66	2.55	2.46	2.40	2.34	2.30	2.23	2.15	2.07	2.03	1.98	1.94	1.89	1.84	1.78
23	4.28	3.42	3.03	2.80	2.64	2.53	2.44	2.37	2.32	2.27	2.20	2.13	2.05	2.01	1.96	1.91	1.86	1.81	1.76
24	4.26	3.40	3.01	2.78	2.62	2.51	2.42	2.36	2.30	2.25	2.18	2.11	2.03	1.98	1.94	1.89	1.84	1.79	1.73
25	4.24	3.39	2.99	2.76	2.60	2.49	2.40	2.34	2.28	2.24	2.16	2.09	2.01	1.96	1.92	1.87	1.82	1.77	1.71
30	4.17	3.32	2.92	2.69	2.53	2.42	2.33	2.27	2.21	2.16	2.09	2.01	1.93	1.89	1.84	1.79	1.74	1.68	1.62
40	4.08	3.23	2.84	2.61	2.45	2.34	2.25	2.18	2.12	2.08	2.00	1.92	1.84	1.79	1.74	1.69	1.64	1.58	1.51
60	4.00	3.15	2.76	2.53	2.37	2.25	2.17	2.10	2.04	1.99	1.92	1.84	1.75	1.70	1.65	1.50	1.43	1.35	1.25
120	3.92	3.07	2.68	2.45	2.29	2.18	2.09	2.02	1.96	1.91	1.83	1.75	1.66	1.61	1.55	1.50	1.43	1.35	1.25
∞	3.84	3.00	2.60	2.37	2.21	2.10	2.01	1.94	1.88	1.83	1.75	1.67	1.57	1.52	1.46	1.39	1.32	1.22	1.00

Volumes of $F_{0.01}$

v_1 = Degrees of freedom for numerator

v_2 = Degrees of freedom for denominator	1	2	3	4	5	6	7	8	9	10	12	15	20	24	30	40	60	120	∞
1	4,052	5,000	5,403	5,625	5,764	5,859	5,928	5,982	6,023	6,056	6,106	6,157	6,209	6,235	6,261	6,287	6,313	6,339	6,366
2	98.50	99.00	99.20	99.20	99.30	99.30	99.40	99.40	99.40	99.40	99.40	99.40	99.40	99.50	99.50	99.50	99.50	99.50	99.50
3	34.10	30.80	29.50	28.70	28.20	27.90	27.70	27.50	27.30	27.20	27.10	26.90	26.70	26.60	26.50	26.40	26.30	26.20	26.10
4	21.20	18.00	16.70	16.00	15.50	15.20	15.00	14.80	14.70	14.50	14.40	14.20	14.00	13.90	13.80	13.70	13.70	13.60	13.50
5	16.30	13.30	12.10	11.40	11.00	10.70	10.50	10.30	10.20	10.10	9.89	9.72	9.55	9.47	9.38	9.29	9.20	9.11	9.02
6	13.70	10.90	9.78	9.15	8.75	8.47	8.26	8.10	7.98	7.87	7.72	7.56	7.40	7.31	7.23	7.14	7.06	6.97	6.88
7	12.20	9.55	8.45	7.85	7.46	7.19	6.99	6.84	6.72	6.62	6.47	6.31	6.16	6.07	5.99	5.91	5.82	5.74	5.65
8	11.30	8.65	7.59	7.01	6.63	6.37	6.18	6.03	5.91	5.81	5.67	5.52	5.36	5.28	5.20	5.12	5.03	4.95	4.83
9	10.60	8.02	6.99	6.42	6.06	5.80	5.61	5.47	5.35	5.26	5.11	4.96	4.81	4.73	4.65	4.57	4.48	4.40	4.31
10	10.00	7.56	6.55	5.99	5.64	5.39	5.20	5.06	4.94	4.85	4.71	4.56	4.41	4.33	4.25	4.17	4.08	4.00	3.91
11	9.65	7.21	5.67	5.32	5.07	4.89	4.74	4.63	4.54	4.40	4.25	4.10	4.02	3.94	3.86	3.78	3.69	3.60	
12	9.33	6.93	5.95	5.41	5.06	4.82	4.64	4.50	4.39	4.30	4.16	4.01	3.86	3.78	3.70	3.62	3.54	3.45	3.36
13	9.07	6.70	5.74	5.21	4.86	4.62	4.44	4.30	4.19	4.10	4.96	4.82	3.66	3.59	3.51	3.43	3.34	3.25	3.17
14	8.86	6.51	5.56	5.04	4.70	4.46	4.28	4.14	4.03	3.94	3.80	3.66	3.51	3.43	3.35	3.27	3.18	3.09	3.00
15	8.68	6.36	5.42	4.89	4.56	4.32	4.14	4.00	3.89	3.80	3.67	3.52	3.37	3.29	3.21	3.13	3.05	2.96	2.87
16	8.53	6.23	5.29	4.77	4.44	4.20	4.03	3.89	3.78	3.69	3.55	3.41	3.26	3.18	3.10	3.02	2.93	2.84	2.75
17	8.40	6.11	5.19	4.67	4.34	4.10	3.93	3.79	3.68	3.59	3.46	3.31	3.16	3.08	3.00	2.92	2.83	2.75	2.65
18	8.29	6.01	5.09	4.58	4.25	4.01	3.84	3.71	3.60	3.51	3.37	3.23	3.08	3.00	2.92	284	2.75	2.66	2.57
19	8.19	5.93	5.01	4.50	4.17	3.94	3.77	3.63	3.52	3.43	3.30	3.15	3.00	2.92	2.84	2.76	2.67	2.58	2.49

20	8.10	5.85	4.94	4.43	4.10	3.87	3.70	3.56	3.46	3.37	3.23	3.09	2.94	2.86	2.78	2.69	2.61	2.52	2.42
21	8.02	5.78	4.87	4.37	4.04	3.81	3.64	3.51	3.40	3.31	3.17	3.03	2.88	2.80	2.72	2.64	2.55	2.46	2.36
22	7.95	5.72	4.82	4.31	3.99	3.76	3.59	3.45	3.35	3.26	3.12	2.98	2.83	2.75	2.67	2.58	2.50	2.40	2.31
23	7.88	5.66	4.76	4.26	3.94	3.71	3.54	3.41	3.30	3.21	3.07	2.93	2.78	2.70	2.62	2.54	2.45	2.35	2.26
24	7.82	5.61	4.72	4.22	3.90	3.67	3.50	3.36	3.26	3.17	3.03	2.89	2.74	2.66	2.58	2.49	2.40	2.31	2.21
25	7.77	5.57	4.68	4.18	3.86	3.63	3.46	3.32	3.22	3.13	2.99	2.85	2.70	2.62	2.53	2.45	2.36	2.27	2.17
30	7.56	5.39	4.51	4.02	3.7s0	3.47	3.30	3.17	3.07	2.98	2.84	2.70	2.55	2.47	2.39	2.30	2.21	2.11	2.01
40	7.31	5.18	4.31	3.83	3.51	3.29	3.12	2.99	2.89	2.80	2.66	2.52	2.37	2.29	2.20	2.11	2.02	1.92	1.80
60	7.08	4.98	4.13	3.65	3.34	3.12	2.95	2.82	2.72	2.63	2.50	2.35	2.20	2.12	2.03	1.94	1.84	1.73	1.60
120	6.85	4.79	3.95	3.48	3.17	2.96	2.79	2.66	2.56	2.47	2.34	2.19	2.03	1.95	1.86	1.76	1.66	1.53	1.38
∞	6.63	4.61	3.78	3.32	3.02	2.80	2.64	2.51	2.41	2.32	2.18	2.04	1.88	1.79	1.70	1.59	1.47	1.32	1.00

Values of χ^2 with Various Values of α and ν

ν \ α	0.99	0.95	0.50	0.10	0.05	0.01
1	0.0002	0.0039	0.455	2.706	3.841	6.635
2	0.0201	0.103	1.386	4.605	5.991	9.210
3	0.115	0.352	2.366	6.251	7.815	11.34
4	0.297	0.711	3.357	7.779	9.488	13.28
5	0.554	1.145	4.351	9.236	11.07	15.09
6	0.872	1.635	5.348	10.64	12.59	16.81
7	1.239	2.167	6.346	12.02	14.07	18.48
8	1.646	2.733	7.344	13.36	15.51	20.09
9	2.088	3.325	8.343	14.68	16.92	21.67
10	2.558	3.940	9.342	15.99	18.31	23.21
11	3.053	4.575	10.34	17.28	19.68	24.72
12	3.571	5.226	11.34	18.55	21.03	26.22
13	4.107	5.892	12.34	19.81	22.36	27.69
14	4.660	6.571	13.34	21.06	23.68	29.14
15	5.229	7.261	14.34	22.31	25.00	30.58
16	5.812	7.962	15.34	23.54	26.30	32.00
17	6.408	8.672	16.34	24.77	27.59	33.41
18	7.015	9.390	17.34	25.99	28.87	34.80
19	7.633	10.12	18.34	27.20	30.14	36.19
20	8.260	10.85	19.34	28.41	31.41	37.57
21	8.897	11.59	20.34	29.62	32.67	38.93
22	9.542	12.34	21.34	30.81	33.92	40.29
23	10.20	13.09	22.34	32.01	35.17	41.64
24	10.86	13.85	23.34	33.20	36.42	42.98
25	11.52	14.61	24.34	34.38	37.65	44.31
26	12.20	15.38	25.34	35.56	38.88	45.64
27	12.88	16.15	26.34	36.74	40.11	46.96
28	13.57	16.93	27.34	37.92	41.34	48.28
29	14.26	17.71	28.34	39.09	42.56	49.59
30	14.95	18.49	29.34	40.26	43.77	50.89

Index

A

Absolute error   2
Additive model   323
Advantages of central difference interpolation formula   169
Alternative hypothesis   347
Analysis of time series   321
Analysis of variance   401
Analysis of variance in one way classification   402
Analysis of variance in two way classification   409
Angle between two lines of regression   306
Applications of time series   322
Approximate numbers   1
Assumptions for f-test   381
Assumptions in the analysis of variance   401
Averaging operator   90

B

Backward differences   89
Bessel's formula   166
Bessel's formula to get the derivative   209
Bisection method or bolzano method   17
Boole's rule   223

C

Central difference interpolation formula   155
Central difference operator   91
Central differences   89
Change of origin and scale   290
Chi-square test   386
Chi-square test to test the goodness of fit   386
Chi-square test to test the independence of attributes   392
Components of time series   322
Computation of missing terms   110
Convergence of newton-raphson method   39
Convergence of the iterative method   27

Critical region 348
Critical value 348
Critical values of f-distribution 382
Cubic spline 197
Curve fitting 288
Cyclical variations 323

D

Differences of a factorial function 109
Differences of a polynomial 93
Different types of operators 90
Differential operator 91
Divided differences 178

E

Error in newton's gregory backward interpolation formula 145
Error in polynomial interpolation 131
Error in newton's gregory forward interpolation formula 131
Errors 2
Errors in numerical computation 5
Euler's method 263
Euler-maclaurin's formula 235
Exact numbers 1

F

Factorial notation 108
Finite differences 87
Finite integration 118
Fitting of a straight line 289
Floating point 9
Forecasting 323
Forecasting methods 324
Forecasting models 323
Forward differences 87
Free hand method or graphical method 325
F-test 380

G

Gauss's backward interpolation formula 158
Gauss's forward interpolation formula 156
Gauss-elimination method 48
Gauss-elimination method with pivoting 52
Gaussian quadrature formula 241
Gauss-jacobi method or jacobi method 71
Gauss-jordan method 54
Gauss-seidel method 77

Geometrical interpretation of newton-raphson method   40

I

Initial and boundary value problems   248
Interpolation   128
Interpolation with unequal intervals   177
Inverse interpolation   195
Iteration method   26

L

Lagrange's interpolation formula   188
Laplace-everett formula   167
Level of significance   348
Lin-bairstow's method   83
Linear regression   303
Linearization   307
Lines of regression   303
Link relative method   338

M

Matrix-inversion method   59
Measurement of cyclical variations   339
Measurement of random or irregular variations   340
Measurement of seasonal variations   332
Measurement of trend   325
Method of least squares   288
Method of least squares   329
Method of moving averages   327
Method of triangularisation or method of factorization   62
Modified Euler's method   267
Montmort's theorem   120
Multiple regression   307
Multiplicative model   324

N

Newton's backward difference formula to get the derivative   205
Newton's divided difference interpolation formula   181
Newton's forward difference formula to get the derivative   203
Newton's-gregory forward interpolation formula   129
Newton's gregory backward interpolation formula   142
Newton-cote's quadrature formula   220
Newton-raphson method   38
Nonlinear regression   307
Normal equations for different forms of curve   290
Normalized floating point numbers   10

432 *Numerical and Statistical Techniques*

Null hypothesis 347
Numerical differentiation 203
Numerical integration 219
Numerical solution of ordinary differential equations 247

O

One tailed and two tailed tests 348
Order and rate of convergence of regula-falsi method 33

P

Paired t-test for difference of means 357
Parameter and statistic 345
Percentage error 2
Picard's method 248
Picard's method for simultaneous first order differential equaiton 254
Polynomial in factorial notation 110
Power of the test 350
Procedure of f-test 381
Properties of divided differences 179
Properties of regression coefficients 304

R

Random or irregular variations 323
Ratio to moving average method 336
Ratio to trend method 334
Reciprocal or negative factorial notation 108
Regression 303
Regula-falsi method or method of false position 31
Relation between bessel's and everett's formula 168
Relative error 2
Romberg's method 233
Rounding off numbers 2
Runge-kutta method 272
Runge-kutta method for simultaneous first order differential equaions 284

S

Sampling distribution 345
Seasonal variations 322
Secular trend 322
Semi average method 325
Shift operator 90
Significant figures 1
Simple average method 332
Simpson's 1/3rd rule 221
Simpson's 3/8th rule 222

Solution of algebraic and transcendental equations   17
Solution of simultaneous linear algebraic equation   47
Standard error   346
Stirling's formula   165
Stirling's interpolation formula to get the derivative   207
Student's t-test   350
Summation of series   119

T

Taylor's method for simultaneous first order differential equation   256
Taylor's series method   256
Test for difference of proportions   364
Test for number of successes   361
Test for single proportion   362
Test of significance   345
Test of significance for attributes   361
Test of significance for difference of means   373
Test of significance for difference of standard deviations   376
Test of significance for single mean   370
Test of significance for variables   370
Time series and forecasting   321
Trapezoidal rule   220
T-test for difference of means   353
T-test for single mean   351
Type-I error and type-II error   349

U

Uses of standard error   346

W

Weddle's rule   224

Y

Yate's correction   394

Z

Z-test   361